SOIL, WATER AND CROP PRODUCTION

PIGEONPEA-SORGHUM
MUNGBEANS
SOLID-PLANTING

SOIL, WATER AND CROP PRODUCTION

D. Wynne Thorne
Emeritus Professor
Utah State University

Marlowe D. Thorne
University of Illinois

AVI PUBLISHING COMPANY, INC.
Westport, Connecticut

Frontispiece courtesy of ICRISAT, Hyderabad, India

Library of Congress Cataloging in Publication Data

Main entry under title:

Soil, water, and crop production.

Includes index.
1. Agricultural ecology. 2. Agriculture.
3. Cropping systems. 4. Agricultural geography.
I. Thorne, David Wynne, 1908- II. Thorne, Marlowe D.
S589.7.S64 630'.27'45 78-21145
ISBN 0-87055-281-3

Printed in the United States of America
by Eastern Graphics, Inc., Old Saybrook, CT.

Preface

The world's major undeveloped arable land resources are in the tropics of Africa and Latin America. Lesser land reserves occur in the temperate zones of North America, Australia and the USSR. Costs of bringing these new lands into production will be high. Also, the major population centers are generally distant from reserves of undeveloped land. With world population still increasing an average of 2% per year, there is reason for concern about our long-range capability to meet food requirements. Studies have further concluded that most of the additional food needed for our growing population must come from increased yields on land now being cultivated rather than from cultivating more land.

Other events are further complicating the food picture, particularly the escalating costs of energy, and shortages and costs of fossil fuels in most countries where food supplies are marginal. Anxiety about food is further enhanced by evidence that improvident use of such production technologies as fertilizers, pesticides, tractors and irrigation water may be injurious to soils, water, and living organisms.

These events emphasize the importance of people in the agricultural production ecosystem. People involved and affected are not only the farmers and those who provide essential supplies to farmers, but all people everywhere who consume food and who live on and look to the earth for sustenance, beauty and enjoyment.

With food requirements rising and the tasks of producing food becoming more complicated, there is need to examine critically present farming systems and to evaluate available and potential technology systems to find the most effective and profitable ways to solve problems. Such is the purpose of this book.

This presentation is divided into three major divisions. First it identifies and examines the largely uncontrollable environmental factors such as soil, climate, and topography which affect the growth of crops. Second, the book examines the various factors that can be managed to improve crop yields while conserving and protecting the basic soil and water and associated plant and animal resources. The third section examines present and potential systems of farming for improved crop production in the major climatic and soil ecosystems of the world.

Scientists were selected to examine alternatives in crop production systems for the various climatic zones because of their experience in particular zones, their specialized knowledge and their past contributions to improved farming in these unique soil and climatic situations.

We recognize the limitations in attempting such a vast undertaking. Only the broad climatic zones and related soils and associated agricultural practices could be examined. The book must, therefore, be considered an introduction and orientation to the major crop production systems of the world, not a definitive treatise. We have attempted, however, to examine the principal crop production systems in the major agro-ecological zones in both the developed and developing nations.

The last chapter considers many of the problems facing operators of small farms in developing countries and indicates some of the most effective answers emerging from agricultural research. We believe those interested in how farmers can improve food production in diverse parts of the world will find in this book many of the important concepts suited to major farming regions.

The presentation is planned for the informed layman. A general knowledge of botany, chemistry, mathematics and agriculture is presumed. However, mathematical and chemical concepts are used in only elementary ways. We believe the book will be helpful to the general public interested in problems facing increased food production in various parts of the world. It should be helpful, also, to advisers working with farmers, to public leaders interested in world food problems, and to students seeking a world-wide orientation in improved crop practices.

The authors acknowledge the help of their many colleagues who have offered advice, criticism, and assistance. The editors gratefully acknowledge the assistance of Jean Roberts who prepared the maps illustrating the various climatic zones and Max Paris who prepared many of the graphs. We are especially indebted to our wives, Alison Comish Thorne and Merle Carter Thorne, who have helped improve the clarity of presentation and who have done much of the final typing. The authors also wish to express their appreciation to Dr. Norman W. Desrosier, Ms. Karen Carter and the AVI Publishing Company for encouragement and assistances in bringing this book into being.

D. Wynne Thorne
Logan, Utah

Marlowe D. Thorne
Urbana, Illinois

October, 1978

Contributors

HAYDEN FERGUSON, Ph.D., Professor of Soils, Dept. of Plant and Soil Science, Montana State University, Bozeman, Montana.

C.A. FRANCIS, Ph.D., Professor, Dept. of Agronomy, University of Nebraska, Lincoln, Nebraska. Formerly Agronomist, Small Farm Systems Program, Centro International de Agricultura Tropical (CIAT), Cali, Colombia.

JOHN J. HASSETT, Ph.D., Associate Professor of Soil Science, Dept. of Agronomy, University of Illinois, Urbana, Illinois.

J. KAMPEN, Ph.D., Agricultural Engineer, Farming Systems Program, International Crops Research Institute for the Semi-Arid Tropics (ICRISAT), Hyderabad, India.

JAMES KRALL, M.S., Professor of Agronomy, Dept. of Plant and Soil Science, Montana State University, Bozeman, Montana.

B.A. KRANTZ, Ph.D., Agronomist, Farming Systems Program, International Crops Research Institute for the Semi-Arid Tropics (ICRISAT), Hyderabad, India.

A. KRISHNAN, Ph.D., Climatologist and Head of Division of Wind Power and Solar Energy Utilization, Central Arid Zone Research Institute (CAZRI), Jodhpur, Rajasthan, India.

H.S. MANN, Ph.D., Agronomist and Director, Central Arid Zone Research Institute (CAZRI), Jodhpur, Rajasthan, India. Also Editor-in-Chief, Annals of the Arid Zone.

JOHN NICHOLAIDES, III, Ph.D., Assistant Professor and Coordinator Tropical Soils Research Program, Dept. of Soil Science, North Carolina State University, Raleigh, North Carolina.

PETER A. ORAM, Ph.D., Senior Research Fellow, International Food Policy Research Institute, Washington, D.C. Formerly Director of the Research Centre, FAO, Rome.

D. WYNNE THORNE, Ph.D., Emeritus Professor, Dept. of Soil Science and Biometeorology, Utah State University, Logan, Utah.

MARLOWE D. THORNE, Ph.D., Professor, Dept. of Agronomy, University of Illinois, Urbana, Illinois.

Contents

1

The Environment and Crop Production Systems

D. Wynne Thorne

The welfare of human beings throughout time has remained closely dependent on renewable natural resources. Until a few thousand years ago this relationship was direct. People were part of broad ecosystems and obtained their food, clothing, shelter, and firewood by gathering and hunting. With increasing population and the development of newer technologies the individual became an exploiter and manager, a planter of seeds, a cultivator of crops, and a herder and manager of animals.

Even today human beings remain tied to climate, soil, water, plant, and animal resources for much of their welfare. But with rapid gains in population and increasing demands of people everywhere for more food, the conservation and wise management of resources to ensure stability and increasing supplies of food and fiber becomes an urgent necessity. In the present state of food production there are two major potentials to meet increasing needs: more land can be placed in crop production and the yields per hectare[1] can be increased.

Recently the University of California staff (Anon. 1974) examined available information on the arable lands of the world and the rates at which present cropland is being expanded (Table 1.1). In Europe the area of cultivated land is declining by about 0.44% per year. The most rapid expansion of tilled cropland is in Latin America (1.6% per year) and Africa (0.9% per year). The average annual world increase in cultivated land is about 0.68%. Meanwhile world population is increasing by about 2% per year and the demand for food on a world basis is increasing by nearly 3% per year. It is apparent that some of the most

[1]Metric units will be used unless otherwise indicated.

TABLE 1.1. ARABLE AND POTENTIALLY ARABLE LAND BY REGION AND PER CAPITA

Area	Arable Land			Potentially Arable (1000 ha)	Annual Growth Rate		Ha/Person 1970
	1960 (1000 ha)	1970 (1000 ha)	1985 (1000 ha)		1960–70 %/year	1970–85 %/year	
World	1426	1461	1610	3190	0.25	0.68	0.40
Europe	153	145	137	174	(0.44)[1]	(0.44)	0.31
USSR	221	233	252	356	0.54	0.54	0.96
Americas	329	355	400	1106	0.79	0.85	0.70
North America	226	236	252		0.44	0.44	1.04
Latin America	103	119	148		1.60	1.60	0.42
Oceania	28	47	95	154	6.70	6.70	3.03
Asia	456	467	484	628	0.24	0.24	0.23
Africa	239	214	242	733	(1.0)	0.90	0.62

Source: Adapted from Anon. (1974).
[1]() denotes negative growth rate.

highly developed areas on earth have essentially exhausted their reserves of arable land. The evidence for Europe and Asia is shown in Table 1.1. But more recently exploited areas of the world may soon face the same dilemma. In 1976 the U.S. Soil Conservation Service (Anon. 1976) reassessed the potentially arable land in the U.S. compared with that being cultivated in 1967. The evidence (Table 1.2) indicated that three-fourths of all land having a good potential for producing crops is already being cultivated. Also their data show that much of the reserve arable lands have problems of susceptibility to erosion, poor drainage, or other hazards.

Clearly the future welfare of people depends more on a steady increase in yields per hectare than on plowing more land. But major

TABLE 1.2. PROBLEMS ASSOCIATED WITH U.S. LANDS HAVING POTENTIAL FOR CONVERSION TO CROPLAND COMPARED WITH 1975 CROPLAND

Problem	Land Class	Potential for Cropland High (1000 ha)	Potential for Cropland Medium (1000 ha)	1975 Cropland
No Problem	1	2,062	139	13,523
Erosion	2E	8,095	1,519	35,476
	3E	6,982	3,485	28,492
	4E	2,376	1,752	12,026
	5 & 6 E	1,297	970	3,562
Subtotal		18,750	7,726	79,556
Excess Water	2W	3,866	838	24,347
	3W	2,439	1,042	12,708
	4W	678	636	1,607
	5 & 6 W	209	868	833
Subtotal		7,192	3,384	39,495
Soil Condition	2S	857	109	8,283
	3S	955	509	4,639
	4S	341	319	2,391
	5 & 6 S	181	431	1,431
Subtotal		2,334	1,368	16,744
Climate	2C	775	295	7,982
	3C	492	109	3,913
	4C	–	137	135
	5 & 6 C	93	105	147
Subtotal		1,360	646	12,177
	7 & 8	–	–	674
TOTAL		31,698	13,286	162,169

Source: Anon. (1976).

increases in crop yields require the integration of a large number of factors into management systems. In only limited areas, however, are crop production systems now designed and carried out based on up-to-date scientific information and technology. Many of the factors involved in or affecting these production systems are only partially under human control. Climate, soil, and water supply are major factors that are controllable to only a limited degree. But because of our inability to predict or control weather, to substantially change soil properties, or to supply water at all places as needed, crop production systems must be kept flexible.

The purpose of this book is to provide a basis for identifying and understanding the practical interrelationships among the various factors important in crop production and to determine how they can be put together in alternative ways to ensure improved yields of crops while conserving soil resources for permanent use.

MAJOR FACTORS CONTROLLING CROP YIELDS

Air, light, temperature, precipitation, and soil are the major factors in crop production that are largely outside of human control. Plant nutrients, crop selection and management, mechanical equipment, and the control of insects and diseases are production factors that are, to an important extent, controllable. These diverse factors responsible for high yields of crop plants are more likely to be brought into optimum combinations if there is an understanding of the principles involved plus the capacity to develop means by which these principles may be applied in specific on-farm production systems.

Crop plants are, in a very literal sense, the product of their natural environment. Without mobility each plant is subject to the diverse and changing aspects of temperature, moisture, oxygen, and carbon dioxide supply, physical and chemical conditions of the soil, and interrelationships with other plants and other living organisms. These microenvironment components vary with latitude, elevation, meteorological conditions, topography, and soil characteristics. The effects of such natural factors can be modified in varying degrees by such practices as irrigation, drainage, windbreaks, tillage, date of planting, chemical treatments, crop selection, and breeding.

Temperature

Growth of higher plants is largely restricted to temperatures between 0° and 60°C. The growing season of more than 150 extensively cultivated crops is confined to a narrow range of temperatures between 10° and 40°C. The optimum growth temperatures for many crops range between 20° and 28°C. However, optimum temperature range data are difficult to interpret because optimum levels for short periods of a few hours or days are often much higher than when such temperatures are maintained for several weeks. Also, the most favorable temperatures may vary with physiological stages of plant development. Lower spring temperatures may favor early stages of plant development such as tillering in wheat. Medium temperatures usually favor stem elongation and fruit development, while warm summer temperatures favor flowering.

Temperature data as reported in climatic records are subject to variations in interpretation. Average daily temperatures, for example, are computed by averaging the maximum and minimum extremes for the day. In some areas, especially in inland arid temperate regions, the daily temperature range during the growing season commonly may exceed 20°C. In humid and more tropical regions, and especially near large bodies of water, the daily range may be less than 5°C. Such daily variations can be significant in crop adaptations. Potatoes, for example, require night temperatures of 10° to 15°C in order to form tubers, while tomatoes set fruit best with night temperatures of 15° to 18°C.

Temperature has extensive effects on soil properties as well as on plant behavior. The activity of many groups of soil microorganisms reaches an optimum between 25° and 30°C and decreases with lower or higher temperatures. Temperature also influences the solubility of plant nutrients in soil solutions and their availability to plants.

Plant growth processes respond to soil temperature variations. Nutrient and water absorption are decreased at low temperatures and relative rates of intake of different nutrients vary with temperature. Release of nitrogen, phosphorus, and sulphur from organic matter and their absorption by higher plants are slow at low temperatures. In one study in Michigan young corn plants absorbed twice as much phosphorus and four times as much zinc at 43°C as at 23°C. Water absorption, transport, and rates of cell division are somewhat proportional to soil temperature. Root growth has been shown in many studies to be influenced by soil temperature. Optimum temperature for root and rhizome production is near 10°C for Canada bluegrass, 15°C for Kentucky bluegrass, and 20°C for orchardgrass and red clover.

Light

Sunlight effects on plant adaptation and growth are only slightly less marked in a geographic sense than temperature. But the full effect of light in controlling plant growth processes is still not fully understood. Although visible light (4000 to 7500 A) makes up only half of the total solar energy, this is the range that is most important for photosynthesis. Most green plants grow normally only when exposed to light having most of the wavelengths of visible light. There are, however, peaks of absorption in photosynthesis in the violet-blue and orange-red regions. Light waves in the ultraviolet and shorter nonvisible range are injurious to plants.

Vegetation screens out light and not only the intensity but the quality of light may be different in shade than in full sunlight. Usually light penetrating leafy vegetation is proportionately lower in blue and violet wavelengths and higher in red. In Texas, in a study of light under a full stand of cotton 122 cm high, the light intensity was reduced by 70% at 55 cm and by 95% at ground level.

Most field crops grow best in open fields under full sunlight. Some plants, however, need shady conditions. Coffee and cacao yield better with only 25 to 50% of full light intensity. Leaves of many plants are smaller and thicker when grown in full sunlight and, for this reason, tea and some tobacco varieties produce better quality products in partial shade.

The proportion of sunlight energy utilized in photosynthesis is usually near 1% or even less. Crops such as sugar cane and maize utilize relatively larger proportions of light but do not usually use more than 2.5%. Recent studies indicate some potentials for increasing light utilization by developing more erect stems and leaves through breeding and selection. Since the capacity to utilize light efficiently in photosynthesis is the basis of agricultural production, such modifications in plants may be important for attaining greater yields.

Carbon Dioxide and Oxygen

Nearly all living organisms obtain most of their energy by using oxygen to oxidize organic materials with the release of carbon dioxide. This is counterbalanced by the photosynthesis of green plants. Large amounts of CO_2 are also produced by combustion processes, both uncontrolled and those that are used to supply heat and energy. There are increasing questions as to whether nature can maintain the traditional balance in atmospheric composition. There is, in fact, evidence of small but significant increases in the percentage of CO_2 in the atmosphere, especially near large cities.

Even though the processes involved in CO_2 and O_2 consumption and production vary with light and dark, with seasonal changes in temperature, and with people's activities, the quantities and proportions in the open atmosphere remain relatively constant, with CO_2 being near 0.03% and O_2 near 21%. The large quantities of organic material produced by plants contrasted with the small percentage of CO_2 in the air has prompted many studies on the adequacies of CO_2 for maximum growth rates and yields of plants. Such investigations have shown that the CO_2 content of air often declines below the atmospheric average in cropped fields or forests when photosynthesis is active. Two factors tend to sustain CO_2 levels in limited areas: the exchange of CO_2 in the air near the crop and that in the more remote atmosphere, and the rather large amounts of CO_2 produced from the soil. Some studies in forests have shown that on warm days with low air movements the proportion of CO_2 in lower layers of the atmosphere may be 6 to 10 times greater than that in the active photosynthetic leaf zone. Air turbulence greatly increases the rate of CO_2 diffusion and assists in keeping the concentration uniform.

A number of studies have shown that under high light intensity the rate of photosynthesis may be limited by the CO_2 supply. In a typical field of an actively growing crop such as maize, the bottom leaves receiving low light intensity may have an adequate CO_2 supply while the upper leaves with high light intensity are limited in photosynthetic rate by lack of CO_2. Under controlled conditions the rate of photosynthesis for various crop plants has been increased from 25% to 400% as a result of elevating the atmospheric levels of CO_2 (Wittwer 1974). Some of the most responsive crops have been cotton, soybeans, sorghum, potatoes, and rice.

Oxygen in the atmosphere above the soil is apparently adequate for all plant activities. Under abnormal soil conditions, such as waterlogging or poor soil structure, O_2 may become limiting for the growth and proper metabolism of plant roots. Under such problem soil conditions the careful management of water and the establishment and maintenance of a good soil structure are important for attainment of satisfactory yields of most crops.

Soils

Soils are the natural medium for plant growth. Although plants can be grown in artificial media such as solution cultures, sand, or artificial exchange resins, and under these conditions can attain equally high yields compared with soil, the technical problems are relatively great and the costs of production are likely to be high. In any event, soils are the most commonly available resource for crop production and offer many unique advantages for practical crop production. Also, plants have evolved with soil as an important part of their natural environment so that, except for special situations, soil will continue as the major resource for gardening and field crop production.

Soils have a wide range of characteristics. In consequence they vary greatly in suitability for the production of high yields of good quality crops. However, plants differ in adaptations to soil conditions, and farming practices can be helpful in modifying many soil conditions that may be unfavorable for most crops such as poor drainage, acidity, or salinity. The intelligent response of farmers and agricultural planners to variations in soil characteristics and to the diverse needs for soil for various types of agriculture has been to first select soils most suitable for the preferred crops or types of farm operations and to gradually extend farming to less suitable soils. A second alternative has been to modify some soil conditions to meet plant requirements. Some soil situations such as waterlogging and low nutrient content often can be modified effectively, while other situations such as steep topography or a high content of clay cannot be corrected economically.

A number of soil characteristics are important to the successful management of soils for crop production. The topography must be relatively level since steep slopes are difficult to cultivate with mechanical equipment and soil erosion is difficult to control. For the successful production of most crops the physical properties of the soil should permit the relatively rapid entry of water, its movement through the soil, and the removal of excess water through natural or artificial drainage channels. Soil particles should remain strongly aggregated so crusting will not prevent emergence of seedlings and so soils can be cultivated as needed for crop management.

Some crops have specialized requirements for soils; rice, for example, needs soils that are only slowly permeable to maintain a continuous level of surface water without excessive losses. Also, in temperate climates long slopes toward the afternoon sun may lengthen the growing season and enhance air drainage, thereby reducing damage from frost.

Manageable Production Factors

Factors and practices that constitute the major variables that require planning, investment, and management by those concerned with improving crop production systems are briefly described here and will be discussed later in greater detail.

Farm Mechanization.—Mechanical equipment such as the plow, the harrow, the mower, the reaper, the thresher, the tractor, and the combine harvester were the ingredients of the first revolution in farming that blossomed during the period of 1820 to 1850 and have been perfected in subsequent years. These engineering developments made possible the cultivation of more land and a more timely conduct of farm operations. Through these innovations the efficiency and effectiveness of farm labor have been greatly increased. However, large investments are required for the purchase of such implements and decisions on the selection and financing of such equipment have become some of the most important problems facing the modern farmer.

Tillage Alternatives.—There are many alternatives relating to soil preparation and management for crop production. The moldboard plow is often replaced on large farms by disk plows, chisel plows, and other devices. Other equipment such as the Graham Hoeme plow and the rod weeder are used on wheat farms in arid regions to keep residues on the soil surface and thereby to reduce soil erosion. In recent years the concept of plowing and cultivating the soil has been challenged by the use of herbicides and limited tillage. Such practices as "limited tillage" and "zero tillage" are gaining in popularity. Direct seeding of soybeans, maize, wheat, and barley following an herbicide treatment is reducing costs of seedbed preparation and soil erosion, and such practices are being widely adopted.

Use of Agricultural Chemicals.—The development of 2, 4-D and other herbicide chemicals was one of the important innovations in agriculture in the 1940s and 1950s. The selective herbicides, in particular, are reducing tillage operations, human labor requirements, and competition of weeds with crop plants.

Growth regulant chemicals are opening exciting new vistas for bringing about specific changes in plant growth and metabolism. Gibberellins are widely used to increase the yield and size of seedless grapes and 2,3,5-tri-iodo-benzoic acid (TIBA) increases the pod set and yield of soybeans and also the yield of alfalfa seed, and brings about earlier maturity of soybeans. A large list of such chemicals is coming into use to increase flowering, to thin fruit sets, and to stimulate or retard plant growth. Such materials and their widespread use seem certain to gain in importance in the future.

Fertilizer Management.—As much as half of the increases in crop yields over the past two decades has been estimated as due to increased and improved use of fertilizers. Major increases have occurred in nitrogen applications, particularly anhydrous ammonia. Phosphorus, potassium, and several other essential elements are now used in small or large amounts where soils are deficient. The proper use of fertilizers in crop production systems is one of the most important factors a farmer can manage for increased yields. World use of primary plant nutrients in 1974-1975 totaled over 90 million tons.

Cropping Systems.—Growing crops in the most favorable sequence or combinations and increasing cropping intensities through multiple and relay cropping are proving to be effective avenues for increasing farm productivity. New cultivars of major crops such as wheat or rice, with shorter than normal growth periods, are adding to potentials for growing two or more crops per year where water supplies are adequate.

Insect and Disease Control.—The use of modern pesticides to control insects has been a major development in reducing damage to crops from both insects and diseases. In many instances improper or excessive uses of such pesticides have been self-defeating in that many helpful insects have been killed. New crop varieties are frequently bred with resistance to specific insects and disease organisms. Where available, such crops provide the least expensive and most effective control. The best practices are now based on the careful selection of crops and the judicious use of insecticides, fungicides, etc., in combination with such other practices as growing crops in the proper sequence, field sanitation to limit insect carry-over between crops, and encouraging the increase of organisms that are parasites on insects harmful to crop plants. The problems are made more complex by the capability of organisms to develop resistance to many insecticides and to circumvent genetic resistance of new crop varieties.

On-Farm Water Management.—Irrigation is highly important in improving the yield and quality of crops in all but the most favorable rainfall zones. Irrigation permits the use of all of the major production improvement factors in farming because with adequate water rights it removes the major hazard of drought that threatens crop production over most of the world. But on-farm water management has come to stand for more than irrigation alone. It includes drainage for soils that become waterlogged. It includes the diversion, harvest, and storage of water during periods of excess rainfall with its potential use for irrigation during dry seasons. It involves making effective use of the water holding capacity of soils to grow crops in dry seasons. New practices or special adaptations of established water management practices are being continually developed to make better use of available water resources. About 80% of the cropland of the world depends on natural precipitation rather than irrigation, but under most conditions, water conservation and management practices are proving to be effective in stabilizing and improving crop yields.

THE PRODUCTION POTENTIALS OF FIELD CROPS

Many attempts have been made to estimate maximum possible yields for various crops under specific conditions. Based on the photosynthetic efficiency of a crop such as maize and an average of 500 cal of light energy per square centimeter per day, Loomis and Williams (1963) estimated that a potential daily production of 71 g of dry plant material could be produced per square meter per day with fully favorable conditions. To this they added an estimated 6 g of nutrients taken from the soil and assumed all dry matter produced during the 35-day grain-development period was deposited in the grain. These estimates resulted in a potential yield of about 27,000 kg of grain per hectare, a yield about 30% higher than has been obtained under record field conditions.

Table 1.3 presents average yields of selected crops for the U.S., the highest recorded yields, and estimates of the yields for the "best farms"

TABLE 1.3. YIELDS OF CROPS IN THE UNITED STATES, KG/HA

Crop	Highest Average 1973–76	Best Farms Estimates	Record
Maize	5,725	12,000	21,200
Wheat	2,250	7,600	14,500
Soybeans	1,910	5,000	7,400
Sorghum	3,685	12,000	20,000
Rice	5,245	6,500	17,650
Barley	2,410	6,500	11,400
Oats	1,725	5,000	10,600
Potatoes	28,800	41,000	72,200
Sweet Potatoes	13,000	30,000	46,400
Sugar Beets	45,000	80,000	121,000

Source: Wittwer (1974).

(an estimated value, and representative of only a small percentage of farms in the most favorable climatic zones). The "best farms" have yields 2 to 4 times higher than the national average, while "record" yields are 50 to 100% higher than yields on the "best farms." Yields obtained on the "best farms" are the product of desirable combinations of near optimum levels of the various growth promoting factors. Since most locations have less than optimum climate, soil, and moisture conditions, national average yields are unlikely to attain such levels unless there are unusual and currently unforeseeable scientific discoveries that will substantially boost yield potentials. However, for large numbers of farms in the U.S., particularly those in favorable rainfall zones, or those with adequate supplies of irrigation water, the "best farm" yields can be considered attainable.

RESPONSES TO CROP PRODUCTION FACTORS

The relationships between a crop yield and increasing quantities of one deficient factor, such as nitrogen or phosphorus, with others present at near optimum levels have been quantitatively studied and yield curves and mathematical functions derived. The most common relationship in the literature is based on the studies of Mitscherlich (Tisdale and Nelson 1975). This relationship assumes that if various units of the deficient factor are added the largest increase in yield will be obtained with the first unit and smaller yield increments will be obtained with the addition of subsequent units of the factor. The resulting curve, Fig. 1.1, follows the general form of the yield response of the general economic law of diminishing returns. While helpful in visualizing relationships of yield to increments of growth factors, the curve does not allow for harmful effects of excessive quantities of a factor beyond the optimum level. In the case of fertilizer salts it is obvious that additions beyond some definite level will be harmful.

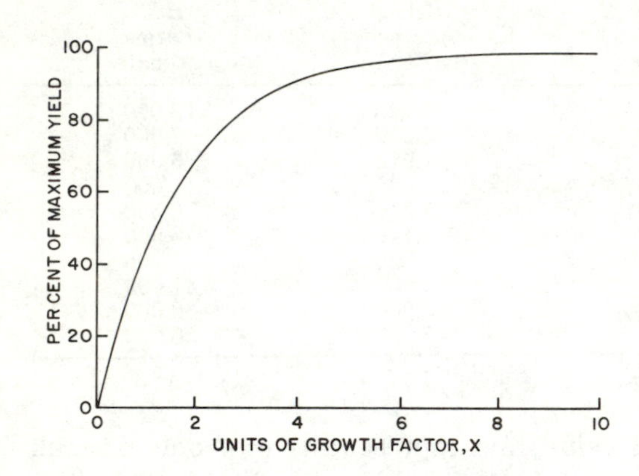

FIG. 1.1. YIELD RESPONSE TO ONE DEFICIENT FACTOR, BASED ON MITSCHERLICH-BAULE'S CONCEPTS

Under field situations optimum combinations of conditions and maximum attainable yields are difficult to define. Also, with multiple factors in varying stages of deficiency the slope of the curve will vary. It is important to note that the general shape of the curve is convex between the initial field conditions and the point attained through application of the various production factors. Important, also, is the capability of the farmer or the planner for identifying the most significant factors responsible for the gap between yields obtained with current practices and the maximum attainable yield.

A primary shortcoming of the above approach of limiting factors is that usually more than one factor is limiting and the combined responses from adding two or more limiting factors are greater than the arithmetic sum of yield responses from the individual factors added separately.

There is an extensive and growing literature on developing mathematical functions and systems models for predicting crop yields from various input factors, particularly available moisture supply and fertilizer increments (NAS-NRC 1961; Pesek 1974). Since this is a specialized field and the consideration of a number of input variables depends on the use of computers and special programs, a detailed discussion is beyond the scope of this book.

A more general approach has been to assign relative weights to different production factors when used alone and in various combinations. One example is the weightings assigned to production factors by a World Bank team in examining the potential for increased crop production in the Indus Valley of Pakistan as summarized in Table 1.4. The principle that appropriate combinations of production practices promote yields greater than the simple additions of the factors taken separately is basic to successful farm management.

TABLE 1.4. ASSUMPTIONS REGARDING CONTRIBUTIONS TO YIELD GROWTH FROM SELECTED YIELD-IMPROVING FACTORS OVER THE PERIOD 1965-2000 IN THE INDUS PLAIN OF PAKISTAN

	Factor Contribution in Isolation %	Cumulative Yield %	Apparent Contribution in Combination %
Present Yield		100	
Factors			
Additional water supplies alone	10	110	10
Removal of waterlogging and salinity	10	121	11
Application of fertilizers	40	169	48
Disease and pest control	15	195	26
Improved seedbed preparation and cultivation practices	20	234	39
Improved varieties	20	281	47

Source: Lieftinck, Sadiri and Creyke (1969).

Essential to an analysis of production systems is selection of the criteria to be used in evaluating the product to be obtained. In most instances this criteria is net economic gain. In such an instance the goal is not maximum yield but the determination of the quantities of combinations of such factors as fertilizer, irrigation water, or pesticides that will each provide the highest increases in income above the costs of the combined factors applied.

MODERN CROP PRODUCTION SYSTEMS

Farm production systems are based on the concept that for each crop or each combination of crops there are certain levels and combinations of production factors that will provide the base for optimum yields. Farming systems have been defined by Krantz (1974) as: "The entire complex of development, management, and allocation of resources as well as decisions and activities which, within an operational farm unit or a combination of such units, results in agricultural production, and the processing and marketing of the products."

This book is primarily a treatise on crop production aspects of farming systems. Most such books to date have been descriptive, concerned primarily with examining past and present farming practices and institutions. Our emphasis is on the future (Duckham and Masefield 1969; Ruthenberg 1976). In our treatments, past and present practices are examined only as a basis for evaluating alternatives for improvement. Improvement in systems is jointly a concern for increased crop production and a concern for conserving and managing resources for a permanently successful agriculture. Emphasis is on improved systems based on the application of established scientific principles, particularly as verified by recent adaptive research under representative conditions with respect to climate, soil, social, and economic situations in different continents.

Systems of farming differ according to the interests and backgrounds of the people involved, the available resources and state of development of these resources, the methods of production, and the crops produced. Farming systems practices include soil and water management technologies; cropping systems; methods, dates, and rates of planting; power equipment; packages for land development and tillage; plant protection; plant nutrient applications; cultivation, harvesting, threshing, drying, processing, and marketing. Credit, markets, transport, and other farm-related services are also involved. Systems of farming differ from each other according to available resources, the state of development of these resources, and the methods of production and the crops produced. Complete systems also include livestock nutrition and management and the handling and marketing of livestock products. In-depth consideration of such broad systems is, however, beyond the scope of this book and so the primary emphasis is restricted to crop production systems.

One important point separates crop production systems from the production of individual crops, and that is that in crop production systems emphasis is placed on the total production or net income throughout the year. This distinction can be important where length of growing season, water supply, and other resources permit cropping throughout all, or much of the year. In such situations crop varieties with shorter maturity dates and somewhat reduced yield potentials may be selected to permit an additional crop to be grown during the year or to favor a succeeding crop in the sequence. Similarly, less than optimum planting dates may be deliberately chosen for one crop in order to provide more favorable growth conditions for another. Consequently, emphasis is placed on the total production in the year rather than on the yields of individual crops. Furthermore, production systems must be deeply concerned with controlling soil erosion, preventing soil deterioration, and with managing water and other resources to maintain the farm in a long-term productive state.

The number of crops produced per year is referred to as the cropping intensity. Studies of traditional crop production systems in various parts of the world indicate that there is opportunity for substantially increasing crop intensities, particularly where intensive use of labor is desirable. In much of the humid tropics, for example, the cropping intensity could be more than doubled. Even in much of the semi-arid tropics two crops can be grown where most farms now produce but one. Some evaluations indicate that even in intensively farmed countries such as Egypt greater crop intensities can be attained. The average cropping intensity there is now about 1.9, whereas there is an estimated potential for an average of 2.5 crops per year.

Improving farm productivity throughout the world through developing more effective and efficient farming systems is essential to the future well being of all people. Essentially all agricultural scientists agree that increased world production of food and fiber over the next two decades will come from the application and extended use of principles and technologies now available. Developing improved farming systems is the systematic way for applying new and old technologies in proper relation to each other.

REFERENCES

ANON. 1974. A hungry world: the challenge to agriculture. Univ. Calif., Berkeley, Div. Agric. Sci.

ANON. 1976. Potential cropland survey, preliminary report. U.S. Dep. Agric. Soil Conserv. Serv. Soil Surv., Washington, D.C.

DUCKHAM, A.N., and MASEFIELD, G.B. 1969. Farming Systems of the World. Praeger Publishers, New York.

KRANTZ, B.A. 1974. Cropping patterns for increasing and stabilizing agricultural production on the semi-arid tropics. *In* International Workshop on

Farming Systems. International Crops Research Institute for the Semi-Arid Tropics, Hyderabad, India.

LIEFTINCK, P., SADIRI, R., and CREYKE, T.C. 1969. Water and Power Resources of West Pakistan. John Hopkins Press, Baltimore.

LOOMIS, R.S., and WILLIAMS, W.A. 1963. Maximum crop productivity: an estimate. Crop Sci. *3*, 67-72.

MCCLOUD, D.E., BULA, R.J., and SHAW, R.H. 1964. Field plant physiology. *In* Advances in Agronomy, Vol. 16. A.G. Norman (Editor). Academic Press, New York.

NAS-NRC. 1961. Status and methods of research in economic and agronomic aspects of fertilizer response and use. National Academy of Sciences-National Research Council Publ. *918*.

NELSON, L.B. 1968. Changing Patterns in Fertilizer Use. Soil Science Society of America, Madison, Wisconsin.

PESEK, J.T. 1974. Crop yield response equations and economic levels of fertilizer use. *In* Fertilizers, Crop Quality, and Economy. V.H. Fernandez (Editor). Elsevier Scientific Publishing Co., New York.

RUSSELL, E.W. 1973. Soil Conditions and Plant Growth, 10th Edition. Longmans, Green, and Co., London.

RUTHENBERG, H. 1976. Farming Systems in the Tropics, 2nd Edition. Clarendon Press, Oxford.

TISDALE, S.L., and NELSON, W.L. 1975. Soil Fertility and Fertilizers, 3rd Edition. Macmillan, New York.

WILSIE, C.P. 1962. Crop Adaptation and Distribution. W.H. Freeman and Co., San Francisco.

WITTWER, S.H. 1974. Maximum production capacity of food crops. BioScience *24*, 216-224.

2

Climate and Crop Production Systems

D. Wynne Thorne

CLIMATE COMPONENTS

Climate is of basic importance in determining the suitability of land areas for agriculture, including suitability for different crops and the relative merits of various types of farming systems. The three most important components of climate from the standpoint of plant growth are temperature, precipitation, and light. Special climatic variables such as wind, hurricanes, and hailstorms may be crucial, however, in specific areas.

Temperature

The major climatic zones of the world usually have been defined by temperature differences. In the tropical zone near the equator and at sea level, temperatures are quite uniform throughout the year, averaging near 25°C for all months. From this region the mean annual temperature decreases with increasing north or south latitudes.

Altitude produces temperature zonations which are highly important in the distribution of natural plant species as well as in crop adaptations. The mean annual temperature decreases about 6°C for each 1000 m of elevation (3.5°F for each 1000 ft). In the tropical latitudes crops requiring high mean temperatures and frost-free climates, such as rubber, cacao, and bananas are replaced at elevations where light frosts occasionally occur by subtropical crops such as rice, citrus, and sugar cane. At elevations as high as 1200 to 1800 m, where temperature ranges resemble those at low elevations at 35° to 45° latitude, temperate zone crops become preferable.

Latitude and altitude affect the length of growing season and the dates of spring plantings. In the central U.S. (35° to 40° N) spring seeding is retarded about 4 days for each increase of 1° in latitude, for each increase of 5° in longitude, and for each 100 m increase in elevation.

Often the length of the frost-free season determines the crops that can be successfully grown. Regions having short growing seasons are handicapped because few crops can mature in the available time. Growing seasons with less than 110 frost-free days are restricted primarily to livestock grazing and to such crops as forages and some cereal grains.

While length of frost-free season gives clues to the kinds of crops to be grown in an area, it tells little of the temperature regimes during the growing season. Information is needed as to ranges in minimum and maximum temperatures. Data on minimum, maximum, and optimum temperatures for each crop provide valuable information as to temperature-plant compatabilities. Although attempts to establish precise cardinal temperatures have been inconclusive, there is an optimum as well as a maximum and minimum range for each crop variety under local sets of environmental conditions.

Precipitation

Under most climatic situations natural precipitation is seldom optimum for the production of farm crops. Even the ideal for a given crop may vary with soil characteristics. A general need is for periodic rains that will fill the soil root zone to near field capacity and come at regular intervals before the soil dries to a point where the growing crop suffers reduced growth from moisture stress. In the classification of climate for agricultural production, therefore, the distribution of rainfall week by week may be as important as the total quantity. Also important are the rates of precipitation. That coming in heavy downpours may be largely lost by runoff and may cause serious soil erosion. There are also periods in which excessive precipitation may be harmful to crops or farm operations, as at harvest time when rains may delay operations or cause damage to the crops.

Frequency of rainfall is usually expressed as the number of precipitation days per month or per year. Various indexes are used to indicate the adequacy of precipitation. Thornthwaite (1948) proposed the concept of potential evapotranspiration (PET), the amount of water that would be evaporated from the soil and transpired from vegetation if it were available, as a basis for climate classification. Thornthwaite used the ratio of precipitation (P) to potential evapotranspiration to calculate the monthly moisture index (Im). If storage of water in the soil remains relatively constant, the formula becomes:

$$Im = 100 \frac{P}{PET} - 1$$

The sum of the monthly Im values gives an annual moisture index. Climatic moisture zones are then differentiated on the basis of the total yearly moisture index, which ranges from 100 for the most humid climate, to 0 to 20 for the moist sub-humid zone, to a −100 for the arid zone.

In most systems of world climate relationships between precipitation and evapotranspiration, such as P−PET, are used to distinguish between the moisture adequacy for different areas. However, where Köppen, Thornthwaite, and others separate zones on the basis of yearly average moisture indexes or total precipitation values, Troll bases his climatic zones on the basis of numbers of humid and arid months per year. According to Kampen and Krantz (1976), Troll's system is better suited to tropical conditions. It is being used extensively for evaluating crop production potentials (Pres. Sci. Advis. Comm. 1967).

Obviously, the variability in soils, precipitation patterns, slope, and plant cover within major climatic zones are great, and so detailed studies of climatic variables and soil characteristics and their interrelationships must be made for each area being considered. Such an analysis will be considered later.

Light

Within climatic zones two aspects of light are important to the growth of plants. These include the effects of clouds in decreasing light intensity and the effects of latitude on the angle of incidence of the sun's rays and on differences in length of daylight. In Southeast Asia there is general acceptance that rice yields for the monsoon season crop are commonly at least 25% less than during the drier season with more sunlight. Sugar plantations near Kailua in Hawaii are reported to receive a heavy rainfall and have cloudy weather and only about 40% as much average light intensity as at Waipio. With similar soil, moisture, and fertility conditions sugar cane yields at Kailua were 146 t per hectare while under the better light conditions at Waipio the yields were 300 t. Experiments with wheat have also shown yields to increase with increasing light intensity until CO_2 in the air becomes a limiting factor.

Because of light absorption by water vapor, light intensity averages less in humid than in arid climates. In areas or seasons where clouds and fogs are prevalent, light may be reduced to as little as 4% of its potential intensity. There is some evidence, however, that in the tropics photosynthesis in many crops increases with light intensity only up to about 30% of full sunlight.

The second light factor, length of day, varies with latitude and season. Daylength on June 21 ranges from 12 hr at the equator to 15 hr at 40° N, to 19 hr at 60° N and 24 hr near the North Pole. At higher latitudes beyond the tropics, length of day increases to the summer solstice and then decreases. Plants respond to differing lengths of day and night. Plants which require short days for flowering are called short-day plants. The dividing point between short- and long-day plants is taken to be 12 to 14 hr. Long-day plants include cereal grains, potato, timothy, and sweet clover. Short-day plants include soybeans, millet, hemp, and some varieties of tobacco. Tomato, early peas, sweet potatoes, pineapple, apple, cotton, and squash are considered day-neutral. Through selective breeding and development, plants have been adapted to differences in daylength. A present emphasis in plant breeding, particularly for wheat and rice, has been to develop varieties relatively neutral to daylength. Such breeding programs have increased variety adaptation to extensive differences in latitude.

Latitude is also important in determining the angle at which solar rays strike the earth. Consequently, light intensity is greatest at the equator and decreases progressively toward the poles. Light is scattered or absorbed by moisture, dust, smoke, and gases in the air and, therefore, light intensity usually increases with elevation. Water vapor absorbs considerable infrared light and ozone layers absorb ultraviolet. Mountain tops at 3000 m above sea level are exposed to about 10,000 lux brighter light than is present at sea level.

Soil and Climate

Soil characteristics are related to broad differences in climate, but are modified by parent materials, native vegetation under which they developed, and by such local conditions as topography and microclimate. For agricultural operations the moisture characteristics of soils, including infiltration rates, permeability, and water holding capacity, are of primary importance. Also of importance are the plant nutrient supplies in forms available to plants.

In an area with similar climate, soil becomes an important factor in determining which farm practices are best suited to good yields of crops. In consequence, a detailed map of soils and a knowledge of their physical and chemical properties constitute a necessary base for devising details of farming systems.

SEASONAL CLIMATES

Most agriculturally oriented classifications of climate (Meigs 1953; Köppen 1931; Troll 1965) identify five major zones based on annual temperature means and annual ranges in temperature. These climatic zones, as described by Troll (1965), include the Polar and Subpolar, the Cold Temperate Boreal (northern), the Cool Temperate, the Warm Temperate Subtropical, and the Tropical. Where agriculture is involved the quantity and distribution of precipitation is also of primary importance. For the purpose of serving as a basis for analyzing the major crop production farming systems of the world, the last three climatic zones are divided on the basis of adequacy of precipitation for plant use during the available growing season. These divisions are made by reapportioning Troll's 33 climatic types among 9 climatic zones instead of the 5 zones used by Troll. Some of the principal characteristics of these resulting nine zones that pertain to their agricultural production characteristics and potentials are described.

The general geographic locations of the various climatic zones are shown in Fig. 2.1. The numbers of hectares of potentially arable land in each zone by continents are given in Table 2.1.

I. Polar and Subpolar Zones

These areas in the vicinity of the two poles and at high elevations in other parts of the world are dominated by inhospitable cold climates. The areas include polar ice caps, ice deserts, polar frost debris, and regions with limited biological productivity such as tundra and subpolar tussock grass and moors. The warmest months in the more favorable areas of the zone have means of 5° to 12°C.

The zones cover about 560 million hectares of land, or about 4.2% of the earth's surface. However, temperature conditions are so severe as to prevent plant growth except for sparse lichens, mosses, and tussock grasses which provide forage for caribou and musk oxen. Consequently, the land in this climate is unsuited for cultivation and none of it is classed as potentially arable.

II. Cold-Temperate Boreal Zone

This zone extends northward to the polar zone from about 50° N in North America and from approximately 55° N in Europe and Asia. The zone is not important in the southern hemisphere. The zone includes three climatic types: oceanic boreal climates, continental boreal cli-

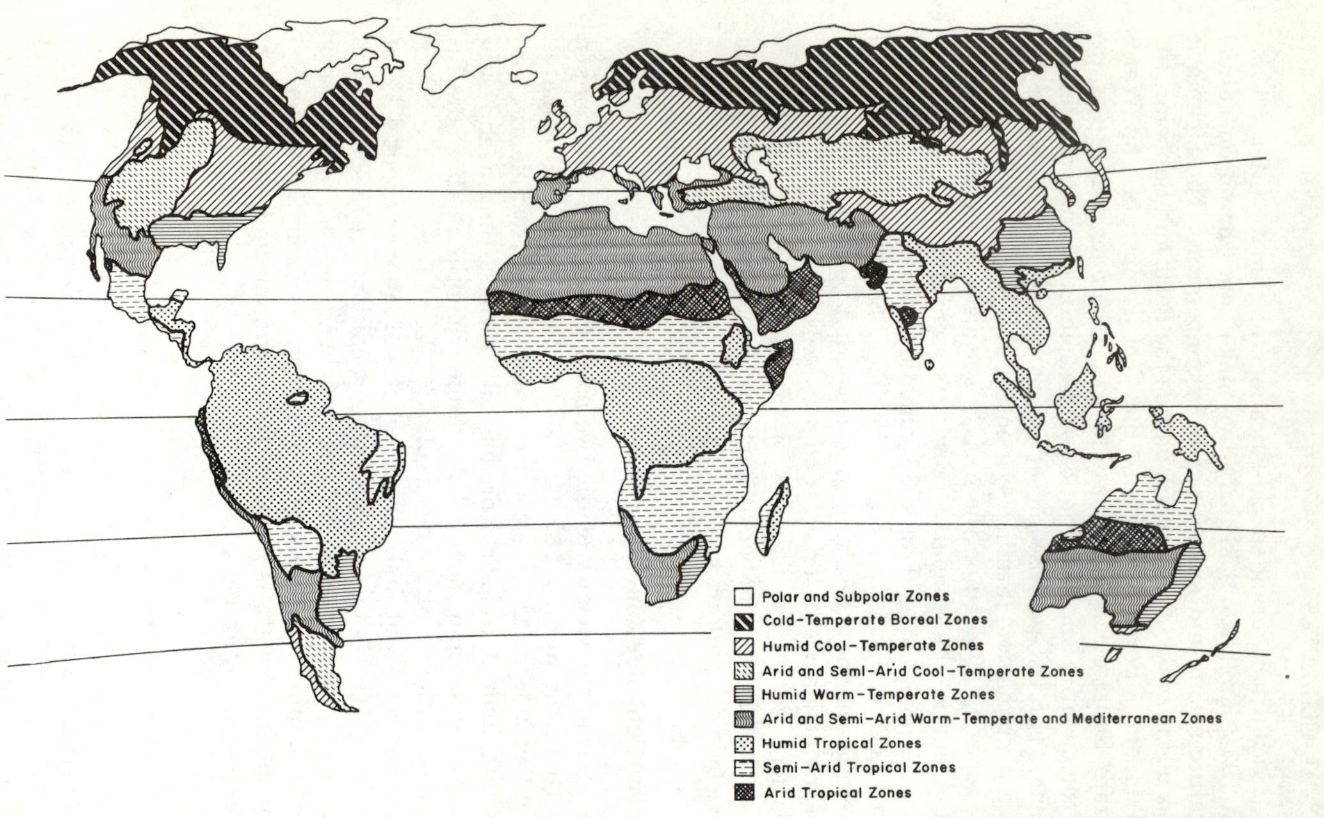

FIG. 2.1. MAJOR CLIMATIC ZONES OF THE WORLD

TABLE 2.1. POTENTIALLY ARABLE LAND IN THE MAJOR CLIMATIC ZONES AND THE DIFFERENT CONTINENTS

		Months of Adequate Climate		Million Ha							
	Climatic Zones	Temperature	Humidity	Africa	Asia	Australia	Europe	North America	South America	USSR	Total
I.	Polar and Subpolar	0	–								00
II.	Cold-temperate Boreal	2–4	–		1.6		20.2	24.3		24.3	52.2
IIIA.	Humid Cool-temperate	4–8	–		109.4	1.2	135.7	189.5	8.5	247.1	691.4
IIIB.	Arid and Semi-arid Cool-temperate	4–6	–		30.4			86.7	12.2	85.1	214.4
IVA.	Humid Warm-temperate	4	6–8	12.2	60.8	40.5	85.1	56.7	56.7		255.3
IVB.	Arid and Semi-arid Warm-temperate	8+	4	56.7	89.1	64.8	34.4	32.4	30.8		308.2
VA.	Humid Tropical	12	7+	332.1	206.6	4.1		28.4	467.4		1038.6
VB.	Semi-arid Tropical	12	2–7	320.0	125.6	36.5		24.3	24.3		530.7
VC.	Arid Tropical	12	2	16.2	8.1	4.1		0.4	1.6		30.4
				737.2	631.6	151.2	172.1	471.1	601.5	356.5	3121.2

Source: Adapted from Pres. Sci. Advis. Comm. (1967); Troll (1965).

mates, and highly continental boreal climates. The zone has fluctuations in mean annual temperatures at different locations ranging from 13°C in the oceanic environment to 40°C under the most continental conditions. The average temperature for the warmest month varies from 10° to 20°C. The native vegetation is primarily coniferous forest.

The Cold Temperate Zone covers about 1968 million hectares, or nearly 15% of the earth's surface. However, except for small coastal areas where oceanic effects modify the temperature, the length of growing season is too short for the production of cultivated crops. Only about 52 million hectares, or about 1.7% of the total for the world, have been classified as potentially arable (Table 2.1). Crop production is limited to some forage crops and cool season food crops during the summer months when the days are long and the intensity of solar radiation is highest.

III. The Cool–Temperate Zones

These major agriculturally important zones (IIIA and IIIB) comprise about 22% of the total land area of the world and about 29% of the potentially arable land. Most of this land is in the northern hemisphere where it occurs largely in a band lying between 35° and 55° N. The zone includes important agricultural production areas in Europe, Asia, North America, and USSR and smaller areas in South America and Australia. Temperature conditions throughout the zones provide a growing season sufficiently long for a wide range of annual and frost-resistant perennial crops. This major region has been divided into two zones with the humid areas comprising the zone receiving adequate precipitation for crop production throughout the growing season, and the semi-arid and arid areas including lands where adjustments to water stress become an important aspect of crop production practices.

IIIA. Humid Cool–Temperate Zones

These zones include eight of Troll's climatic types. The growing season ranges from 4 to 8 months, with the 8-month season being restricted to a limited region with an oceanic environment. Precipitation makes crop production possible throughout the growing season, although periodic gaps in rainfall create frequent moisture stress problems in most climatic types. The native vegetation is primarily woodland of deciduous and mixed types. The zones include about 691 million hectares of potentially arable land. A major portion of this is now being cropped. In some parts of the region multiple cropping is feasible by growing a cool season crop and a warm season crop in sequence.

These zones have considerable potential for increased crop production through improved technology on land now being cropped. The soils range from neutral to highly acid in reaction. Lime treatments are needed in much of the zone for acid-sensitive crops. Also generally needed are liberal fertilizer applications for high levels of crop production. The principal order of soils in the zone include Spodosols (highly leached podzolized soils), Alfisols (less weathered podzolic and associated soils), some traditional Ultisols (includes red and yellow podzolic soils), associated Entisols (azonal group), and Histosols (bog soils).

IIIB. Semi-Arid and Arid Cool-Temperate Zones

These zones include primarily soils that have developed under native steppe and desert shrub types of vegetation. Included in these areas are 214 million hectares of potentially arable land with the most extensive parts being in the USSR and the northwestern part of North America. The soil orders occurring in the zones include the Mollisols (includes Chernozem, Chestnut, and associated solonetz soils), Aridisols (desert soils), and associated Entisols. These soils are usually well supplied with many of the essential mineral elements and are usually neutral to alkaline in reaction. Salt accumulations are common hazards to successful farming. Nitrogen is generally inadequate in these soils for high yields of crops and the supply of available phosphorus is often low.

The region is seasonally dry during part to all of the growing season. Lack of sufficient moisture during the growing season restricts the choice of crops or limits land use to grazing or to irrigation agriculture where water supplies are available. Wheat is an extensively planted crop under rain-fed conditions. Water conservation practices become an important consideration in all parts of the zone.

IV. Warm-Temperate Zones

These major zones include about 21% of the world's land area and 18% of its potentially arable land. They dominate the two large land areas lying between 20° and 35° N and 20° and 35° S. Temperatures are sufficiently high to permit vegetative growth throughout the entire year in all but the high latitude margins of the zone and at higher elevations. The frost-free period usually exceeds 200 days and the winters are mild. The zones are often referred to as subtropical.

The amount and seasonal distribution of precipitation are the dominant climatic factors that determine agricultural productivity, the crops that can be grown, and the general types of farming systems. Accordingly, the region is divided into two zones, one being humid in climate and the other including arid and semi-arid climates where precipitation limits cropping practices, except where irrigation is developed.

IVA. Humid Warm-Temperate Zones

These zones are broadly represented in all of the continents except Europe and USSR. Included in the zones are about 255 million hectares of potentially arable land, much of which is now cultivated. The zones include three of Troll's climatic types. The native vegetation includes monsoon woods, wooded steppe, subtropical high grassland, and subtropical humid forests of laurel and conifers. Precipitation is adequate for crop production for six or more months of the year and usually permits a cool season or winter crop and a warm season summer crop. Rather intensive multiple cropping is possible in many locations.

The zones include a range of soils that are neutral to acid in reaction and represent such orders as Mollisols, Ultisols, Entisols, and Vertisols. Fertilizers and lime are commonly needed for high crop yields. Erosion control and drainage frequently need attention in farming systems.

IVB. Semi-Arid and Arid Warm-Temperate Zones

These zones include more than 300 million hectares of potentially arable soils, or nearly 10% of the world total. Included are such soil orders as Mollisols, Aridisols, and associated Entisols. The soils are usually neutral to alkaline in reaction and in the more arid and irrigated areas salinity and sodium accumulations are common problems. The great deserts of Africa, Asia, Australia, North America, the Middle East, and South America are wholly or in part in these zones.

Wheat and grain sorghum are grown rather extensively under the more favorable rainfed conditions. Irrigation is required for full crop production during the growing season and for essentially all crop production in the arid areas. The zones include much of the dry summer Mediterranean area where winter and early summer grain and seed legumes are frequently grown.

V. The Tropical Zones

This large area contains nearly half of the world's potentially arable land and a total of nearly 5 billion hectares of land, amounting to 38% of the land on earth. The three tropical zones also contain appreciably more than half of the potentially arable land of the world that has not been developed for cropping. The region also includes much of the traditional shifting cultivation, small-farm type of crop production. The three tropical zones include a substantial part of the current world areas of food deficit, but they also include the areas that have the greatest potential for increasing food production. Unfortunately, however, the human population is concentrated in Asia while the undeveloped land is largely in Africa and in Latin America.

The zone occurs largely between 20° N and 20° S. In such latitudes temperatures show little or no seasonal variation and are favorable for continuous crop growth throughout the year. The agricultural productivity of the region is determined, not by temperature constraints, but by moisture supply and soil characteristics. The wettest zone is generally within 5° N and 5° S and rainfall tends to decrease with further distances from the equator.

VA. The Humid Tropical Zone

The humid tropics include a narrow belt of land near the equator and adjacent areas where there is in excess of 1500 mm of rainfall per year and in which 7 or more months are classed as humid. However, in much of the zone there are distinctive dry seasons of up to five months. This large area contains an estimated one billion hectares of potentially arable land, or about one-third of the world's total. Less than half of this is now being cultivated. With good water supply and an abundance of arable land, reinforced by a full 12-month growing season, this is a region of great potential for increased food production.

The soils of the region are highly oxidized and leached of many plant nutrients. In decreasing order of weathering the major soil orders are Oxisols, Ultisols, and Inceptisols. Since the soils are commonly highly leached and oxidized when placed under crop production they become rapidly depleted of supplies of available plant nutrients. Farming systems in the humid tropics must give careful attention to improving soil fertility, the control of soil erosion, and managing water to reduce problems of drought in dry seasons and controlling excesses in the rainy seasons. Some parts of the zone have periods of thick cloud cover in which light may be a limiting factor in plant growth.

VB. Semi-Arid Tropical Zone

This climatic zone occurs in a wide belt across parts of South Asia, India, Africa, Northern Argentina, Northeast Brazil, and in Central Mexico. These areas contain about 530 million hectares of potentially arable land, or about 17% of the total in the world. Average yearly precipitation ranges from about 500 to 1500 mm. The humid months range from 2 to 7. About 95% of the annual precipitation occurs during the rainy season and the periods of rain during the rainy season are often highly variable. Also, the mean annual rainfall intensities are high, averaging 2 to 4 times greater than in temperate regions. In consequence, water intake rates by soils are often too slow to absorb precipitation and runoff water and soil erosion are general problems.

The principal soil orders in the semi-arid tropics are the Alfisols which are the most weathered; the Vertisols, often occurring as high-swelling black clay soils; and Entisols, or undeveloped soils. Managing the soils and crops to make effective use of the variable precipitation is one of the central focuses of farming systems. Also, the rather low state of fertility of many soils requires soil-building practices and a careful use of fertilizers.

VC. The Arid Tropical Zone

This zone is relatively small in total area, aggregating only 30 million hectares of potential arable land or about 1% of the world's total. Effective crop production requires irrigation, using water imported from moderately distant areas with more plentiful precipitation. The limited quantity of precipitation, less than 500 mm per year, can support but one crop under the most favorable conditions. In India alone, however, this zone covers 15% of the geographic area and has a human population of over 20 million and a livestock population of 23 million.

The soil orders in the zone are primarily of the Aridisols, or dry soils, with some Alfisols and Mollisols (soils with humus-rich surface horizons).

Crop production under rainfed conditions is limited to the more favorable precipitation and soil sites and is intensively oriented toward conservation and efficient use of water. In some parts of this zone some interesting moisture conservation and crop management practices are being developed. These water conservation practices are closely similar in the arid tropics and the arid warm temperate areas.

REFERENCES

COMM. ON CLIMATE AND WEATHER. 1976. Climate and Food. National Academy of Sciences-National Research Council, Washington, D.C.

COMM. ON TROP. SOILS. 1972. Soils of the Humid Tropics. National Academy of Sciences-National Research Council, Washington, D.C.

CRITCHFIELD, H.J. 1974. General Climatology. Prentice-Hall, Englewood Cliffs, New Jersey.

KAMPEN, J., and KRANTZ, B.A. 1976. The farming systems research program. International Crops Research Institute for the Semi-arid Tropics, Hyderabad, India.

KÖPPEN, W. 1931. Outline of Climatology. Walter de Gryter Co., Berlin.

MEIGS, P. 1953. World distribution of arid and semi-arid humoclimates. *In* Review of Research on Arid Zone Hydrology. UNESCO, Paris.

PRES. SCI. ADVIS. COMM. 1967. The World Food Problem, Vol. II. The White House, Washington, D.C.

SOIL SURV. STAFF. 1975. Soil taxonomy. U.S. Dep. Agric., Agric. Handb. 436.

THORNTHWAITE, C.W. 1948. An approach to a rational classification of climate. Geogr. Rev. 38, 55-94.

TROLL, C. 1965. Seasonal climates of the earth. *In* World Maps of Climatology, 2nd Edition. H. E. Landsberg *et al.* (Editors). Springer-Verlag, New York.

TROLL, C., and PAFFEN, K.H. 1965. Seasonal climates of the earth. *In* World Maps of Climatology, 2nd Edition. H. E. Landsberg *et al.* (Editors). Springer-Verlag, New York.

WIESIE, C. P. 1962. Crop Adaptation and Distribution. W.H. Freeman and Co., San Francisco.

3

Soils

John J. Hassett

The success or failure of crop production is in large part determined by the general make-up of the soil. The soil provides for many of the physiological needs of the plant. Water stored in the soil is taken up by the plant in large quantities in response to demands created by transpiration and to a lesser extent by the metabolic needs of the plant. Hence, soil factors which affect the water holding capacity of the soil as well as factors which affect root growth need to be examined as they affect crop production.

The essential nutrients needed by the plant to grow and reproduce come predominantly from the soil. Soils that are highly weathered have considerably different fertilizer needs than younger, less highly weathered soils. The soil also provides mechanical support that allows the plant to stand upright. Soil temperature and aeration affect crop production, mainly through their effects on seed germination and root growth.

It is generally accepted that most of the water and nutrients used by the plant come from that portion of the soil that the plant roots are in direct contact with; very little water or nutrients move into the root zone. The implication of this is that a plant with a well developed root system has a larger supply of water and nutrients to draw from than a plant with a shallow or weakly developed root system. Therefore, the soil factors which determine the suitability of the soil as an environment for plant roots are major factors in determining the success or failure of crop production.

The recently developed Comprehensive Soil Classification System and the Cooperative Soil Survey reports provide valuable information about the differences between soils and their suitabilities for crop production. In the classification system ten major soil orders are recognized, soil orders being the broadest category. Soils in a given order have similar morphology and often have formed under similar circumstances. As an example, soils developed under prairie vegetation, where accumulation of organic matter in the surface horizons (epipedons)[1] is a common property, are usually classified in the order Mollisol.

Orders are subdivided into suborders by major differences in climatic environment, wetness of profile or other major profile characteristics. Suborders are divided into great groups by similarities in profile characteristics. Great groups are further divided into subgroups and then into families and eventually into series.

In the classification system each succeeding name in going from order to subgroup is formed by taking the name or a portion of the name of the higher level and adding a descriptive formative element to it. An example of this nomenclature is as follows:

Order—Mollisol
Suborder—Aquoll
Great group—Haplaquoll
Subgroup—Typic Haplaquoll
Family—Fine-silty, mixed, noncalcareous, mexic
Series—Drummer

The resulting names provide useful information about the major properties of the soil. The family name provides textural, clay mineralogy, pH and climatic information about soil. As an example, fine-silty, mixed, noncalcareous, mexic implies that the soil is fairly fine textured, has a mixture of different types of clay minerals present, is free of carbonates, hence pH is less than 7, and occurs in the central U.S. The series name is equivalent to the common name of the soil. It provides no direct information about the soil, but it is the most commonly used name by farmers, soil scientists, and other individuals.

Texture is one of the more important physical properties of a soil. Texture refers to the relative proportion of sand (particles >0.05 mm in diameter), silt (0.002 to 0.05 mm in diameter) and clay (<0.002 mm in diameter) in the soil. As the particles in the soil become smaller the ratio of the particle's surface area to its mass becomes much larger. Since the adsorption of water and nutrients is primarily on mineral surfaces, soils which have a predominance of small particles (clay) have much greater water and nutrient holding capacities than sandy soils where large particles predominate.

---

[1]Epipe = surface, pedon = soil; hence, surface soil.

SOIL ORDERS

Entisols

Entisols are mineral soils with only the beginnings of horizon development. These soils have only weakly developed surface horizons with no subsurface horizon development. They are found in a wide variety of climatic conditions, generally on recently formed or exposed parent materials or on parent materials such as sand dunes which are very resistant to soil development. The order name is formed by adding "-ent" from rec<u>ent</u> to "sol" which means soil.

Inceptisols

These are mineral soils which are more highly developed than the Entisols. These soils have thin, light-colored surface horizons with weakly developed subsurface horizons (cambic). Also included in this order are soils which are not highly developed due to the extreme coldness of their climates. Inceptisols are found in all climatic regions and represent a degree of profile development that is slightly more advanced than the Entisols.

Alfisols

Alfisols are mineral soils which have a surface horizon which is too thin or too light-colored (Ochric epipedon) to be classified as a Mollic epipedon (thick dark-colored—not highly leached surface horizon). These soils have a highly developed subsoil horizon (argillic or natric) in which there has been substantial clay movement out of the surface into the subsurface horizon. The subsoil horizons of these soils have been leached but are not as highly leached as in the order Ultisol. These soils are typical of the forest soils of the central U.S., although not all Alfisols are forest soils.

Ultisols

Ultisols are mineral soils which are more highly weathered than the order Alfisol. These soils have either the thin light-colored surface horizon of the Alfisol (Ochric epipedon) or a highly leached dark-colored surface horizon (Umbric epipedon). The main distinguishing characteristic is an Argillic horizon which is more highly leached and hence more acidic than the Argillic horizons of the order Alfisol. The Argillic horizon of the order Ultisol has less than 35% base saturation while the Argillic horizon of the order Alfisol has greater than 35% base saturation. This illustrates the more intensely weathered profile of the Ultisol, due to its warmer and more humid climate. These soils are typical of the southeastern part of the U.S. Ultisols are prominent in subtropical areas of the world.

Spodisols

These are mineral soils which represent an extreme amount of weathering in a cool moist climate. These soils are characterized by a subsoil accumulation of organic matter and oxides of aluminum with or without iron oxides (spodic horizon). This order of soils is formed mainly on coarse-textured, highly leached acid parent materials under forest vegetation. These soils are commonly found in north central and northeastern U.S. The zone (albic-A2) directly above the spodic horizon generally is quite light in color due to the extreme amount of leaching and weathering to which these soils have been subjected.

Oxisols

These are mineral soils of tropical regions which have been subjected to intense weathering and leaching due to their presence in the hot, wet tropical climates. These soils have a subsoil accumulation of resistant minerals (oxic horizon) that is the result of the less resistant minerals being removed by extreme weathering. These are the most highly weathered soils of the ten Orders. Depth of weathering in the Oxisols may be 50 or more feet deep. These soils tend to occupy old landscapes and occur mainly in the tropics.

Aridisols

Aridisols are mineral soils formed in dry climates, where dryness of the climate has restricted normal soil development. These soils have an ochric epipedon with some subsoil development and are distinguished from the orders Entisols and Inceptisols by their presence in a dry climate which restricts normal soil development. These soils may have accumulations of soluble salts (salic), gypsum (gypsic), or calcium carbonate (calcic) due to the restricted amount of leaching in the dry climates. These soils are found in most of the arid regions of the world (deserts and near-deserts).

Vertisols

Vertisols are mineral soils which are characterized by a high content of swelling-type clay minerals (Montmorillonites). In dry periods these soils develop wide, deep cracks. The surface material can and does fall into the cracks, causing the surface few feet to invert or turn over after a number of years. These soils tend to form on old marine deposits that are easily weathered to swelling-type clays.

Mollisols

These are mineral soils which have a thick dark-colored, not highly leached surface horizon (mollic epipedon). These soils generally form under prairie vegetation. Mollisols may have no subsoil development, weak subsoil development (cambic), or a substantial amount of subsoil development (argillic). Soils of this order are characterized by a thick dark-colored, surface horizon with neither the surface nor subsurface horizons showing the strong acidity due to leaching such as found with the order Ultisol.

Histisols

Histisols are organic soils that differ from the other nine orders in that the properties of the soil are due primarily to the organic matter, not the mineral matter. These soils form under conditions that favor the accumulation rather than the decomposition of organic matter. Conditions that favor organic matter accumulation are wet, boggy, or swampy conditions.

Relationships Among and Within Soil Orders

Figure 3.1 shows the general relationship of several of the orders, proceeding from little soil development (no subsoil horizons) to a high degree of soil development (argillic or oxic subsoil horizons).

The orders Mollisol and Aridisol can have a variety of subsoil developments ranging from no subsoil to cambic subsoil development for Aridisols and from no subsoil to argillic subsoil development for the Mollisols. The orders Vertisol and Histisol are orders which show unique parent materials, swelling clays, or high organic matter contents without differences in subsoil development. Hence, these orders do not fall nicely into a development sequence.

Suborders of the ten orders show either the specific macro climate the soil occurs in, if the order occurs in widely different climates, or show major profile characteristics. To illustrate this let's examine the orders Alfisol and Mollisol.

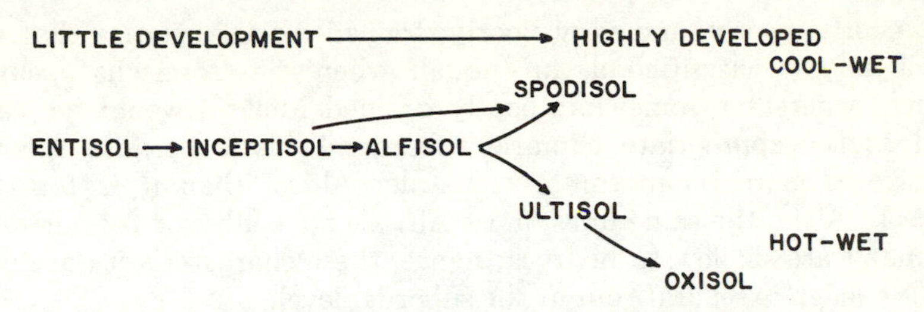

FIG. 3.1. RELATIONSHIP OF SELECTED ORDERS TO MACRO CLIMATE AND DEGREE OF SOIL DEVELOPMENT

Order	*Suborder*
1. Mollisol	1.1 Alboll
	1.2 Aquoll
	1.3 Rendoll
	1.4 Boroll
	1.5 Udoll
	1.6 Ustoll
	1.7 Xeroll
2. Alfisol	2.1 Aqualf
	2.2 Boralf
	2.3 Udalf
	2.4 Ustalf
	2.5 Xeralf

In both orders four climatic formative elements are used:

(1) Bor = a cool climate
(2) Ud = a humid climate
(3) Ust = a drier climate than Ud, usually hot in summer
(4) Xer = a dry climate with an annual dry season.

In the order Mollisol three additional non-climatic suborders are possible and these take preference over the climatic suborders. Hence, a poorly or very poorly drained Mollisol is classified as an Aquoll regardless of the macro climate it exists in. A Mollisol with an albic subsoil horizon (one from which clay and free iron oxides have been removed), which is a unique condition for Mollisols, is classified as an Alboll. A Mollisol that is shallow to limestone with large unweathered limestone chunks within the profile is classified as a Rendoll. In the order Alfisol only one non-climatic suborder is possible and that is Aqualfs, somewhat poorly, poorly, and very poorly drained Alfisols. Notice that an Alfisol may be classified as an Aqualf when it is somewhat poorly drained; whereas a somewhat poorly drained Mollisol would be classified as the appropriate climatic suborder. This is because "good" drainage is more the norm of the order Alfisol than it is for the Mollisols. Along the same lines, most Alfisols have albic or A2 horizons and many are shallow to bedrock; hence, these characteristics are not unique enough to separate out at the suborder level.

Interpreting Soil Characteristics from Names

To illustrate how the suborder, great group, and subgroup names can be used to supply useful information about specific soils consider the following examples.

Example 1. Bloomfield - Psammentic Hapludalf

alf =alfisol
ud =humid climate, drainage better than somewhat poorly drained
hapl =a weakly developed alfisol-minimal subsoil development
Psammentic =Psamm stands for a sandy soil, entic for entisol-like. The whole term implies a weakly developed entisol-like soil due to resistance of the parent material, sand.

Example 2. Flanagan-Aquic Argiudoll

oll =Mollisol
ud =humid climate, not poorly or very poorly drained, not shallow to limestone, not containing an albic (A2) horizon
argi =a well developed subsoil horizon in which substantial clay has moved out of the surface into the subsurface in contrast to a cambic or no subsoil horizon
aquic =shows depressed drainage but not sufficient for aqu suborder. Hence, this soil is somewhat poorly drained. A well drained or moderately well drained soil would be typic.

Example 3. Wabash-Vertic Haplaquoll

oll =Mollisol

aqu = poorly or very poorly drained
hapl = weakly developed subsoil horizon (cambic)
vertic = a high concentration of swelling type clay minerals (Vertisol-like). This high concentration of swelling type clay could be either the cause of the poor drainage or at least may hinder any attempt to establish internal drainage.

The discussion of the comprehensive soil classification system in this chapter has been kept general and fairly free of specific terminology and definitions. The discussion is intended to introduce the reader to the existence of the system and its usefulness. A more complete discussion can be found in specific soil texts and in Soil Surv. Staff (1975).

An additional source of valuable information about soils and their potential use for crop production are the county soil surveys published by the United States Department of Agriculture Soil Conservation Service, usually in cooperation with the State Agricultural Experiment Stations.

A modern soil survey usually consists of several parts: (1) detailed maps of the county showing the individual soil series that occur in a given area of land; (2) a discussion of the general nature of the county, climate, physiography, relief, and drainage; (3) a discussion of how the survey was made; (4) a general soil map showing the location of the important soil associations (groups of related soils) in the county; (5) a detailed description of all of the soil series occurring in the county; (6) a discussion of the formation and classification of the soils; and (7) a discussion of the use and management of the soils.

SOIL CAPABILITY GROUPINGS

A portion of the soil survey report, in addition to the classification of the soil that is useful for judging the soil's general usefulness for crop production, is its capability grouping. Capability groupings give in a general way the suitability of soils for various kinds of row and forage crops. The soils are grouped according to their restrictions when used to produce crops, the risk that cropping has for the soil from damage such as excessive erosion, and the way that the soils respond to various soil treatments.

The capability grouping of a soil does not take into account major and often expensive soil treatments such as land forming that would change the slope, depth, or other important restrictive characteristics of the soil. It does allow for soil change due to normal agricultural practices such as liming to correct for excess soil acidity or fertilization that generally improves the tilth of the soil. The grouping also does not take into consideration possible but unlikely major reclamation projects such as the correction of a Natric (Sodic) horizon and does not apply to crops which require special management such as rice or other wetland or horticultural crops.

In the capability grouping system, the kinds of soils are grouped at two or more levels: the capability class and the subclass. Capability classes are designated by Roman numerals I through VIII. The Roman numerals indicate increasing limitations and fewer choices for practical field crop use:

Class I —Soils that have few limitations to restrict their use.

Class II —Soils that have moderate limitations that reduce the choice of plants or that require moderate conservation practices.

Class III —Soils which have severe limitations that reduce the choice of plants, require special conservation practices, or both.

Class IV —Soils that have very severe limitations that reduce the choice of plants, require very careful management, or both.

Class V —Soils which are subject to little or no erosion but have other limitations, impractical to remove, that limit their use largely to pasture, range, woodland, or wildlife habitat.

Class VI —Soils that have severe limitations that make them generally unsuited to cultivation and limit their use largely to pasture or range, woodland, or wildlife habitat.

Class VII —Soils that have very severe limitations that make them unsuited to cultivation and restrict their use largely to pasture or range, woodland, or wildlife habitat.

Class VIII —Soils and landforms which have limitations that preclude their use for crop production and restrict their use to recreation, wildlife habitat, water supply, or for aesthetic purposes.

Capability classes can be subdivided into subclasses within each class based on special limitations. Subclasses are designated by adding a small letter c, e, s or w to the Roman numeral; for example, Vw shows a class V soil in which the main restriction is wetness. Limitations of the subclasses are as follows:

c— shows that the chief limitation is climate that is too cold or too dry.

e— main limitation is erosion unless close growing plant cover is maintained.

s— soils that are limited mainly because it is shallow, droughty, or stony.

w— shows that excess water in or on the soil interferes with plant growth or cultivation; in some soils the wetness can be partly corrected by artificial drainage.

There are many problems which face modern crop production, but among the most common and often the most severe are those problems which are related to the inherent water and/or nutrient supplying capability of the soil. There are two main conditions which lead to soil having a low water supplying capability. First, the soil may have a low plant available water holding capacity due to its textural class. As an example, the soil may be sandy and have a low total as well as a low plant available water holding capacity. In the classification system the modifier "psamm" is often used to show the sandy nature of these soils. The second general type of condition that leads to a droughty soil is where some soil property restricts root development and hence restricts the amount of plant available water the plant roots are in contact with.

The chemical and/or the physical properties of the subsoil can restrict root development. In the southeastern portion of the U.S. and in much of the subtropical and tropical regions of the world where the orders Ultisol and Oxisol predominate, subsoil acidity can greatly reduce root growth. In these soils subsoil pH values are often less than 5.0. This strong acidity and the associated high levels of soluble aluminum greatly reduce root growth of many crops. In droughty years when subsoil moisture is critically needed by the crops, the restricted root growth and the subsequently reduced water uptake is manifested in reduced yields or crop failures. In some highly weathered soils, low subsoil fertility can also result in poor root growth and have the same effect on water and nutrient uptake and hence on yield.

Physical as well as chemical properties of the subsoil can reduce root growth. Physical conditions such as high bulk density subsoils or the presence of cemented zones or pans in the subsoil can actually prevent root penetration into the subsoil. The presence of cemented pans is often shown in the classification of the soil through use of formative elements such as: frag—showing the presence of a fragipan or dur—showing the presence of a durapan or plac—showing the presence of a thin pan.

Even the presence of large amounts of clay in the subsoil can restrict root growth. This is usually due to the poorly aerated conditions that are created by the slow movement of excess water through the high clay zone (clay pan). The suborder Aqualf often has its poor drainage due to the existence of such a clay pan.

REFERENCES

SOIL SURV. STAFF. 1975. Soil taxonomy. U.S. Dep. Agric., Agric Handb. *436*.

4

Soil–Water–Plant Relations

Marlowe D. Thorne

Soil water is one of the most important factors affecting crop production. The success or failure of a particular cropping system may well depend on the water management program followed. Water must be available in the soil to replenish that lost by evaporation from the soil and by transpiration of the crop throughout the season. Soil water also carries nutrients in solution for the growing crop and has significant effects on aeration and temperature conditions in the soil. The soil must be able to provide water as it is needed if maximum yield is to be obtained.

Seldom is the water or moisture content of a soil at the optimum value for maximum crop production. Usually there is either a deficiency or a surplus of moisture and crop production is reduced. It is difficult to determine the exact soil moisture content which is optimum and it is even more difficult to maintain a soil at a specified moisture content over a period of time. The ranges of soil moisture content over which many crops achieve highest yields are fairly well known. More is being learned about this from innumerable studies throughout the world.

MOISTURE RELATIONS OF SOILS

Figure 4.1 shows the distribution of moisture in an uncropped dry soil when water is applied to the surface. During and immediately after the application, the surface layer is at or near the saturation point. The major part of the wetted soil is at a moisture content called the *field capacity.* Below the zone at field capacity, moisture content decreases markedly with depth until the original soil moisture content is reached. A day or two after the application ends, the moisture content at the surface has dropped below saturation and the field capacity zone extends to a slightly greater depth.

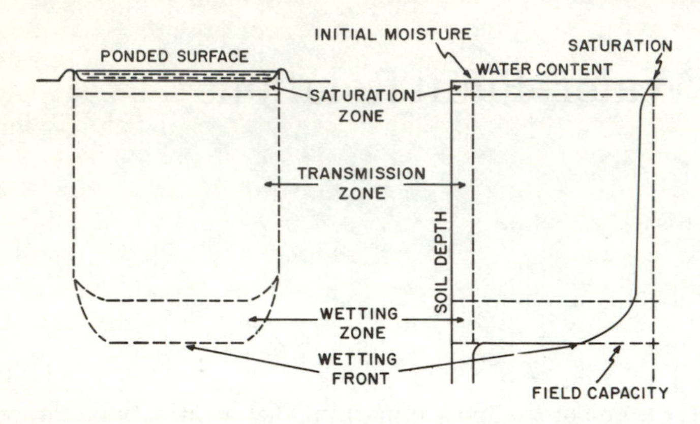

Adapted from Hillel (1971)

FIG. 4.1. THE INFILTRATION MOISTURE PROFILE

Left, a schematic section of the profile; Right, the water content vs depth curve.

If plants are growing in the soil they will extract moisture, and reduce the moisture content of the soil. Eventually the soil moisture drops so low that roots can no longer obtain the water and the plant wilts. The moisture content in the root zone at this point is termed the permanent *wilting point* or wilting percentage. The immediate soil surface may be at a moisture content less than the wilting point because of moisture loss by evaporation.

Energy Concept of Soil Moisture

Since soils vary widely in their ability to hold moisture, it is impossible to establish a direct relationship between plant response and the percentage of moisture in soil which applies to soils of different textures. For example, maize may wilt in a clay soil having 15% moisture, but a sandy soil may not be able to hold 15% moisture.

If the moisture content of a soil is expressed in terms of the energy status of that moisture, we have a measurement system in which we can more easily compare the availability of the moisture in soils of different textures. As water is squeezed from a sponge, increasingly greater effort must be expended to get the next bit of water out. While there are many differences between a soil and a sponge, they are similar in that as more and more water is removed (as by plants growing in the soil), increasingly greater effort is required to extract the next increment.

Procedures have been devised for determining the energy status of soil water and various units of measurement are used. The most commonly accepted unit at present is bars of suction. Suction is a negative pressure, so the higher the numerical value of suction, the lower the energy status of the water. When most soils are at field capacity, the soil moisture has an energy content between 0.1 and 0.3 bars of suction. When soils are at wilting point, the moisture has an energy content of 15 bars of suction.

Thus, if we know the energy content of the moisture in a soil, we immediately have valuable information about the availability of that moisture to plants. The relationship of suction to percentage of water in soils with different textures may be shown as in Fig. 4.2.

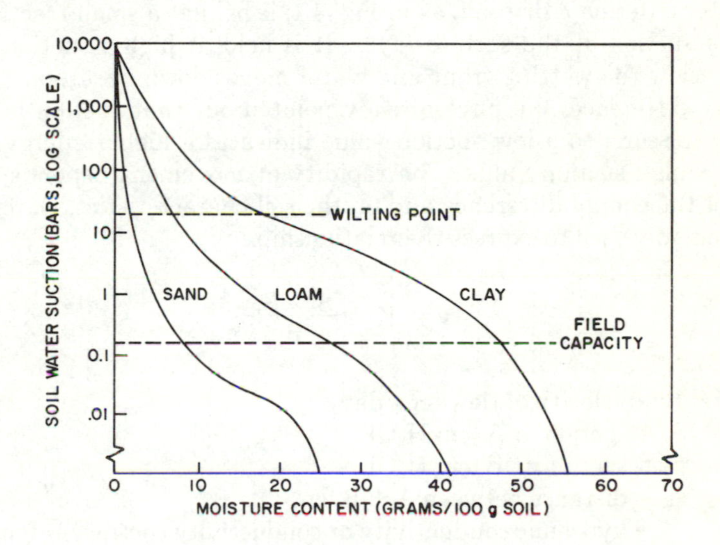

FIG. 4.2. RELATIONSHIP BETWEEN SOIL WATER SUCTION AND PERCENTAGE MOISTURE FOR THREE SOIL TEXTURES

Dotted lines show approximate suction values for two common moisture constants.

We can calculate the equivalent depth of available water per unit depth of soil from the following relationship:

$$\Theta_A = \frac{d_B \times l_S \times \Theta_S}{d_W}$$

Where: Θ_A = available soil moisture (cm/cm)
d_B = bulk density of soil (g/cm^3)

d_W = density of water (g/cm³)
l_S = depth of soil (cm)
Θ_S = percentage of soil moisture (g water/100g dry soil)

Water Movement in Soils

Water will move in a soil from one point to another if the water at the first point has higher energy status than the water at the second. This may be compared to water moving through a pipe from a point of high pressure toward a point of lower pressure. In the pipe the pressures are normally greater than atmospheric pressure. (We usually use atmospheric pressure as our zero point.) In unsaturated soils, however, the pressure is less than atmospheric and the energy has a negative value compared to our zero point.

Water entering a dry soil, as in Fig. 4.1, is held at a small fraction of a bar of suction at the surface layer. It is held at higher suction in the zone below the wetting front and water moves down in response to the energy difference. We have already pointed out that suction is a negative pressure, so a low suction value indicates a higher energy status than a high suction value. The rapidity of movement depends on the size of the energy difference and on the soil characteristics. Darcy's law is commonly used to express the relationship.

$$v = K \frac{S_A - S_B}{d}$$

Where: v = velocity of flow (cm/day)
S_A = suction at A (cm H_2O)
S_B = suction at B (cm H_2O)
d = distance between A + B (cm)
K = hydraulic conductivity or conductivity coefficient (cm/day)

Since S_A and S_B are both really negative values, the numerator becomes positive when S_B is numerically greater than S_A. The value of K in unsaturated soils changes as the suction changes. As soil moisture content decreases and suction values increase in numerical size, the value of K decreases. This relationship is shown in Fig. 4.3. At high suction values (low soil moisture contents), K may be so low that water movement is of little practical value in supplying water to a growing crop.

If water is applied to the surface of a soil (by rain or irrigation) much faster than it can enter the soil and be transmitted downward, the

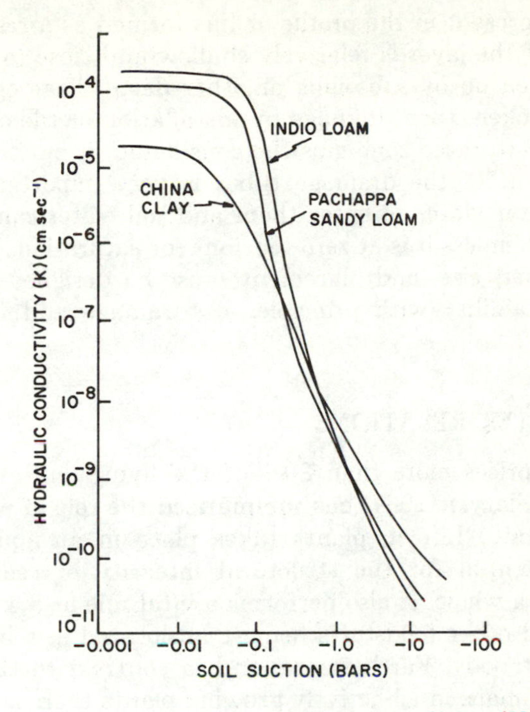

Adapted from Slatyer (1967)

FIG. 4.3. RELATIONSHIP BETWEEN HYDRAULIC CONDUCTIVITY, K, AND SOIL SUCTION FOR THREE SOILS WITH DIFFERENT TEXTURES

excess water accumulates on the surface. It will, of course, move laterally to a lower elevation if it is possible to do so. If the slope of the land is great, erosion will likely result unless the soil surface is extremely stable and/or well protected by plant residues or other materials.

Drainage

The detrimental effects of excess water on soils and plants may arise from (a) water being added to the surface at a rate greater than it can infiltrate into the soil or (b) water moving downward through the soil to a profile layer faster than it can be transmitted through that layer. If, in the case of (a), the water can move laterally across the surface, the accumulation will not be to an appreciable depth, but water may be lost and erosion may occur. Land shaping can be of immense benefit in alleviating surface ponding and any drainage program should begin with this operation.

The situation in (b) can arise when a layer with limited water permeability is present in the profile or has formed as a result of cultural operations. If the layer is relatively shallow and close to the surface, it may be broken up by subsoiling or other deep tillage operations. If it cannot be broken, then drainage by installation of tile or plastic drain tubes in the saturated zone must be considered. It must be emphasized that placement of the drainage tubes is very important. They have perforations for water to enter them and soil water can not flow into such openings unless it is at zero tension—or saturation. The spacing of the tubes, their size, and placement must be designed by competent technicians familiar with principles of drainage and their field application.

PLANT-WATER RELATIONS

Water comprises more than 80% of the living and growing cells of most plants. Slatyer (1967) has summarized the role of water very well in stating that, "Life in plants takes place in an aqueous medium. Water is essential for the structural integrity of cells, tissues and organisms as a whole. It also performs a vital role as a solvent, mineral nutrients and other foodstuffs being translocated in solution throughout the plant body. Furthermore, and in contrast to the situation in terrestrial animals, in all actively growing plants there is a liquid phase continuity from the water in the soil through the plant to the liquid-gas interface at evaporation sites in the leaves. The proliferation of roots in the soil provides an extensive absorbing surface across which passes virtually all the water and mineral nutrients utilized by plants."

Growing plants require large amounts of water. Plants absorb from the soil, transport, and lose to the atmosphere quantities of water many times greater than they use in their own metabolic processes. This loss of water, called *transpiration*, occurs in response to the evaporative demand of the atmosphere. Plants may grow in a saturated or nearly saturated atmosphere with little transpirational loss, but in dry climates the weight of lost water may be hundreds or thousands of times the dry weight of the plant.

The water loss from plants occurs through the stomates or pores in the leaf surface. The stomates must be open to take in carbon dioxide which is necessary for photosynthesis. If the evaporative demand of the atmosphere is high, the stomates may close or partially close to reduce water loss. Photosynthesis will then also be reduced because carbon dioxide entry is limited. Also, cell enlargement and division is greatly reduced at high suction. Plant water suction may be so high during daytime hours that little growth takes place. Thus the crop may make a large share of its growth at night when plant water suction is lower.

The relative water requirement of different crops is usually expressed as the *transpiration ratio.* This is defined as the number of units of water required to produce each unit of dry matter—usually exclusive of roots. This may also be called the water requirement of crops. The inverse of this ratio is sometimes referred to as the *water use efficiency.*

The transpiration ratio for various crops may vary from below 200 to over 1000. The value for any given crop will fluctuate widely with light intensity, relative humidity of air, temperature, wind velocity, water availability, nutrient availability, and other factors. Transpiration ratio was at one time used extensively in characterizing the water relationships of crops. However, it is now evident that other characteristics may be more useful in crop management decisions. Transpiration losses appear to be similar for a wide range of crops so long as the crop completely covers the soil surface and the plants are in a turgid condition. Transpiration rates have a fairly constant relationship to the rate of evaporation of water from a free water surface under similar climatic conditions. Much of the previously found differences in transpiration ratios of different crops appears to be due to differences in the proportion of the ground covered by actively growing plants and by the differences in weather conditions when the measurements are made.

Crop characteristics which are important in moisture management decisions include the ability of plants to adjust to high moisture stress, the time and length of stages in plant growth and development when moisture is most critical, and reaction of plants to adverse temperature or to abnormal solar radiation periods when moisture stress is not excessive. These factors may be associated with the probability that soil moisture deficiencies will occur at various times, and the probability of adverse temperature and/or solar radiation conditions. Sometimes it is possible to alter the planting date sufficiently to avoid unfavorable moisture or temperature conditions at critical periods in the plant's growth cycle. Or it may be possible to use a variety or hybrid with a different period of time between planting and flowering or between planting and harvesting.

Availability of Soil Water to Plants

As a plant transpires water, a water suction develops in the cell walls of the plant. This suction is transmitted throughout the plant unless water can move into the roots and up through the plant fast enough to prevent it. Water moves into the plant whenever suction in the water in the plant is greater than that in the water in the soil. The rate of movement into the plant depends upon the suction difference and the ability of the soil to allow movement of the water to the plant roots. As previously noted, most plants can withdraw water from soils until the soil moisture suction reaches about 15 bars.

Many soil properties affect the availability of soil moisture to plant roots and the rate of movement of water in the soil. One of the most important of these properties is soil texture. Figure 4.3 shows the relation between the amount of water in the soil and the soil suction for soils of different texture. Fine-textured soils hold more water than sands at field capacity and can supply more water as suction increases (e.g., from 0.1 atm to 1 atm). This helps to explain why fine-textured soils are less droughty and can support plants in good moisture status for longer periods after a rain or an irrigation.

Water Requirements of Crop Plants

An excellent report dealing with the water requirement of crop plants has been prepared by a committee of the American Society of Civil Engineers (Jensen 1973). This states: "There is confusion as to the meaning of terms employed in reports on use of water. Such terms as consumptive use, evapotranspiration, transpiration, total evaporation, stream-flow depletion, water requirements, irrigation requirements and duty of water have been interchanged." That report considers the terms "consumptive use" and "evapotranspiration" as being synonymous and we follow that usage here.

Potential evapotranspiration is defined as "the rate at which water if available would be removed from the soil and plant surface expressed as the rate of latent heat transfer per square centimeter or depth of water. For comparative purposes in this report potential evapotranspiration refers to a well-watered crop like alfalfa (lucerne) with 30 to 50 cm of top growth and about 100 m of fetch under given climatic conditions . . . " (Jensen 1973).

The ratio of evapotranspiration (ET) under a given set of conditions and the potential evapotranspiration (PET) under those conditions is a useful value. It is called relative evapotranspiration or crop coefficient (K_{co}). Figure 4.4 shows how the crop coefficient for grain sorghum varies with stage of growth.

The relationship between yield of winterwheat and ET at Akron, Colorado has been shown by Hanks *et al.* (1969). Average annual precipitation is about 42 cm at this location. Additional water (and increased ET) increased yields until water was no longer the limiting factor. Energy is required to evaporate water from the soil and to cause plants to transpire. Most of the energy comes from solar radiation; some comes from warm winds. The amount of radiation received at any

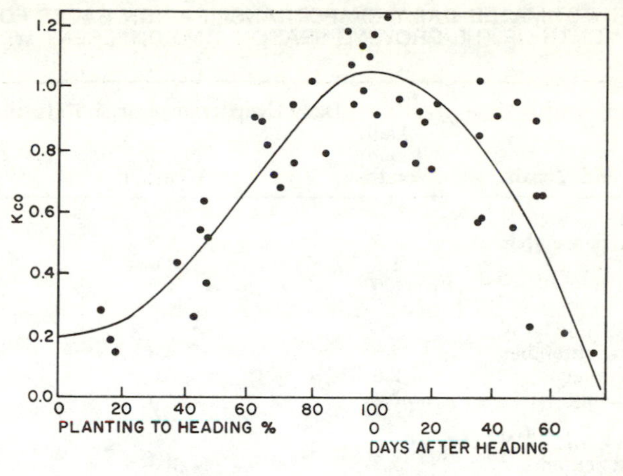

Adapted from Jensen (1966)

FIG. 4.4. CROP COEFFICIENTS, K_{co}, FOR GRAIN SORGHUM

locality is limited and the amount which a growing crop can utilize is only about 1 to 2% of that received. Thus the utilization of energy may become the next limiting factor when moisture is adequate and good cropping practices are followed. Some years there is unusually high solar radiation at the crop surface and high yields and high ET result if moisture is available to the crop.

Many workers have proposed equations for computing evapotranspiration rates. Jensen (1973) has a good presentation of most of these equations and discussion of their value. As previously indicated, transpiration losses under a given climatic situation are similar for a wide range of crops if the crop completely covers the soil surface and plants are in a turgid condition.

Jamison and Beale (1959) estimated daily moisture losses for maize during various months and with 3 weather conditions at 3 latitude ranges, as given in Table 4.1. Values at latitudes nearer the equator could be expected to be somewhat higher than these. Peak use rates over twice as high as those in Table 4.1 have been reported. These, of course, do not usually occur many days in succession and must not be confused with average rates over a period of time. It is readily apparent that total water use per month can amount to 180 mm or more. Seasonal ET for well-watered common crops in the U.S. and Canada are given in the ASCE report previously referred to (Jensen 1973). Doorenbos and Pruitt (1977) also present useful information on crop water requirements and Table 8.1 presents some of their data. Many crops have critical stages of growth when a water deficit will cause an unusually large reduction in yield of the marketable product (see Chapter 8).

TABLE 4.1. ESTIMATED DAILY EVAPOTRANSPIRATION RATES FOR CORN, BY LATITUDE, MONTH OF THE GROWING SEASON, AND DIFFERENT WEATHER CONDITIONS

Latitude and Month	Daily Evapotranspiration Rate (mm)		
	Dull Cloudy Weather	Normal Weather	Bright Weather
Between 48° and 40°N			
April and September	1.5	2.3	3.3
May and August	1.8	3.0	4.6
June and July	3.0	4.2	5.6
Between 40° and 34°N			
April and September	2.0	2.8	3.6
May and August	2.8	3.6	4.8
June and July	3.6	4.3	5.8
Between 34° and 30°N			
April and September	2.3	3.3	4.1
May and August	3.3	4.1	5.6
June and July	3.6	4.3	5.8

Source: adapted from Jamison and Beale (1959).

CONTROLLING AND MEASURING SOIL MOISTURE

It appears that maximum crop production would be attained if soil moisture suction could be held at a value low enough that the energy exerted by the plant in absorbing water would be minimal, yet high enough to avoid saturation and consequent poor aeration. In actual practice, moisture content (and suction) may vary from the ideal to extremes of excesses and deficiencies. Good management of cultural practices can help reduce the variation.

Instruments of many types have been developed and marketed for use in measuring soil moisture. All are based on a measurement of some property of the soil which changes with a change in soil moisture content. Instruments which measure the electrical resistance of soils have been widely publicized. Some workers report good satisfaction with instruments of this type and others report they are of little help. They must be used with caution and interpretations should be made by someone having an understanding of factors affecting the readings. Some sellers of instruments claim that a certain resistance value corresponds to the same availability of water in all soils. This claim should be viewed with caution and some calibration of instrument readings with moisture availability is advisable.

Resistance measurements commonly are made with some sort of blocks buried in the soil, with wires extending to the surface for attaching to the electrical meter. The blocks may have plaster of paris (gypsum), nylon, fiberglass, alundum, or other material between the electrodes to absorb moisture from the soil. The electrical properties of this material change with moisture content and this is what is measured. Nylon, fiberglass, and alundum blocks show greater sensitivity at higher moisture contents, but are usually more susceptible to errors due to soluble salts, and they usually cost more. The electrical meter required for resistance measurements may be a moderately expensive investment for the small farmer.

The neutron probe is a more sophisticated and expensive instrument which has thus far found use mainly in research. It has a source of neutrons and a detector which measures the reflection of neutrons by hydrogen atoms. The reading obtained is quite sensitive to changes in soil moisture content since each water molecule contains hydrogen atoms and this is the main source of hydrogen atoms in the soil. Access tubes are permanently installed in the soil profile to the desired depth and the compact neutron source and detector are lowered into the tube when measurement is to be obtained. Three large irrigation districts in the lower Colorado river region of the U.S. are reported to be using this instrument for soil moisture measurements.

Tensiometers are instruments which may be used fairly successfully on some soils. These instruments have a ceramic cup connected by a water-filled tube to a vacuum gage. The unit is inserted into an auger hole so the cup is in close contact with the soil and the gage is above ground. At suctions greater than about 0.8 bar, a tensiometer will fail to operate because the water column breaks and air enters the cup. Since sands hold a larger fraction of their available water with suction less than 0.8 bar than do loams or clays (Fig. 4.2), tensiometers are more useful for moisture measurements in sands. Tensiometers are relatively expensive for small farmers, if secured in sufficient numbers to give measurements in many locations. Generally, they must be left in place during the growing season and not moved from place to place. Tensiometers may be placed with cups at one or more depths below the soil surface and used to determine when to irrigate. They may also be used in determining whether moisture is moving vertically and, if so, the direction of movement. This may be important to know when salt accumulation is a problem or when salt removal is being attempted.

The removal of a soil sample from the appropriate depth and determination of moisture content by drying and weighing can give reliable information with reasonable capital expenditure. Since soil is heterogeneous, sampling to give reliable results is always a problem. A composite sample can easily be taken from shallow depths and one gravimetric determination made. This procedure does not give a continuous appraisal of the soil moisture situation and there may be disturbance of the growing crop. Gravimetrically determined moisture content does not immediately give one information about moisture availability and a calibration procedure of some sort must be followed.

Some soils have distinct color changes as they wet or dry. With such soils, an experienced worker can learn to identify moisture availability rather well. The "feel" of most soils changes with moisture content and some can calibrate their sense of touch to give quite an accurate measure of available moisture.

Perhaps the simplest method is to observe both crop and soil closely for signs of moisture stress. Some plants show early signs of moisture depletion and can indicate dangerous levels of moisture stress. The curling or rolling of maize leaves is one such plant indication which may be used. Some workers have reported that moisture stress in cotton and some other crops can be detected by feeling the temperature of the leaves during the daytime.

A bookkeeping procedure may be used to indicate soil moisture status if rainfall and evaporation can be measured or estimated accurately enough. This system uses daily rainfall and/or irrigation amounts as credit entries and daily evapotranspiration as debit entries. Rainfall can be measured at the farm or obtained from a nearby weather station. Evapotranspiration is either estimated from open-pan evaporation, obtained from weather station, or computed from weather data. Sometimes average loss rates for an entire month can be used with fairly good success.

In some areas, high speed computers are being used to compute water loss by using meteorological data and the information is reported to the farmer at scheduled intervals. Generally this service is provided by a consultant or a service company for a fee and is of most interest to farmers who need the information for scheduling irrigation applications.

REFERENCES

DOORENBOS, J., and PRUITT, W.O. 1977. Crop water requirements. Irrigation Drainage Pap. *24*. FAO, U.N., Rome.

HANKS, R.J., GARDNER, H.R., and FLORIAN, R.L. 1969. Plant growth-evapotranspiration relations for several crops in the central Great Plains. Agron. J. *61*, 30-34.

HILLEL, D. 1971. Soil and Water, Physical Principles and Processes. Academic Press, New York.

JAMISON, V.C., and BEALE, O.W. 1959. Irrigating corn in humid regions. U.S. Dep. Agric., Farmers' Bull. *2143*.

JENSEN, M.E. 1966. Empirical methods of estimating or predicting evapotranspiration using radiation. Proceedings, Conference on Evapotranspiration and Its Role. Chicago, December. ASAE, St. Joseph, Michigan.

JENSEN, M.E. 1973. Consumptive use of water and irrigation water requirements. Report of Committee on Irrigation Water Requirements. Am. Soc. Civ. Eng., New York.

SLATYER, R.O. 1967. Plant Water Relationships. Academic Press, New York.

5

Plant Nutrients and Fertilizers

D. Wynne Thorne

Fertilizers accounted for about half of the increase in crop yields in the U.S. in the 15-year period from 1941 to 1955. Most developed countries have found fertilizers similarly important, and increased use of fertilizers along with related improved technologies are a major hope for future gains in food production in developing nations.

There is an extensive literature and an abundance of data about the response of crops to field applications of plant nutrients in various forms and quantities. Extensive references are available, including Russell 1973; Tisdale and Nelson 1975; Fernandez 1974; Nelson 1965; and Olson *et al.* 1971.

Approximately 20 elements are known to be necessary for the normal growth of crop plants. Twelve of these frequently are deficient in agricultural soils. Nitrogen, phosphorus, and potassium are used so extensively to enhance crop yields that they are known as the fertilizer elements. Calcium and magnesium are constituents of lime, which is frequently applied to soils to correct acidity. Sulfur, another major element, is a common constituent of many fertilizers. The remaining elements are required in relatively small quantities for plant growth and are known as the micronutrients.

GENERAL INFORMATION ABOUT FERTILIZERS

Effective selection and use of fertilizers requires a knowledge of terminology and principles. The guarantee of plant nutrient content of fertilizers is expressed in the percentage of total nitrogen (N), percentage of available phosphorus, expressed as phosphorus pentoxide (P_2O_5), and percentage of water soluble potassium, expressed as potassium oxide (K_2O). In a complete fertilizer the guarantee, or fertilizer formula, is written in 3 figures such as 10-20-10, referring respectively to the percentage of N, P_2O_5, and K_2O.

Acidity and Basicity of Fertilizers

The residual effects of fertilizers on soil reaction are important. To the extent that fertilizers substantially increase soil acidity they increase exchangeable and toxic aluminum. Potassium fertilizers are essentially neutral. Phosphate fertilizers are considered neutral except for the ammonium phosphates which have acidic effects. All fertilizers containing ammonium compounds have acidic residual effects on soils because of the oxidation of ammonia to nitric acid. In fertilizer mixing plants dolomitic limestone is often used in mixed fertilizers as a neutralizer of potential acids. The magnesium also helps improve availability of the fertilizer phosphate.

Acidic fertilizers are generally considered desirable in irrigated alkaline calcareous soils that tend to accumulate lime when irrigated.

Costs of Fertilizers

In choosing among fertilizers the grade or guarantee is of major importance. If different mixed or single nutrient carrier fertilizers are available to choose from, and all are of approximately equal quality in terms of nutrient availability, cost becomes important. For single nutrient carriers, the cost per unit of the element can be estimated by dividing the percentage of the element into the cost per ton. Estimates can also be made by comparing the cost of materials in single carriers with costs in mixed materials, by multiplying the cost per unit of each element by the percentage or number of such units in each ton.

FERTILIZER ELEMENTS AND MATERIALS

Nitrogen

Nitrogen is the most extensively deficient and the most widely applied plant nutrient to cropped soils. Its availability in soils is influenced by microbial activity, moisture, temperature, and other factors. Whether applied in organic or inorganic form, under favorable soil, temperature, and moisture conditions, the nitrogen in such materials is converted rapidly to the nitrate form. As nitrate it is subject to leaching from the soil and it can also be denitrified by soil microorganisms and lost to the atmosphere.

The supply of potentially available nitrogen in soil is contained primarily in the organic matter fraction. The quantities and activities of organic matter and microbes in soil are important to the nitrogen nutrition and general well-being of plants. A special discussion about nitrogen is given in Chapter 7.

Plants absorb nitrogen primarily as the ammonium (NH_4^+) or nitrate (NO_3^-) ions. While there are some plant preferences, ammonia nitrogen in well aerated, fertile soils is normally converted so rapidly to nitrate that, except for rice, this is the most abundant form absorbed. In the nitrification process ammonium ions are converted to nitric acid, so ammonium fertilizers have an acidifying effect on soils.

Nitrogen Fertilizers

Nitrogen fertilizers are produced primarily through a catalytic process of combining relatively pure nitrogen and hydrogen gases to form ammonia. Natural hydrocarbon gas is used to produce the hydrogen and to furnish energy for the reaction. With the current and increasing shortages of natural gas and oil, the price of nitrogen fertilizers has risen rapidly. The responses of improved crop varieties to nitrogen fertilizers are often dramatic compared with traditional varieties. This is shown in Fig. 5.1 in which varying amounts of nitrogen fertilizer were applied to a local unimproved rice variety and to an improved one, IR.8, in India.

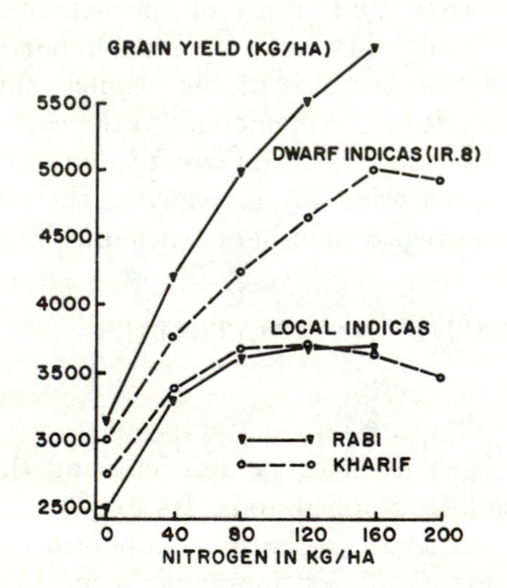

Adapted from Ram (1975)

FIG. 5.1. YIELD RESPONSE OF TWO VARIETAL GROUPS OF RICE TO NITROGEN DURING KHARIF AND RABI SEASONS IN INDIA

The principal nitrogen fertilizers used for general field crop production are as follows:

Anhydrous Ammonia.—This is the most concentrated (82% N) of the nitrogen fertilizers. It is a gas under normal pressures and temperatures but for storage and transport it is converted to liquid form and stored under pressure. Ammonia is usually the least expensive nitrogen fertilizer and is convenient to use where bulk handling and other appropriate equipment are available to inject the gas into the soil.

It is the most extensively used of the various nitrogen fertilizers in the developed countries where much of it is applied directly to the farmers' fields by fertilizer companies. Anhydrous ammonia is not widely used in developing nations for a number of reasons. They usually lack the infrastructure to handle such specialized materials. Handling of high pressure tanks presents problems in transportation and in safety. The empty tanks must be shipped back, thus increasing costs where land transportation is expensive.

Urea.—Urea is gaining rapidly in popularity in much of the world. It is relatively low in cost, has a high content of nitrogen (46%), has a granular form that is easily handled, and is readily available to plants. Many plants can absorb urea directly. When incorporated in moist soils, microbial action converts urea nitrogen rapidly to the ammonium form which is retained by the soil and is subject to nitrification or it may be in part lost by utilization by soil bacteria. Urea can be applied either directly to the soil or sprayed on the leaves of plants when dissolved in appropriate concentrations. Urea should, therefore, be worked into the soil as soon as possible and managed much the same as ammonium fertilizers. Urea currently accounts for over one-fourth of the world supply of nitrogen fertilizer and the proportion is increasing rapidly.

Ammonium Nitrate.—Ammonium nitrate is a desirable, relatively concentrated (34% N) nitrogen fertilizer. Half of the nitrogen is in the ammonium form and half is nitrate. Consequently it has a moderate acidifying effect on the soil. Ammonium nitrate is currently selling on the international market at a higher price per unit of nitrogen than either urea or anhydrous ammonia. It currently constitutes about one-fourth of the nitrogen fertilizer on the world market but a decreasing proportion of the trade is projected for the future.

Ammonium Phosphates.—The economies in transporting, handling, and storing high analysis fertilizers containing both nitrogen and phosphorus have led to the production of a number of different ammonium phosphate materials with mono-ammonium phosphate and diammonium phosphates as the principal ingredients. Common grades of ammonium phosphate fertilizers include 11-48-0, 13-39-0, 16-20-0, 16-

48-0, and 18-46-0. These materials are usually thought of as phosphate fertilizers more than as nitrogen fertilizers. The production of these and related materials is increasing.

Ammonium Sulfate.—This has been one of the important fertilizers but since 1955 it has declined from about 30% of the world nitrogen fertilizer supply to less than 10%. When added to soil much of the ammonium component is oxidized to nitrate and the net effect is the release of equivalent quantities of nitric acid and sulfuric acids. Because of its strong acidifying effect ammonium sulfate may have injurious effects on poorly buffered or acidic soils. In addition to the acid residues, the nitrogen content is only 21%, which is less than half of that in urea, and so when shipped some distance the marketing costs may be relatively high.

Nitric Phosphates.—These fertilizers are manufactured by reacting nitric acid with rock phosphate. In some preparations the water solubility is low and in such cases use on acid soils is recommended. Nitric phosphates are currently used primarily in Europe but have a potentially broader market. A wide range of fertilizer grades and rates can be produced and production is increasing in many countries.

Phosphorus

Phosphorus is normally present in soils in small proportions of less than 0.11%. It is retained in soils in compounds of relatively low solubility, and where high yields of crops are sought phosphorus deficiencies are one of the most common limitations. Good phosphorus nutrition of plants encourages vigorous root development.

Phosphorus Fertilizers

The value of phosphate materials for fertilizer is based on its relative solubility. The most used tests involve solubility in a neutral ammonium citrate solution or in water. For most neutral or alkaline soils or for attainment of high crop yields, highest value is placed on water soluble forms. For highly acid soils less soluble forms of phosphate are at least moderately suitable; even finely ground rock phosphate becomes slowly available.

Ordinary Superphosphate.—This is the longest used phosphate fertilizer, made by treating finely ground rock phosphate with sulfuric acid. The product is a mixture of monocalcium phosphate and gypsum. Depending on impurities in the rock phosphate, the phosphorus pentoxide content (P_2O_5) amounts to 16 to 22% (7 to 9% P), which is about 90% water soluble. This is a high quality fertilizer and constitutes an important but declining portion of the fertilizer used. Because of the low content of phosphorus, handling and shipping are rather expensive. But because of simplicity of manufacture it can be readily produced and used where the acid and rock phosphate are available.

Phosphoric Acid.—Most of the phosphoric acid used in the fertilizer trade is produced by treating rock phosphate with sulfuric acid. The resulting wet process acid contains about 55% P_2O_5 (24% P). This may be concentrated by evaporation to a 72% P_2O_5 product. Much of the phosphoric acid is used in producing various liquid and mixed fertilizers. It is reacted with ammonia to produce various ammonium phosphates. Phosphoric acid is also being used in increasing quantities for direct soil treatment, particularly with calcareous soils and is suited to application in irrigation water.

Concentrated Superphosphate.—This is one of the extensively used phosphate fertilizers, accounting for about 15% of the world's supply. Made by treating rock phosphate with phosphoric acid, it contains about 44 to 52% P_2O_5 (19 to 23% P). The phosphorus contained is 95 to 98% water soluble. It is used for direct soil application and in mixed fertilizers.

Ammonium Phosphates.—Several such materials are produced by reactions between ammonia and phosphoric acid or with a mixture of phosphoric acid and sulfuric acid. Ammonium phosphates are completely water soluble and are marketed in both dry granular form and in solutions. Because of their high content of plant nutrients the shipping and handling costs are relatively low.

Phosphate Fixation and Availability

Phosphorus is usually absorbed by plants as the $H_2PO_4^-$ and HPO_4^{2-} ions. Most phosphorus-containing minerals in soils are of low solubility. In some soils much of the phosphate is present in the organic matter fraction, from which it is gradually released on decomposition.

When fertilizers or other soluble sources of phosphate are added to soils the phosphate is often fixed in low soluble forms. In acid mineral soils, aluminum and iron are absorbed on clay and react quickly with soluble phosphate to produce compounds that decrease in solubility with time. In less acid soils, particularly those above pH 5.0, phosphate may react directly with silicate clays. Clays with low silica sesquioxide ratios, such as in Ultisols and Oxisols, generally have high phosphate-fixing capacities. In neutral or alkaline soils, phosphate fixation is associated with calcium dominated clays and calcium and magnesium carbonates. Phosphate may be absorbed on the surface of lime particles or it may react to form such compounds as hydroxy or carbonate apatites.

In most situations the combinations formed after fertilizers are added to soils will continue to release small concentrations of available phosphate into the soil solution for some time. Soils between pH 6.0 and 8.5 are usually the most favorable for phosphate nutrition of plants. Even under the relatively unfavorable conditions of the tropics, however, phosphate fixation may be only a partial barrier to the nutrition of crops and small applications have been followed by residual benefits for several years. This is shown in data in Table 5.1, in which 19.5 and 39 kg of phosphorus were applied followed by 6 years of cropping (Boswinkle 1961). The experimental site at Nakaveti, Kenya had limited rainfall and only in 1954 of the test period was there enough moisture for two crops. The one 39 kg treatment resulted in increased yields of crops in each of the 6 years. The soil is not highly oxidized and is probably related to the Alfisols.

TABLE 5.1. THE RESIDUAL EFFECT OF A SINGLE DRESSING OF SUPERPHOSPHATE ON A BASEMENT COMPLEX OF SOIL AT MAKAVETA, KENYA

Crop	1952 Millet[1] kg/ha	1953 Beans[2] kg/ha	1954 Maize kg/ha	1954 Millet kg/ha	1955 Millet kg/ha	1956 Millet kg/ha	1957 Beans kg/ha
No Phosphate	405	755	1920	410	1290	790	785
Response to Phosphorus							
19.5 kg/ha	525	160	585	120	105	85	40
39.0 kg/ha	720	250	785	175	190	120	125

Source: Boswinkle (1961).
[1]Pearl millet: *Pennisetum typhoideum.*
[2]Beans: *Phaseolus vulgaris.*

Potassium

Potassium is absorbed by plants in larger quantities than most mineral elements. Potassium is also frequently present in soils in large quantities but it is low in coarse-textured soils and soils that are highly leached and oxidized such as the Oxisols. The readily available potassium in soils includes that which is dissolved in the soil solution and that occurring in the exchangeable form on soil colloids. It is absorbed as the K^+ ion. Soil minerals such as mica and feldspars gradually weather and release substantial amounts of potassium that become available to plants. Even though relatively abundant in soils, potassium deficiencies are widespread. Potassium deficiencies often occur because of the large and rapid uptake by many crop plants as farmers create conditions favorable for high yields. Under such conditions some soils are unable to release potassium rapidly enough to supply plant needs during high demand periods.

Potassium Fertilizers

The principal potassium fertilizer materials are potassium chloride (50 to 60% K_2O) and potassium sulfate (45 to 50% K_2O). These two materials make up the bulk of the potassium fertilizer trade because they are high quality fertilizers, the potassium is readily available to plants, and the materials are easily handled. Potassium sulfate is, of course, also a source of sulfur for areas where this element is deficient.

Potassium Fixation and Availability

In many soils having clays with 2:1 types of lattice, particularly illite, the exchangeable potassium is often in part converted to nonexchangeable forms within the lattice of clay minerals. Such fixed potassium is largely unavailable to plants and cannot be removed by leaching the soil with the usual salt solutions. Ammonium ions undergo similar types of reactions. Additions of ammonium-type fertilizers may slow the release of the potassium that has been thus absorbed. Swelling and contracting of clay particles as a result of cycles of wetting and drying and freezing and thawing encourage a gradual release of fixed potassium or ammonium ions. With continued applications of potassium fertilizers the fixation sites become tied up so that with subsequent applications there is an increase in the proportion of potassium that remains available. In strongly acid soils having aluminum as the dominant exchangeable element, potassium is not firmly held as an exchangeable ion. Under such conditions losses from leaching are enhanced. Liming such soils leads to replacement of the exchangeable aluminum by calcium, and potassium is retained better in calcium-dominated soils than in acid soils.

CALCIUM AND MAGNESIUM AND LIMING OF SOILS

Calcium and magnesium are essential plant nutrients. They also play an important role in maintaining a soil environment favorable to major cultivated crops. Both elements originate in soil from the weathering of rocks and minerals and both are important exchangeable cations. Calcium is usually more abundant than magnesium in soils and helps maintain the soil in a good physical condition and the pH in the slightly acid to slightly alkaline range. Magnesium deficiencies are fairly extensive, probably more common than direct deficiencies of calcium. Ammonium nitrogen and potassium absorption by plants tend to reduce the uptake of magnesium.

Acid Soils

Soils become acid as exchangeable basic elements such as calcium, magnesium, and potassium are leached with replacement by hydrogen, followed by hydrogen being replaced by aluminum, iron, or manganese ions. A high content of exchangeable aluminum on soil clay and toxic quantities of aluminum ions in the soil solution are the most common ingredients of acid soils that are harmful to plants. Exchangeable aluminum and iron are difficult to replace by potassium, calcium, and magnesium ions in the soil solution. Consequently acid soils do not retain these nutrients well against leaching.

In acid soils, phosphate ions react strongly with exchangeable aluminum and iron, reducing the effectiveness of phosphate fertilizers. A lack of available phosphorus and calcium and the presence of toxic aluminum all contribute to reduced root development in acid soils.

Lime Treatments

Treatments of acid soils with lime materials have the multiple purposes of reducing acidity, furnishing a supply of calcium and magnesium, and reducing exchangeable aluminum, iron, and manganese. Other indirect effects are improved physical properties of the soil, improved availability of phosphorus, and an improved soil capability for retaining and releasing to plants other essential nutrients such as potassium, zinc, and molybdenum.

There are many different materials for soil liming treatments. Most important are finely ground limestone; ground dolomitic limestone; burnt lime, or calcium oxide; slaked lime, or calcium hydroxide; and various slags from industrial operations which contain calcium silicate.

The neutralizing capacities of lime materials are usually expressed in terms of calcium carbonate equivalents. Thus, pure calcite would have a neutralizing value of 100 and calcium hydroxide would have a value of 136. Degree of impurities may greatly alter the neutralizing values of various liming materials. Fineness of grinding affects the rate of action. Lime materials are usually evaluated on the basis of neutralizing value, fineness of grinding, and the content of magnesium.

Lime materials are usually broadcast on the soil surface and plowed under, obtaining as good a mix with the soil as possible. Application some time before planting an acid sensitive crop is desirable. Applications are usually made at intervals of several years.

The quantity of lime needed to raise the pH of the soil to a desired value as deep as plow depth is termed the lime requirement. The desired pH and depth of plowing may vary with the crop and the nature of the soil. For oxidized and leached soils, such as Oxisols and Ultisols, and with general field crops a pH value of 5.0 to 6.0 is commonly sought. With less weathered soils, such as Mollisols, best crop growth may be obtained with a soil pH range of 6.0 to 6.4. Most soil testing laboratories are equipped to determine the lime requirement.

In many tropical humid regions with highly weathered and leached soils, local sources of lime are often not available and the cost of shipping in materials for full soil treatment are considered excessive. However, general benefits have not been reported from applying lime to oxidized soils of the tropics. The evidence summarized by Russell (1974) indicates no general benefits have resulted from applying lime to bring the soil pH above 5.0.

MICRONUTRIENTS

Micronutrient elements are required by plants in small quantities, ranging from a few hundred grams to a few kilograms per acre. The micronutrients of principal importance in crop production are listed and some general information about each is summarized in Table 5.2.

Treatments for micronutrient deficiencies usually involve either inclusion of small quantities of the appropriate carrier of the element in mixed fertilizer, or application as foliar sprays. In addition, however, there is growing information about genetic differences of plants in susceptibility to specific micronutrient deficiencies. Available local information should be secured, both as to materials and methods for soil and foliar treatments and as to availability of seed for good varieties with resistance to the nutrient deficiency being encountered.

ESTIMATING FERTILIZER NEEDS

Plant appearance and growth rate, together with markings on the leaves, are good indications of specific nutrient deficiencies (Bear *et al.* 1949). Observing deficiency symptoms during the year may help in developing a fertilizer program for subsequent years. But in the same year, by the time plant symptoms are evident, it is usually too late to gain full value from fertilizer applications. Also, plants may not show definite nutrient deficiency symptoms even though yields are being reduced due to inadequate supply of one or more nutrients.

TABLE 5.2. GENERAL INFORMATION ABOUT MICRONUTRIENTS

Plant Nutrient	Principal Forms in Soil	Favorable Soil Conditions	Forms Assimilated by Plants	Fertilizer Compounds	Methods of Application
Iron	Clay Oxides	pH below 8.0	Fe^{2+}	$FeSO_4$ Chelate	$FeSO_4$ or chelated Fe foliar sprays Chelated iron to soil
Manganese	Clay, O.M. Oxides	pH below 7.5	Mn^{2+} Chelated	$MnSO_4$	Foliar spray
Zinc	Clay, O.M. Phosphates	pH below 8.0 Good structure	Zn^{2+} Chelated	Zinc oxides $ZnSO_4$ Chelate	Foliar spray Soil
Copper	Clay O.M.	pH below 8.0	Cu^{2+} Chelated	$CuSO_4$ Chelate Glass frits	Foliar spray Soil
Boron	Soil minerals, e.g., Tourmaline	pH below 8.0	$B_4O_7^{2-}$, $H_2BO_3^-$	Borax Boric acid Glass frits	Soil Foliar spray
Molybdenum	Absorbed on ferric oxides	pH above 6.0	MoO_4^{2-} $HMnO_4^-$	Sodium or ammonium molybdate	In mixed fertilizers Foliar spray

Soil Tests

Greatest benefits from fertilizers are commonly obtained by applying them before or at planting time so that plants can make a healthy growth throughout their productive lives. For this reason soil analysis has become the most extensively used and most advantageous means for diagnosing fertilizer needs. Soil fertility tests typically consist of extracting a soil sample with a weak acid or a buffered salt solution, or, as in the case of the heavy metals, a dilute solution of a chelating agent. Such tests are usually made in chemical testing laboratories that serve an agricultural area.

One essential in the making of soil fertility tests is the securing of representative samples of soil. A soil tube or auger is usually used to obtain samples because such an instrument removes a rather uniform slice of soil from the surface to the depth of insertion. In dry or gravelly soils a spade or trowel may be necessary. A soil sample of about 0.5 kg is usually taken from an area of 2 to 10 ha that is relatively uniform in soil, crop appearance, cropping and fertilizer history, and in topography. Each sample should be a composite of 15 to 20 sampling sites taken at random over the area. Usually sampling is to the normal depth of plowing which is about 15 to 20 cm if fertilizer needs or lime requirement is the primary purpose. If salt or other special soil conditions are a potential problem, deeper samples to a depth of a meter or more may be advisable.

Plant Analysis

The nutrient content of specific parts of plants is also useful in determining needs for fertilizers. Two types of plant analysis have been used. One is a fresh tissue test, usually made in the field. The other is a total plant or tissue analysis performed with accuracy in a laboratory. Since such tests must be compared with other similar analyses for valid interpretations, directions for selection of plant materials, preparation, and methods of analysis should be followed carefully. More detailed procedures are given in such references as Donahue *et al.* 1977 and Tisdale and Nelson 1975.

MIXED FERTILIZERS

Since in crop production more than one nutrient element is often needed at the same time, it is common practice to mix materials to provide a fertilizer with 2 or 3 elements and even other nutrients. At one time many U.S. farmers purchased separate materials and mixed their own. However, as farming has become more specialized, the trend has been toward commercial mixing plants. The use of mixed fertilizers is also increasing in many countries other than the U.S. In areas where large commercial farms are dominant, there are bulk blending or mixing plants.

Mixed fertilizers are produced in various ways from dry, gaseous, and liquid materials. But in recent years the trend has been toward the use of nitrogen solutions, ammoniation, and granulation. For local areas there has been an increase in small bulk blending plants which mix dry, solid fertilizer materials. The materials are usually handled in bulk and are delivered to the farmer and applied to the soil without hand lifting or bagging. The saving to the farmer approximates $8 to $12 per ton of nutrients. Fluid mixed fertilizers rather than dry granulated products are popular in California, Texas, and parts of the Midwest in the U.S.

For small farms in developing countries, bagged and granular fertilizer materials are easiest to handle. These may be either mixed or single nutrient carriers. While farmers in developing countries lack the broad choice of fertilizer materials and grades, the added labor for applying different materials in more than one operation is also less of a limitation.

As a general rule in the U.S. and Europe, nitrogen is applied as a separate material, at least for a portion of the total applied. This is because of the increasingly large quantities of nitrogen being used and its application on particular crops at special times during the crop growing season.

FERTILIZER APPLICATION

In actual practice the application of fertilizers is usually a compromise between the ideal and the practical. Ideally, fertilizers should be applied when needed by plants and placed in a strategic location in the root zone. In practice, the costs of extra or special field applications and of special equipment are weighed against the differences in fertilizer effectiveness.

The appropriate time to apply fertilizer depends on the climate, soil, crop, and nutrients to be applied. The amount of precipitation usually occurring between time of fertilizer application and utilization is important. If precipitation is normally high, if the soil is coarse in texture, and if the nutrients are subject to leaching, then special split applications are usually advisable with part being side-dressed during the crop season. With loam or finer textured soils and when the soil is maintained in a relatively high state of fertility, fertilizers are usually applied and worked into the soil before the crop is planted. Later side dressing of fertilizer will ordinarily be limited to supplemental treatments with nitrogen carriers.

Where potassium and phosphorus are deficient, usually applications can be made before planting crops that are most responsive to these elements. Less demanding crops may be adequately supported with the residual quantities remaining in the soil.

In planning fertilizer programs for a cropping system a good nitrogen-fixing legume crop is commonly followed by a crop with high nitrogen requirements. Nitrogen fertilizers are usually planned for each crop, whereas phosphate and potash applications are frequently made on selected crops in the cropping sequence. Such considerations are especially common under conditions where modest quantities of fertilizer are used and in which high yields of all crops are not given top priority.

Method of Application

Fried (1974) has reported the percentage of nitrogen in maize derived from isotopically labelled ammonium sulfate as related to different methods of application. His findings are given in Table 5.3. The data indicate a more efficient utilization of fertilizer nitrogen when it is applied during periods of high plant demand near to the tasseling growth stage. With favorable soil moisture the side-dressed material was more effective than the banded application.

Fertilizer placed in bands a few centimeters distance from the seed protects germinating seeds from salt effects of the fertilizer. Such applications also make the nutrients positionally available to plant roots during the early growth stages. Banding and side dressing are advantageous also where soils are very low in fertility and only limited quantities of fertilizer are used. With small applications of potassium or phosphate fertilizers, banded applications usually give better crop responses than when mixed with the soil.

Broadcast applications are commonly used when moderately large amounts of fertilizer are applied and when soils are maintained in high states of fertility. Phosphorus and potassium fertilizers, some micronutrients, and lime are usually applied broadcast and plowed under or disked into the soil. Nitrogen fertilizers can also be broadcast and worked into the soil except with crops or soil conditions favoring split applications, and then part is usually side-dressed during the cropping season. Anhydrous ammonia and nitrogen solutions are usually injected into the soil with special equipment. Such applications are often made by fertilizer dealers or through special contracts with those equipped to handle such materials.

Broadcast fertilizer applications are commonly made on established

TABLE 5.3. EFFECT OF PLACEMENT AND TIME OF APPLICATION OF AMMONIUM SULFATE ON THE PERCENTAGE OF NITROGEN IN THE GRAIN OF MAIZE DERIVED FROM FERTILIZER

			Side-dressed[1]			
Location	Plow-down	Banded[2]	50 cm	½ bet	T	T + 10
Argentina	20	23	16	16	10	4
Brazil	27	28	34	39	33	24
Columbia[3]	19	25	38	50	45	34
Peru[3]	24	38	46	43	46	32
Egypt[3]	20	18	36	42	46	37

Source: Fried (1974).
[1]Refers to stage of growth: 50 cm = 50 cm high; ½ bet = halfway in time between 50 cm high and tasseling; T = at tasseling; T + 10 = 10 days after tasseling.
[2]15 kg P/ha applied in band.
[3]Irrigated.

perennial sods or legume crops. Such applications are also used where no-till planting is practiced. But in such systems plowing every 4 or 5 years is helpful in distributing the accumulated fertilizer nutrients through a greater proportion of the root zone.

REFERENCES

BEAR, F.E. *et al.* 1949. Hunger Signs in Crops. American Society of Agronomy and National Fertilizer Association, Washington, D.C.

BOSWINKLE, E. 1961. Residual effects of phosphorus in Kenya. Emp. J. Exp. Agric. *29*, 131-142.

DONAHUE, R.L., MILLER, R.W., and SCHICKLUNA, J.C. 1977. An Introduction to Soils and Plant Growth. Prentice-Hall, Englewood Cliffs, New Jersey.

FERNANDEZ, V.H. 1974. Fertilizers, Crop Quality, and Economy. Elsevier Scientific Publishing Co., New York.

FITTS, J.W. 1974. Proper soil fertility evaluation as an important key to increased crop yields. *In* Fertilizers, Crop Quality, and Economy. V.H. Fernandez (Editor). Elsevier Scientific Publishing Co., New York.

FRIED, M. 1974. The effect of cultural practices on efficiency of fertilizer use determined by direct measure in field experiments using isotopically labelled fertilizers. *In* Fertilizers, Crop Quality, and Economy. V.H. Fernandez (Editor). Elsevier Scientific Publishing Co., New York.

HUBER, D.M., WARREN, H.L., NELSON, D.W., and TSAIDE, C.Y. 1977. Nitrification inhibitors—new tools for food production. BioScience *27*, 523-529.

NELSON, L.B. 1965. Advances in fertilizers. *In* Advances in Agronomy, Vol. 17. A.G. Norman (Editor). Academic Press, New York.

OLSON, R.A., ARMY, T.J., HANWAY, J.J., and KILMER, V.J. 1971. Fertilizer Technology and Use, 2nd Edition. Soil Science Society of America, Madison, Wisconsin.

RAM, M. 1975. Ten years of dwarf rice in India. World Crops *27*(11) 33-36.

RUSSELL, E.W. 1973. Soil Conditions and Plant Growth, 10th Edition. Longmans, Green and Co., London.

RUSSELL, E.W. 1974. The role of fertilizers in African Agriculture. *In* Fertilizers, Crop Quality, and Economy. V.H. Fernandez (Editor). Elsevier Scientific Publishing Co., New York.

SOIL IMPROVEMENT COMM., CALIF. FERT. ASSOC. 1975. Western Fertilizer Handbook, 5th Edition. Interstate Printers and Publishers, Danville, Illinois.

TISDALE, S.L., and NELSON, W.L. 1975. Soil Fertility and Fertilizers, 3rd Edition. Macmillan, New York.

6

Physical Properties of Soils

Marlowe D. Thorne

Two basic physical properties of soils are *texture* and *structure*. Soil texture refers to the relative proportions of the various sizes of soil particles in a particular soil. It may change over a long period of time, but for the purposes of soil and crop management programs, texture is an inherent property of the soil not generally subject to manipulation. Soil structure refers to the arrangement of the particles into groups or aggregates and is greatly affected by management. Soil texture and structure together have strong influence on nutrient supplying ability, water relationships, aeration, root growth and health, soil microbial population and vigor, and many other factors which affect the success of a cropping system.

SOIL TEXTURE

Soil texture has been discussed in Chapter 3. It should be recalled that the texture of the surface horizon of the soil is indicated by the class term in the soil name, e.g., Egbeda sandy loam or Motatua clay. The textural name for a given soil horizon is determined by a laboratory analysis of a sample from that horizon. Suitable chemical treatment is given so that individual soil particles are separated from each other and the proportions of different size groups are measured. The U.S. Department of Agriculture's scheme for naming the textural class according to relative properties of sand, silt, and clay is generally used. The triangular diagram showing the textural relationships is given in most elementary soils texts.

Since a given mass of soil material will have increasingly greater surface area as it is subdivided into increasingly smaller particles, properties related to surface area are more pronounced in finest textured soils. These properties include heat of wetting, plasticity, cohesion, swelling, adsorption for gases, soluble salts, and water. Clays have greater water holding capacity than silts and water movement is slower. They become sticky when wet and can be cloddy on drying. They are often called "heavy" soils. Sands are called "light" soils—they do not become sticky and hard to till. A sandy soil with little clay or silt may often be tilled immediately after a rain or irrigation.

The nutrient holding ability as well as the physical properties of a soil varies with the textural class and with the type of clay mineral in the colloidal fraction. While there are many clay minerals which are found in soils, the three alumino-silicates, *kaolinite*, *illite*, and *montmorillonite* are considered to have primary influence on soil properties. Such properties as swelling, shrinking, high cohesion, and plasticity are more predominant as montmorillonite content increases. Illite content has less influence than montmorillonite, and kaolinite clays show these properties to the least marked extent.

SOIL STRUCTURE

The principal forms of soil structure generally recognized are platy, prismatic, columnar, blocky, and granular. Sometimes additional forms, such as subangular blocky and crumb, are also listed. A single grain condition, as occurs in sands, and massive, cloddy, or puddled conditions are usually regarded as lack of structure or as an artificially induced state of structure.

Soil structure is markedly affected by tillage operations. Plowing, chiseling and cultivating are usually done at least partly to loosen soil and to make a more favorable environment for growth of roots of the desired crop. Tillage, also, can cause deterioration of soil structure with resultant detrimental effects on crop roots. Tillage will generally hasten decomposition of organic matter. When land is brought under cultivation for the first time, organic matter content generally decreases. If adequate organic residues are not returned to the soil or if temperature and moisture conditions are unfavorable for organic matter maintenance, the decline in organic matter content can seriously reduce productivity.

Measurement of Soil Structure

Research workers have long sought a physical measurement which would give an overall evaluation of a soil's structure. We know that the degree of aggregation and the arrangement and stability of aggregates affect many other soil characteristics which influence aeration, water flow, water retention, temperature, and impedance to roots. As yet, however, we do not have any system in which a given value on the scale tells us as much about soil structure as the water potential value tells about availability of the soil moisture to plants grown in that soil. Thus we must measure various physical properties and attempt to make inferences about the structural suitability of the soil.

Pore Space

Pore space is that part of the soil volume not occupied by solid particles but filled with air or water. The size-distribution of individual pores gives us valuable information about soil aeration and soil moisture characteristics. Analysis of the moisture retention curve of a soil (Fig. 4.2) can give us a classification of soil pores according to their effective sizes. The capillary rise equation, $h = 2T/rdg$, relates the height of capillary rise, h, to the effective pore diameter, r. Thus, h is directly related to the suction in the soil water. It is possible to calculate from the release curve the volume of water drained at different pressures and thus to determine the volume of pores having a given range of sizes. Pore size distribution is strongly influenced by the aggregation of soil particles as well as the soil texture.

Bulk Density

Bulk density is one measurement frequently made to evaluate structure. This refers to the weight of a given volume of soil. The higher the bulk density the more compacted the soil and the lower the pore space. Bulk density values for surface layers of fine textured soils generally range from 1.0 to 1.6 g per cc. Sands or sandy loams may run as high as 1.8 because sand particles tend to remain in close contact. They do not aggregate together leaving larger voids between aggregates, as do finer textured soils. Subsoil layers generally have higher bulk densities. They may have less organic matter, be less well aggregated and may be compacted from heavy equipment. Tillage operations may not be deep enough to disturb them. Some soils have compact subsurface layers created during the soil formation period. For example, there may be movement of soil constituents downward and a redeposition at the lower depth. Some of these are thin enough that they may be broken up by specialized tillage operations.

SOIL AERATION

Soil aeration is a property which relates to the ability to provide air of suitable composition to plant roots and to organisms growing in the soil. Good aeration depends on adequate exchange of air in the soil with air from the atmosphere. If a soil is well-aerated, the composition of the soil air will not be greatly different from that in the atmosphere. If aeration is impeded, the soil air will be higher in carbon dioxide and lower in oxygen than the atmosphere above the soil. Plant roots and soil organisms use oxygen and release carbon dioxide so lack of free interchange with the atmosphere may result in appreciably altered composition of the soil air. Diffusion of air through soils seems to be much more directly dependent on the volume of air-filled pores than on pore sizes.

INFILTRATION RATE

Infiltration rate or intake rate is the rate (usually expressed in units such as millimeters per hour) at which rain or irrigation water enters a soil. The rate depends upon characteristics of both the surface soil and the subsurface horizons. The rate is at maximum value soon after water application begins, then falls to a somewhat lower value which will usually be maintained without appreciable further decrease for a considerable period of time. Figure 6.1 illustrates this decrease in infiltration rate as well as the influence of initial moisture content on the infiltration rate. In Fig. 6.1, Θ_{vi} is used to indicate the percentage of soil volume which is occupied by water initially as wetting begins. Infiltration rate is seen to be inversely proportional to Θ_{vi}.

The condition of the soil surface (e.g., degree and stability of aggregation, or protection by means of crop residues or growing crop) has significant influence on the infiltration rate of a soil. The organic matter content and soil texture may also have appreciable influence. Sandy soils generally have high infiltration rates and the higher organic matter content contributes to stability of structural units while wet.

The movement of water downward in a soil was shown in Fig. 4.1. The permeability of the soil when it is not saturated was discussed in Chapter 4 and illustrated in Fig. 4.3. The longer the time of water application the greater the transmission zone between the surface and the wetting front, with a resultant continued decrease in infiltration rate.

The success of a cropping system may be greatly influenced by the ability to capture and hold limited rainfall. Practices which keep a growing crop on the land as high a proportion of the year as possible and which provide crop residues to protect otherwise bare soil may be favored.

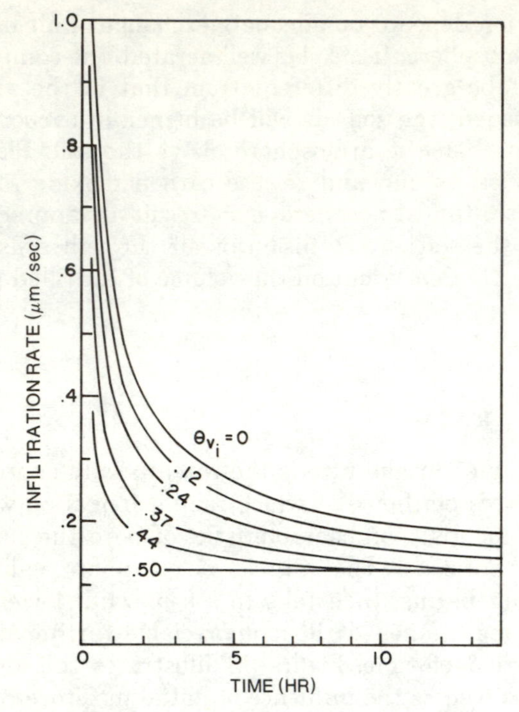

Adapted from Taylor and Ashcroft (1972)

FIG. 6.1. VARIATION OF INFILTRATION RATE WITH TIME FOR YOLO LIGHT CLAY AT DIFFERENT INITIAL WATER RATIOS

SOIL TEMPERATURE

Soil temperature affects the biological processes in the soil, such as microbiological activity, seed germination, and root growth. Organic matter decomposition with resultant release of nutrients increases with increasing temperatures. Seed germination is slowed by low soil temperature and planting must be timed to avoid the unfavorable temperatures.

Soil temperature may vary appreciably with depth below the surface and surface temperatures may undergo a wide range of fluctuation at different seasons of the year. The amount of radiant energy reaching the soil surface and the thermal properties of the soil are factors responsible for the temperature variations. Dark colored soils absorb a higher proportion of the radiant energy. Radiant energy will be absorbed in greater proportion when it strikes the surface at nearly a right-angle as compared to when it strikes at a smaller angle.

Energy absorbed by the soil surface may be re-radiated to the atmosphere as longwave radiation, it may heat the air above the soil by convection, it may raise the surface soil temperature, or it may be conducted downward and raise the temperature of lower layers.

Fluctuations in soil temperature are influenced by both the heat capacity and the thermal conductivity of the soil. These, in turn, are affected by the proportions of the soil volume filled with solid, liquid, or gaseous constituents. Since the thermal conductivity of water is greater than that of soil particles and of air, the rate of heat flow increases with increasing moisture content. However, the amount of heat necessary to change the temperature of a volume of soil increases as moisture content increases since water has a greater heat capacity than the solid or gaseous components. Thus the surface layer of a wet soil warms slower than that of a dry soil because heat is conducted more rapidly to lower layers and it takes more heat to change the temperature a given number of degrees.

The variation in temperature of surface layers during a sunny day may be considerably different for a bare soil as compared to one with a vegetative cover. Bare soil is heated by the sun's rays and early in the morning it reaches a temperature higher than the air above the soil. During the hottest part of the day, the surface layer of soil may be many degrees hotter than the atmosphere. During the night, soil surface temperature may fall below air temperature. A soil with a dense vegetative cover may never reach a temperature higher than the atmosphere above the soil. The presence of plant residues may also keep surface soil temperatures from rising above air temperatures.

The shading effect of a growing crop can be beneficial in avoiding excessive soil temperatures. The practice of interseeding a crop in the previously grown crop has distinct advantages in avoiding excessively high soil temperatures in tropical areas. Seeding a new crop in the residue of the previous crop may have similar advantages. Both these practices can conserve soil moisture as well as reduce maximum soil temperatures.

Germination of seeds or growth of plant roots may be affected more by soil temperature than by any other physical factor except moisture. In temperate climates, soil temperatures may be too low for rapid germination and if adequate moisture is present, the seed or seedling may fall victim of diseases because of low vigor. In tropical climates, soil temperatures, particularly near the surface, may be too high for the seedling to get off to a vigorous start. Some seeds are particularly susceptible to high temperatures and germination and seedling vigor are reduced if the seeds must be stored at above-optimum temperatures.

In general, roots are more susceptible to injury by high temperatures than are the above-ground portions of the plant. High soil temperatures are, of course, usually accompanied by intense sunshine and low soil moisture content. Soil temperatures lower than optimum for a crop can result in reduced root growth and higher incidence of some soil-borne diseases.

TILLAGE PRACTICES

Soils are tilled for three general reasons: to change soil structure, to manage crop residues, and to control weeds. Changes in soil structure may be desired to make it easier to plant seeds and to provide a good seedbed for germination of seeds and development of roots. Improvement of moisture intake, storage, and transmission, and of air interchange and temperature relationships may also be objectives of tillage operations.

Crop residues must be managed to facilitate planting of the new crop and to expedite its proper development and maturation. In some cases it may be desirable to cover all or most of the crop residues; in other cases the maximum amount of residues may be desired to remain on the surface.

Weed control is, of course, needed to minimize competition of the weeds for light, moisture, and nutrients. Weeds have requirements for growth which are about the same as those for crop plants. Practices must be followed to keep weeds from germinating and emerging at the same time as the desired crop, or chemicals must be applied which affect the weeds and not the crop. If neither of these practices is feasible, hand-weeding will be required.

Larson (1962) proposed a division of the seed bed into two sub-zones: (1) the seedling environment zone, which is immediately around the seed, and (2) the water-management or inter-row zone. This general concept had been recognized in various tillage systems which tilled only the seedling environment zone and left the water management zone untilled or roughly tilled. The water management zone may be left quite unfavorable for germination of weed seeds but favorable for water intake and resistance to erosion from water and wind.

A desirable seedbed is difficult to describe in terms of soil physical parameters. Most farmers would describe it in terms somewhat as follows:

> It is one from which the fine grain-like granules or aggregates trickle through your fingers when you pick up a handful of the soil. It feels loose and mellow in your hand but not too loose for good germination of the seed. It is well supplied with readily available organic material. It contains enough large pores to allow rapid moisture movement into it during periods of heavy rainfall. It also allows rapid movement of water into

> drain tiles. It contains enough air so that the millions of soil micro-organisms are actively decomposing an abundant supply of readily available organic material, liberating nutrient materials for the plant. The granules and aggregates are in small clumps and not in large clods. The granules are stable and resist the beating action of raindrops. It has these characteristics for the surface 15 cm layer that is tilled, but in addition it also has favorable air water and nutrient relationships throughout the entire zone that the roots penetrate (Browning 1950).

A soil which has these desirable properties is said to have good tilth.

It is now generally thought that "conventional" tillage practices have been excessive and the tendency is to reduce the number and intensity of tillage operations. Various schemes are being tried and promoted and they are known by such names as no-till, zero-till, conservation tillage, minimum tillage, wheel-track planting, and plow-plant. All these systems involve fewer operations across the field with tillage implements. There are obvious savings in cost of operation of the equipment (including fuel costs) and time of the operator. There may also be appreciable saving of soil and water (Siemens and Oschwald 1976).

In developed countries the trend has been to tractors of increasingly large horsepower and with four wheel drive. In 1875, it took 150 man-hours of labor on the farm to produce 2700 kg of maize. In 1975 it took 4 man-hours. A farmer in North Dakota has stated that he can now plow 130 ha per day; his own father initially plowed only ¼ ha per day on the same farm. This larger equipment with better traction permits tillage and other operations to be done nearer the optimum time. Most farmers in midwestern U.S. can now plant their entire corn acreage in one week. This may be extremely important in years when rains are unusually frequent during the time for tillage and planting. While it may be thought that these large machines consume excessive amounts of fuel, one study has shown that farms smaller than 40 ha use 90% more fuel per hectare for field operations than do farms larger than 105 ha (Nelson *et al.* 1975).

An excellent presentation of tillage operations in relation to multiple cropping systems has recently been published by Unger and Stewart (1976). The reader is referred to that article and to one by Raney and Zingg (1957) for further information on the subject.

SOIL EROSION

It has long been recognized that a soil with desirable tilth is one which is resistant to both wind and water erosion. Undesirable tilth usually, though not always, leads to excessive losses from erosion. Tillage operations performed at optimum soil moisture content contribute to formation and maintenance of good tilth. Tillage when too dry or too wet can result in excessive cloddiness which may or may not be resistant to wind and water. The reader is referred to a recent article by Siddoway and Barnett (1976) for an excellent review of erosion in relation to multiple cropping practices.

Water Erosion

A universal soil loss equation has been developed by U.S. Department of Agriculture scientists working in cooperation with state agricultural experiment stations (Wischmeier and Smith 1965). It allows the computation of average annual soil loss as the product of 2 quantitative factors (soil-erodibility and rainfall-erosivity) and 4 qualitative factors (Wischmeier 1976). The equation is:

$$A = R\ K\ L\ S\ C\ P \quad \text{where}$$

A — is the average soil loss, in metric tons per hectare, for the time period used for factor R (usually average annual).

R — is the rainfall and runoff erosivity index. It reflects the combined potential of raindrop impact and the turbulence of overland flow carrying the dislodged soil particles from a field. It is directly proportional to total kinetic energy of a storm times its maximum 30-minute rainfall intensity.

K — is the soil erodibility factor. It is the average soil loss in tons per hectare per unit of R, for a given soil on a "unit plot" which is defined as 23.1 m long, with 9% slope, continuously fallowed, and tilled parallel to the land slope. It is related to soil structure texture, type of clay mineral, organic matter content, and permeability.

L — is the slope-length factor and is the ratio of soil loss from a given length of slope to that from a 23.1 m length with all other conditions identical.

S — is the slope-steepness factor. It is the ratio of soil loss from a given percentage of slope to that from a 9% slope with all other conditions identical. (In practice, factors L and S are usually combined in a single topographic factor denoted by LS.)

C — is the cover and management factor and is the ratio of the soil loss with specified cover and agronomic practices to that from the fallow condition on which factor K is evaluated.

P is a factor relating to erosion control practices and is applicable only on cropland. It is the ratio of soil loss with supporting practices such as contouring or strip-cropping to that with straight row farming up and down the slope.

The values of factors R, K, L, and S are essentially fixed for a particular field and, together, they determine that field's erosion potential. Factors C and P are highly responsive to management practices followed. The effect of good management and erosion control practices in reducing loss of soil from water erosion is shown primarily through these two factors (Wischmeier 1976).

The universal soil loss equation was developed for cropland east of the Rocky Mountains in the U.S. It has been shown to estimate soil loss quite well for that area. C and P values are available for conventional practices. New practices can be evaluated for their effectiveness in controlling water erosion and C and P values determined for them. Caution should be observed in using the equation for regions in which it has not been evaluated. Tropical regions differ from eastern U.S. in most of the factors included in the soil loss equation, so it would be surprising to find accurate soil loss predictions in tropical areas from this equation without modification.

A recently published two-volume report (USDA 1976A, 1976B) gives an excellent presentation of the effect of soil and water management practices on control of water erosion. It is pointed out that erosion-control practices rely primarily on five means of reducing erosion: (1) vegetation, (2) plant residues, (3) improved tillage methods, (4) residual effects of crops in rotation, particularly systems that include grass and legume meadow, and (5) mechanical supporting practices.

A combination of crop canopy and plant residues can provide year-round protection of the soil surface from raindrop impact and significantly reduce soil loss. With multiple cropping, the soil may have adequate protection without following practices which leave abundant plant residues on the surface. With single-crop systems, protection may be provided by surface residues during the part of the year between harvest and the development of a full crop canopy by the next crop.

No-till systems utilize plant residues quite fully. The residue of the previous crop is disturbed only in the narrow band in which the next crop is planted. No-till planting in chemically killed sod can reduce soil loss by 95%, and uniformly distributed cornstalk residue at the rate of 6 metric tons per hectare can reduce soil loss by 85% on moderate slopes (USDA 1976A). These practices cannot be used on all soils, however. High amounts of surface residue delay soil warming and drying and may unduly delay planting on poorly drained soils. Weed control and proper fertilizer placement may be difficult to achieve.

Tillage practices which involve less tillage than moldboard plowing but more disturbance than results from no-till are termed conservation tillage practices. They usually involve some kind of non-inversion tillage and retain part of the crop residue on the surface. Conservation tillage practices generally result in less soil loss from water erosion than comes from moldboard plowing and associated secondary tillage operations. They are applicable to many soils where no-till cannot be advantageously practiced. Fertilizer placement, effective weed control, and timely planting may be more easily achieved in poorly drained soils with conservation tillage than with no-till.

Wind Erosion

Land is most vulnerable to wind erosion when vegetative cover is sparse and the soil surface is dry. Thus arid and semi-arid climates usually are associated with the most severe wind erosion. Soils in humid climates can dry rapidly when there is little cover, especially sandy soils. Serious erosion can be experienced on these soils during the winter time if they are fall-plowed and in the spring before the crop is well established.

A wind erosion equation has been proposed which seems to have considerable value (Woodruff and Siddoway 1965). This is:

$$E = f\,(I\ K\ C\ L\ V) \text{ where}$$

E is potential annual soil loss; I is soil roughness; C is a climatic index based on wind speed and surface soil moisture; L relates to field size; and V to the equivalent quantity of vegetation which takes into account the kind and orientation of plant material.

Adequate cover of the soil surface with a growing crop or with crop residues affords good protection from wind erosion as well as from water erosion. Roughness of the soil surface provides protection from wind erosion also. Where the surface is left for long periods without protection by plant materials, tillage operations may be performed specifically to provide a rough surface. Where it is not feasible to grow more than one harvestable crop during the year, a second crop may be grown primarily for the protection it provides from wind erosion. The secondary crop may be plowed under or may be left standing so the crop residue can provide protection to the newly planted crop being grown for harvest.

REFERENCES

BROWNING, G.M. 1950. Principles of soil physics in relation to tillage. Agric. Eng. *31*, 341-344.

LARSON, W.E. 1962. Tillage requirements for corn. Soil Water Conserv. *17*, 3-7.

NELSON, L.F., BURROWS, W.C., and STICKLER, F.C. 1975. Recognizing productive energy-efficient agriculture in the complex U.S. food system. ASAE Pap. *75-7505.* St. Joseph, Michigan.

RANEY, W.A., and ZINGG, A.W. 1957. Principles of tillage. *In* U.S. Department of Agriculture Yearbook. Washington, D.C.

SIDDOWAY, F.H., and BARNETT, A.P. 1976. Water and wind erosion control aspects of multiple cropping. *In* Multiple Cropping. R.I. Papendick, P.A. Sanchez and G.B. Triplett (Editors). Spec. Publ. *27.* American Society of Agronomy, Madison, Wisconsin.

SIEMENS, J.C., and OSCHWALD, W.R. 1976. Corn-soybean tillage systems: erosion control, effects on crop production, costs. ASAE Pap. *76-2552.* St. Joseph, Michigan.

TAYLOR, S.A., and ASHCROFT, G.L. 1972. Physical Edaphology. W.H. Freeman and Co., San Francisco.

UNGER, P.W., and STEWART, B.A. 1976. Land preparation and seedling establishment practices in multiple cropping systems. *In* Multiple Cropping. R.I. Papendick, P.A. Sanchez and G.B. Triplett (Editors). Spec. Publ. *27.* American Society of Agronomy, Madison, Wisconsin.

USDA. 1976A. Control of Water Pollution From Cropland, Volume I. U.S. Dep. Agric. Rep. *ARS-H-5-1.* Washington, D.C.

USDA. 1976B. Control of Water Pollution From Cropland, Volume II. U.S. Dep. Agric. Rep. *ARS*-H-*5-2.* Washington, D.C.

WISCHMEIER, W.H. 1976. Cropland erosion and sedimentation. *In* Control of Water Pollution from Cropland, Volume II. U.S. Dep. Agric. Rep. *ARS-H-5-2.* Washington, D.C.

WISCHMEIER, W.H., and SMITH, D.D. 1965. Predicting rainfall-erosion losses from cropland east of the Rocky Mountains. U.S. Dep. Agric. Handb. *282.*

WOODRUFF, N.P., and SIDDOWAY, F.H. 1965. A wind erosion equation. Soil Sci. Soc. Am. Proc. *29*, 602-608.

7

Soil Organic Matter, Microorganisms and Crop Production

D. Wynne Thorne

Soil organic matter is essential to productive soils. It promotes desirable physical and chemical properties of soil; it is the life base of a diverse population of microorganisms; and it acts as a storehouse of many plant nutrients.

Soil organic matter includes a range of freshly supplied materials and humus that is both intermediate decomposition products and finely divided residual compounds resistant to further microbial decomposition. Humus constituents include lignin residues of decomposition combined in part with proteins and related materials from the bodies of microorganisms that have been involved in decomposition. It also includes simple to complex carbohydrates, tannins, waxes, and complex polymers of various carbonaceous and protein materials. Soil humus is amorphous and is closely associated with the clay fraction. The proportion of soil organic matter that is bonded to clay particles, in all but the very sandy soils, ranges from 50 to over 90% (Allison 1973).

THE ROLE OF ORGANIC MATTER IN SOIL FERTILITY

The two primary functions of organic matter in soils are the improvement of soil physical properties and the general enhancement of microbial activities.

Physical Benefits

Organic materials left as crop residues on the surface of soils have a significant role in soil management. One pronounced benefit from such surface mulches is control of soil erosion. If there is a modest quantity to cover much of the soil surface, wind erosion is reduced and water is trapped and helped in entering the soil. Such surface materials can also reduce evaporation losses.

Under mechanized farming, leaving crop residues on the soil surface requires special tillage, planting, and weed control equipment and practices. In the U.S. Great Plains region, where wheat is the principal crop, tillage with sweeps or blades and control of weeds with revolving rod weeders are referred to as stubble mulching. A related and widely used system of maintaining residues on the soil surface is that of zero, reduced, or conservation tillage. In all of these systems herbicides are the primary agents for weed control. Crop residues may be chopped or left in place. Periodic disking or plowing may be practiced to reduce the volume of surface materials and incorporate some of the materials in the soil. Special equipment is usually required to cut through the surface residues and place seeds uniformly in moist soil.

Benefits Associated with Organic Matter Decomposition

Plant Nutrients' Availability.—For non-fertilized soils almost all of the nitrogen is tied up with the organic matter fraction. Less than 1% is ordinarily present at any given time in such available forms as ammonia or nitrate. Phosphorus is commonly present in organic combinations to the extent of 20 to 50% of the total, but this can be as high as 75%. The proportion of sulfur in the organic fraction varies appreciably, but in productive, non-saline soils half or more is likely to be in such combinations.

Plant nutrient elements other than nitrogen, phosphorus, and sulfur are also contained in, or associated with, soil organic matter. Potassium, calcium, magnesium, copper, zinc, manganese, and iron are in part associated with the organic complex.

Decomposition of organic matter makes mineral elements available to plants. Essentially all soils harbor a diverse and active population of microorganisms. Farm operators have a role in creating and maintaining a good state of activity of these organisms by maintaining a good supply of moisture in the soil, providing drainage as needed to make soils well aerated, and replenishing the organic matter supply periodically to keep the decomposition process active.

Cation Exchange Capacity.—Soil humus has a high capacity to absorb and exchange cations. The cation exchange capacity of humus increases with pH and varies appreciably with different cations, being much lower with potassium than with calcium or barium. The cation exchange capacity of humus usually amounts to 150 to 300 meq per 100 g. The proportion of the cation exchange capacity of productive mineral soils that is due to organic matter varies greatly but a range of

10 to 50% is quite common. In oxidized tropical soils, having primarily kaolinite types of clay mineral with low exchange capacities, most of the exchangeable cations are held by the organic fraction.

The common ranges of cation exchange capacity of humus and some of the principal clay minerals are shown in Table 7.1. Russell (1974) reported that organic manures increased the cation exchange capacity of soils in a tropical area in Nigeria. In a long term field experiment 15 annual applications of 5 t per hectare of farmyard manure to a sandy clay loam soil increased the cation exchange of the soil from 2.17 to 2.83, raised the soil pH from 5.42 to 5.82, and substantially decreased the exchangeable aluminum and increased exchangeable calcium, magnesium, and potassium.

TABLE 7.1. THE CATION EXCHANGE CAPACITY OF CLAY MINERALS AND ORGANIC MATTER

	Cation Exchange Capacity meq/100 g
Humus	150–300
Vermiculite	100–150
Montmorillonite	80–150
Illite	10– 40
Kaolinite	3– 15

Soil Structure Stability.—Products of organic matter decomposition are important agents for holding soil particles together in aggregates that are stable in water. Organic materials shown to be effective in holding stable aggregates together are primarily the products of microbial action such as polysaccharides. While moderately resistant to decomposition, these polysaccharide materials are gradually attacked by microorganisms. In consequence, maintaining soils in good physical condition is enhanced by periodic additions of organic materials. Oxides of iron and manganese also act as effective soil binding agents, and soils rich in such minerals usually have favorable granular structures.

The addition of decomposable organic matter to soils in poor physical condition often results in slight improvement. Studies have shown that in order to improve soil structure there must be a binding agent and an aggregating force or agent. The most desirable granular structures are usually found in soils just plowed out of grass covers. The fine network of grass roots furnishes one of the best aggregating agents and the decomposing root residues supply needed binding substances thoroughly distributed through the soil. Other plant roots are also beneficial. Other aggregating forces include some types of tillage such as the moldboard plow, wetting and drying, and freezing and thawing.

Growth Promoting and Phytotoxic Substances.—When organic matter decomposes, a wide variety of products is formed. Some of these have pronounced effects on plants. Growth promoting substances such as auxin, gibberellins, and β-indolyl acetic acid have been identified in decomposing organic matter in soils. Under some conditions these may stimulate root development, but the general practical value of such hormones in soil and crop management has not been adequately evaluated.

Small quantities of antibiotics have been identified in soils by some investigators following applications of fresh plant materials. The phytotoxins identified and found toxic to some plants include 3-indolyl acetic acid, coumarin, penicillin, citrinin, glutinosin, and a number of phenolic compounds.

Possibly related to the effects of phytotoxins on plants are the instances in which the turning under of certain crop residues injured the germination and growth of crops. Accumulations of wheat straw, for example, following several years of stubble mulch tillage were sometimes found to retard the development of maize. Guenzi and McCalla (1962) found that water soluble extracts of decomposing residues of several crops were toxic to the germination and seedling development of maize, wheat, and sorghum. The possibilities of phytotoxic effects are well established and should be avoided by planting after the early flush of rapid decomposition of plant materials, recently worked into a soil, has taken place.

ORGANIC MATTER CONTENT OF SOILS

When virgin soils are placed under cultivation there is usually a rapid decline in the organic matter content. The rate of decline gradually slackens and after long cultivation a new equilibrium is reached. Where drainage is poor, organic matter accumulates irrespective of climate. Virgin soils of arid regions are typically low in organic matter and nitrogen. Often when such soils are placed under irrigation and cropped, the content of organic matter increases.

The organic matter content of soils is the balance between rates of addition and rates of decomposition. Because of high rates of organic matter production, under many tropical conditions the organic matter is unusually high in relation to the generally accepted relationships with temperature and moisture. Such high organic matter situations have been observed in Colombia, Costa Rica, and Puerto Rico (Broadbent 1953).

Organic matter retention in soils is influenced by both the quantity and the type of clay present. Soil humus is adsorbed most vigorously by montmorillonite types of clays, less by illite types, and least by kaolinite types. Adsorption of organic materials on clay, particularly montmorillonite types, has been shown to decrease the rates of microbial decomposition. Workers are uncertain as to whether clay-humus combinations affect the amount of humus retained in soils.

No recommendation or formula can be given as to a desirable level of organic matter in soil. Europe has a tradition for including key crops in rotations and for conserving and applying available organic manures. In consequence, soils of many European countries are generally well supplied with decomposable organic matter. Even productive soils with good tilth benefit from frequent additions of organic matter. Therefore, crop residues should be returned to the soil where possible and animal manures should also be used to the extent possible.

MAINTAINING ORGANIC MATTER IN SOILS

The management of productive soils requires frequent applications of organic materials to replace those removed by decomposition. The principal materials available for the improvement of cultivated soils are crop residues, green manures, farmyard manure, composts, and urban and industrial wastes.

Crop Residues

Crop residues are the most important source of organic materials for soils and in large areas of the world the only important source. For most crops the above ground unharvested materials and the roots are distinguishable in management practices and in alternative uses. In all but the root crops, the roots are normally left in the soil. Where monoculture systems are practiced with cereal crops and in many other types of farming the straw is also returned to the soil. However, in many countries most of the straw of cereal crops is harvested for either livestock feed or fuel.

The quantities of crop residues that are available from cereal crops, such as wheat, barley, oats, and rye, are shown in Table 7.2. The actual quantities and the proportions of grain, straw, and roots vary considerably depending on the environment, the levels of soil fertility, and the supply of available moisture. According to these data and estimates of a 2% loss in soil organic matter per year, roots and stubble alone often will not provide all the organic matter needed to maintain a favorable soil balance.

If the straw or stalks are returned to the soil, or if harvested and used for livestock feed, and if the farmyard manure is returned to the soil, the soil organic matter supply is at least modestly sustained. Often where the straw is used for livestock feed, the animal manure deposited in the farmyard is used for fuel and the stubble may be closely grazed, compounding the organic matter deficit.

TABLE 7.2. RANGES IN YIELD AND NITROGEN AND PHOSPHATE CONTENT OF CEREAL CROPS

	Grain	Straw	Roots and Stubble
Yield, t/ha	1.6–2.0	2.5–3.0	2.0
Nitrogen, kg/ha	40–50	15–24	12–15
Phosphate, P_2O_5, kg/ha	10–16	12–15	12–15

Source: Adapted from Konova (1966).

The root systems of many grasses and some of the legume crops used for forage or grazing are substantial. Russell (1973) reported that roots in the upper 10 cm of soil in England totaled 6 t per hectare for *Agropyron cristatum*; 2.44 for *Agropyron pauciflorum*; 5.5 t for *Poa pratensis*; and 2.4 for alfalfa.

Green Manures

Green manure crops are grown primarily when they can be fitted into a cropping pattern without interfering with production of regularly harvested crops. A green manure crop is also often combined with a cover crop and with livestock grazing. The costs of seed and land preparation generally make the production of a crop, solely to plow under as a green manure, unprofitable.

Since the major response to a green manure type of crop is in added nutrients provided for subsequent crops, best results are usually obtained from the use of leguminous crops.

Farmyard Manure

After examining data for various types of livestock and differing management practices in the U.S., Salter and Schollenberger (1939) concluded that acceptable values for the recovery of fertilizer constituents in livestock excrement are approximately: nitrogen, 75%; phosphorus, 80%; and potassium, 85%. The recovery of organic matter in the feed averages about 40%. Their data also indicate that of the fertilizer nutrients in the excrement, 40% of the nitrogen, 2% of the

TABLE 7.3. ESTIMATED DAILY MANURE PRODUCTION BY VARIOUS FARM ANIMALS AND THE CONTENT OF FERTILIZER ELEMENTS

Animal	Raw Manure Production kg	Composition of Manures, Dry Basis			
		Moisture %	N %	P %	K %
Man	0.15	77			
Beef cattle	24.5	82	3.5	1.0	2.3
Dairy cattle	46.6	87	2.7	0.5	2.4
Chicken	0.15	73	4.3	1.6	1.6
Sheep	1.7	73	4.0	0.6	2.9
Swine	3.2	84	2.0	0.6	1.5
Horse	20.1	60			
Turkey	0.4	68			
Duck	0.34	73			

Source: Adapted from Meek *et al.* (1975).

phosphorus, and 75% of the potassium are in the urine. The quantities of manure voided and the content of fertilizer elements for various animals are shown in Table 7.3.

Where animals are grazing and the manure is directly voided on the soil, the nutrients are well conserved but the poor distribution results in less benefit to crops than would be expected based on the quantities involved. In the case of commercial dairy farms, manure is usually voided on concrete floors and hauled to the field daily or at short intervals. This is one of the most effective ways of conserving the value of manure. Where the manure and litter are allowed to accumulate, partially decompose, and to dry out or to be leached, large losses of nutrients are likely to occur. Ammonia nitrogen may be volatilized on drying and nitrate, nitrogen, and potassium are readily leached.

Value of Farm Manure.—Ordinary cow manure from the farmyard is generally considered to contain about 5 kg of nitrogen, 2.5 kg of phosphoric acid, and 5 kg of potash per ton. At 40¢, 35¢, and 18¢ per kg the nutrient value of one ton of manure would be about $3.77. Poultry manure may contain more than double the nutrients of cattle manure and would have proportionately higher value.

In the U.S. and in other countries with highly developed commercial farming, farmyard manure is considered primarily as a nitrogen fertilizer material, and as being too low in readily available phosphorus to be of benefit. However, in some tropical areas primary benefits from manure treatments have been attributed to the phosphate content. Russell (1974) reported that Heathcote found that maize in Samaru, Nigeria responded better to the ashes of farmyard manure than to the fresh manure plowed under. He also reported that at Kano, Nigeria the benefits of farmyard manure were only slightly less than for equivalent quantities of phosphorus supplied as superphosphate.

Numerous field experiments in the U.S. and elsewhere have shown that the value of farmyard manure for improving crop production may not exceed the benefits derived from equivalent quantities of fertilizer applied in mineral fertilizers. In consequence, many operators of commercial farms consider cattle and pig manure to be too low in plant nutrients to justify hauling and spreading (Allison 1973). Recent increases in fertilizer cost and differences in labor supply and cost in many other countries justify good management and field application of manures. In either situation, farm manures must be disposed of rather than being allowed to accumulate. Cultivated soils usually provide the best available place for disposal, and under most conditions, benefits from manure in increased crop yields will at least repay the costs of handling.

Urban Wastes

Cities and many food processing industries are encountering increasing difficulties in disposing of waste materials. In the U.S. the production of sewage sludge currently amounts to about 6 million tons per year on a dry weight basis. Based on average composition this contains about 200,000 t of nitrogen, 138,000 t of phosphorus, and 6000 t of potassium. About 40% of this sludge is now disposed of in landfills, 20% is applied to land, 25% is incinerated, and 15% is discharged into lakes and oceans. Some of the major plant nutrient elements and other mineral elements in sewage sludge, based on analysis of samples from 150 different sewage treatment plants in 6 North Central states are give in Table 7.4.

The nutrient elements in treated sludge are variable and depend on the source of the sewage and method of treatment. Nitrogen and phosphorus are the principal nutrients of general value that are present in substantial amounts. The micronutrients may be valuable under appropriate conditions. Considerable emphasis has been placed on hazards from the heavy metals present, particularly zinc, lead, cadmium, copper, and nickel. These are strongly adsorbed on clay particles and accumulate with repeated applications. Obviously precautions are needed to assure that soils are not made unfavorable for plants or that toxic concentrations of metals enter the food chain.

Nitrogen in an average sludge would consist of about 40% in ammonia forms and 60% in complex organic compounds. Unless injected into the soil or worked under soon after application, much of the ammonia nitrogen may be lost by volatilization. Under good management about half of the ammonia nitrogen would be lost and the re-

TABLE 7.4. PLANT NUTRIENTS AND METALLIC ELEMENTS IN SEWAGE SLUDGE FROM 150 TREATMENT PLANTS IN 6 NORTH CENTRAL STATES

	Concentration Dry Weight Basis	
	Range	Median
	(%)	(%)
Nitrogen	0.1−17.6	3.3
Phosphorus	0.1−14.3	2.3
Potassium	0.02−2.64	0.3
	(ppm)	(ppm)
Manganese	18−7,100	260
Boron	4−760	33
Cobalt	1−18	4
Molybdenum	5−39	30
Zinc	101−27,800	1,740
Copper	84−10,400	850
Mercury	0.5−10,600	5
Arsenic	6−230	10
Lead	13−19,700	500
Cadmium	3−3,410	16
Nickel	2−3,500	82
Chromium	10−99,000	890

Source: Sommers (1977).

mainder could be considered immediately available to plants. Also about 20% of the organic nitrogen would become available the year of application. About 3% of the residual nitrogen would become available each of the 3 succeeding years.

In some instances excess available phosphorus may also build up in soil with repeated sludge treatments. This should be checked with soil tests. The criteria used to prevent metal injury from sludge application on land are based upon the total amount of lead, copper, zinc, nickel, and cadmium added. Whereas nitrogen commonly limits the annual application rate of sludge, its metal content will determine the length of time a given area can receive sludge. The upper limit for metal additions is given in Table 7.5. One further precaution is that annual additions of cadmium should not exceed 2.2 kg per hectare. A soil pH above 6.5 must be maintained on sludge treated soils to reduce toxicity and plant uptake of the potentially toxic heavy metals.

Sewage sludges have proved effective in reclaiming such land types as mine spoil banks, dune and dredged sands, and abandoned garbage dumps and sanitary land fills.

TABLE 7.5. TOTAL AMOUNT OF SLUDGE METALS ALLOWED ON AGRICULTURAL LAND

	Soil Cation Exchange Capacity (meq/100 g)[1]		
	0–5	5–15	15
Lead	500	1000	2000
Zinc	250	500	1000
Copper	125	250	500
Nickel	50	100	200
Cadmium	5	10	20

Source: Sommers and Nelson (1976).
[1]Determined by the ammonium acetate procedure.

MANAGING MICROBIAL PROCESSES

The diverse population of microorganisms in most productive soils totals several hundred million per gram. Many of the activities carried out by these organisms are important and often essential to the production of high yields of good quality crops. These microbial processes include organic matter decomposition, the conversion of elements in organic matter to available forms, and nitrogen fixation. There are also microbial activities considered undesirable, such as denitrification and the diseases of plants.

Organic Matter Decomposition

A modest rate of decomposition that maintains an active organic matter base that will contribute to the cation exchange capacity of the soil, to a granular soil structure, and to a gradual release of available plant nutrients is desirable. There are seldom problems of organic matter decomposition in productive agricultural soils. This is because the conditions favoring organic matter decomposition are about the same as those favoring good growth of most field crops. These conditions under which microorganisms decompose organic materials effectively include a supply of available moisture, a well aerated soil, ample supplies of available plant nutrients, and a warm temperature (20° to 35° C).

To attain these favorable conditions soils should be well drained. Where waterlogging occurs soils are poorly aerated and organic matter decomposition is retarded. Drainage is one solution. Cultivation of soils usually improves aeration and speeds organic matter decomposition.

Maintaining a favorable carbon:nitrogen (C:N) ratio is one of the important factors in avoiding problems of unbalanced plant nutrition in soils that have recently received a fresh supply of organic matter. If the C:N ratio of freshly added organic matter is much wider than 20:1 the microorganisms involved in the decomposition process are unable to obtain sufficient nitrogen from the organic material to metabolize the carbon and they must draw on available supplies of nitrogen in their environment. Crop plants fare badly in such competition and if there is not adequate available nitrogen in the soil the crop will suffer from nitrogen deficiency.

The straw of cereal crops and the stalks of maize and millet have C:N ratios ranging from 40:1 to 80:1. In consequence, incorporating these materials in soils should be accompanied by addition of supplemental nitrogen fertilizer, or time should be allowed for decomposition before a crop is planted.

Nitrogen Transformations

Since nitrogen is often inadequate for high crop yields and since its availability in soils is closely associated with the activities of soil microorganisms and with organic matter, the changes in form of soil nitrogen and the activities of the microorganisms are important to farm management. Plants absorb most of their nitrogen as either NH_4^+ or as NO_3^- ions, so these two forms receive special attention.

Ammonification.—This is the process through which ammonia is formed by the action of microorganisms on organic nitrogen compounds. Ammonia is formed as a by-product of organic matter decomposition, with production becoming active as the C:N ratio approaches 10:1. The released ammonia forms compounds such as NH_4HCO_3 or the ammonium ion is adsorbed on the soil colloids. Only in very sandy soils is ammonia lost to any appreciable extent by leaching, but near the soil surface it may be volatilized to the air as the soil or organic matter dries. Ammonification takes place under as broad a range of soil conditions as will support the growth of crop plants.

Nitrification.—Although a number of microorganisms have been shown to participate in the oxidation of ammonia nitrogen to nitrate nitrogen, there is a general consensus (Allison 1973) that in agricultural soils the primary agents are two groups of autotrophic bacteria: the

Nitrosomonas group that oxidizes ammonia to nitrite and the Nitrobacter group that oxidizes nitrite to nitrate. These are collectively referred to as nitrifying bacteria.

The conditions favoring the nitrification process in soils are more restricted than those required for most soil biological processes. Favorable conditions include a good supply of organic matter with a gradual release of ammonia, good soil aeration, adequate available soil moisture, a temperature between 20° and 30°C, and a pH between 6.0 and 9.0.

Denitrification.—Denitrification is the biological reduction of nitrate or nitrite to gaseous nitrogen. This is one of the most important causes of loss of available nitrogen from soils. The process is brought about by numerous species of bacteria that normally use oxygen from air, but which can use the oxygen in nitrate or nitrite combinations under anaerobic conditions. The nitrogen is then released as either nitrogen gas or as gases of nitrous oxide. The process can proceed rapidly when there is a supply of nitrate or nitrite in the presence of decomposable organic matter and conditions become anaerobic. Denitrification starts when oxygen in the air is reduced below about 6%. Denitrification often occurs in and under manure piled on the soil surface as well as in soils.

Water may leach nitrate and nitrite ions from upper well aerated parts of the soil or from a pile of manure to lower and more moist layers where conditions are anaerobic. At pH values below 6.0 most of the nitrogen losses occur as nitrous oxides; at pH 7.3 to 7.9 the losses are mostly as nitrogen gas. Alternate aerobic and anaerobic conditions, usually brought about by conditions of limited and of excess water, favor the formation of nitrate and then its reduction and loss.

Nitrogen Fixation

For about 100 years it has been known that certain groups of soil microorganisms, both free-living and in close association with the roots or leaves of higher plants, can fix atmospheric nitrogen and thereby enrich the soil or supply nitrogen to plants. The inoculation and growth of leguminous plants has been the major practical use of this knowledge. The symbiotic relationship between legumes and Rhizobium strains of bacteria is now widely utilized. Although there have been many reports of inoculating soils with free-living nitrogen-fixing microorganisms, there is little evidence of substantial benefits from such practices. Recently the discovery that nitrogen may be fixed by certain bacteria living in close association with the roots of a number of tropical grasses has attracted interest, renewing hope for further control and use of the biological nitrogen fixation process.

Nitrogen-Fixing Microorganisms

The major groups of microorganisms that have been found to fix atmospheric nitrogen include free-living species in the Azotobacter, Beijerinkia, Clostridium, and blue-green algae groups. Other organisms require close association with higher plants to accomplish nitrogen fixation. These include Rhizobium bacteria living in nodules on the roots of leguminous plants. Some actinomycete-like organisms in nodules on the roots or leaves of non-legumes such as Alnus, Casuarina, Ceanothus, and Eleagnus plant genera also have this capability. Some blue-green algae seem to be active nitrogen fixers in close association with the leaves of a small fern, Azolla. Evidence is also accumulating that certain species of *Spirillum* and other bacteria are active nitrogen fixers when in close association with the roots of selected species of the grass family.

REFERENCES

ALLISON, F.E. 1973. Soil Organic Matter and Its Role in Crop Production. Elsevier Scientific Publishing Co., New York.

BARTHOLOMEW, W.V. 1972 Soil nitrogen and organic matter. *In* Soils of the Humid Tropics. National Academy of Science-National Research Council, Washington, D.C.

BROADBENT, F.E. 1953. The soil organic fraction. *In* Advances in Agronomy, Vol. 5. A.G. Norman (Editor). Academic Press, New York.

GUENZI, W.D., and MCCALLA, T.M. 1962. Inhibition of germination and seedling development by crop residues. Soil Sci. Soc. Am. Proc. *26*, 456-458.

KONOVA, M.M. 1966. Soil Organic Matter. Pergamon Press, Oxford. (Russian)

MCCALLA, T.M., and ARMY, T.J. 1961. Stubble mulch farming. *In* Advances in Agronomy, Vol. 13. A.G. Norman (Editor). Academic Press, New York.

MEEK, B.L. *et al.* 1975. Guidelines for manure use and disposal in the western region. Wash. Res. Cent. Bull. *814*.

MOORE, A.W. 1969. Azolla: biology and agronomic significance. Bot. Rev. *35*, 17-34.

NEYRA, C.A., and DOBEREINER, J. 1977. Nitrogen fixation in grasses. *In* Advances in Agronomy, Vol. 29. N.C. Brady (Editor). Academic Press, New York.

QUISPEL, A. 1974. The Biology of Nitrogen Fixation. North Holland Publishing Co., Amsterdam.

RAPPAPORT, L., and SACHS, R.M. 1976. Physiology of Cultivated Plants. Rappaport and Sachs, Davis, California.

RUSSELL, E.W. 1973. Soil Conditions and Plant Growth, 10th Edition. Longmans, Green and Co., London.

RUSSELL, E.W. 1974. The role of fertilizers in African agriculture. *In* Fertilizers, Crop Quality, and Economy. V.J. Fernandez (Editor). Elsevier Scientific Publishing Co., New York.

SALTER, R.M., and SCHOLLENBERGER, C.J. 1939. Farm Manure. Ohio Agric. Exp. Stn. Bull. *605*. Wooster.

SOMMERS, L.E. 1977. Chemical composition of sewage sludges and analysis of their potential use as fertilizers. J. Environ. Qual. *6*, 225-232.

SOMMERS, L.E., and NELSON, D.W. 1976. Analyses and their interpretation. *In* Application of Sludges and Wastewaters on Agricultural Land: A Planning and Educational Guide. B.D. Knezek and R.H. Miller (Editors). Res. Bull. *1090*. Ohio Agricultural Research Development Center, Wooster.

WARSI, A.S., and WRIGHT, B.C. 1973. Influence of nitrogen on the root growth of grain sorghum. Indian J. Agric. Sci. *43*, 142-147.

8

Irrigation and Crop Production

D. Wynne Thorne

Crop plants require much larger quantities of water for growth than of any other growth factor. A range of 200 to 1000 kg of water may be absorbed and largely transpired for each kg of dry matter produced. Most irrigation waters contain sufficient soluble salts to require special practices to prevent accumulations in soils. Although irrigation has the potential for greatly increasing crop yields in arid regions, the costs of storing, transporting, pumping, and applying water to soils are often great and the costs and benefits need to be carefully examined before initiating an irrigation development program.

In view of the costs of water and the facilities and labor required for effective application, the adoption of relatively high levels of technology (good seeds, high plant populations, fertilizers, pest control, land levelling) is often needed to make irrigated farming profitable.

CLIMATE IN RELATION TO IRRIGATION

Irrigation is potentially justifiable wherever periods of moisture stress occur that would substantially reduce crop yields and where a supply of good quality water is available at reasonable cost. In consequence, irrigation is often profitable in climatic zones in which total precipitation is relatively high, but in which periods of drought frequently reduce crop yields. Since the available moisture that can be retained in soils varies, irrigation may be more essential for high crop yields on sandy or shallow soils than on relatively deep loam soils. Deep-rooted crops can tap soil moisture reserves and continue growth longer than shallow-rooted crops during drought periods; hence, the latter may have greater need for irrigation.

Evapotranspiration (ET) (see Chapter 4) is important in determining time and frequency of irrigation. Doorenbos and Pruitt (1977) have assembled crop coefficients (compared with grass rather than alfalfa) for a number of crops. These are shown in Table 8.1. The magnitude of ET values shown varies not only with the physiology of the crop, but also with the length of the growing season. Some of the tropical fruits, for example, have a 12-month growth period in frost-free areas.

TABLE 8.1. APPROXIMATE RANGE OF SEASONAL ET (CROP) IN COMPARISON WITH ET (GRASS)

Crop	Seasonal Crop ET	
	mm	Grass (%)
Alfalfa	600–1500	90–105
Avocados	650–1000	65–75
Bananas	700–1700	90–105
Beans	250–400	20–25
Cocoa	800–1200	95–110
Coffee	800–1200	95–110
Cotton	550–950	50–65
Dates	900–1300	35–100
Deciduous Trees	700–1050	60–70
Flax	450–900	55–70
Grains (small)	300–450	25–30
Grapefruit	650–1000	70–85
Maize	400–700	30–45
Oil Seeds	300–600	25–40
Onions	350–600	25–40
Oranges	600–950	60–75
Potatoes	350–625	25–40
Rice	500–950	45–65
Sisal	550–800	65–75
Sorghum	300–650	30–45
Soybeans	450–825	30–45
Sugarbeets	450–850	50–65
Sugar Cane	1000–1500	105–120
Sweet Potatoes	400–675	30–45
Tobacco	300–500	30–35
Tomatoes	300–600	30–45
Vegetables	250–500	15–30
Vineyards	450–900	30–55
Walnuts	700–1000	65–75

Source: Doorenbos and Pruitt (1977).

Irrigation systems have been developed in different countries with differing goals. In the U.S. and in most developed countries systems have been designed to provide sufficient water for near maximum yields of most crops. In India and many other developing countries, and especially where irrigation projects evolved under the philosophy of the British Empire, emphasis has been to supply water to ensure modest yields over as large an area as possible. Under the latter concept the command area of an irrigation project often includes 2 to 3 times as much farm land as can be irrigated for maximum crop yields.

WATER SUPPLY AND QUALITY

The quantity and periods of availability of irrigation water limit irrigation practices. Rights to irrigation water vary greatly. In some instances rights to water are tied to land. In others, rights are established by prior use or by purchase and ownership. Laws governing water rights vary among countries and even among states within countries. For each farm the nature of the water right and quantity and dependability of water supply should be determined before investments are made in irrigation facilities. If, for example, water comes by direct diversion from a naturally flowing stream, the quantity may fluctuate greatly. Also, water is often available only at periodic intervals such as every 10, 15, or 20 days. For any situation, estimates can be made of water requirements for specific crops and soils and some adjustments can be made to use available water as effectively as possible. If the quantity of irrigation water available for the crop season is known, this can be added to water stored in the soil at planting time and to anticipated precipitation. These data are a basis for planning cropping practices and also indicate whether the potential for high crop yields is great enough to justify purchase of fertilizers and other inputs.

Water Quality

The quality of irrigation water depends primarily on its content of silt and salts. Silt may add some fertility to soils, but it tends to clog canals and may be deposited on limited parts of fields, necessitating periodic land levelling.

Among the most important salt factors that determine water quality are total concentration, the proportion of sodium to other cations, and the presence of special toxic ions such as borate, or, for some crops, chloride, sodium, or bicarbonate. The classification system usually used for rating the quality of irrigation waters is that developed by the U.S. Salinity Laboratory staff. Total salt concentration is measured as electrical conductivity. For each increase of 1.0 millimhos per centimeter the salt content of the water increases by about 640 ppm. As salt concentration increases, problems of drainage and leaching to prevent excessive accumulation become increasingly important.

The hazard of sodium being absorbed on soil clay can be estimated from the concentrations of sodium, calcium, and magnesium in the water using the convenient nomogram shown in Fig. 8.1. With medium and fine textured soils exchangeable sodium on soil clay in excess of 12 to 15% causes swelling and reduces permeability. Also, the balance of available plant nutrients may be disturbed.

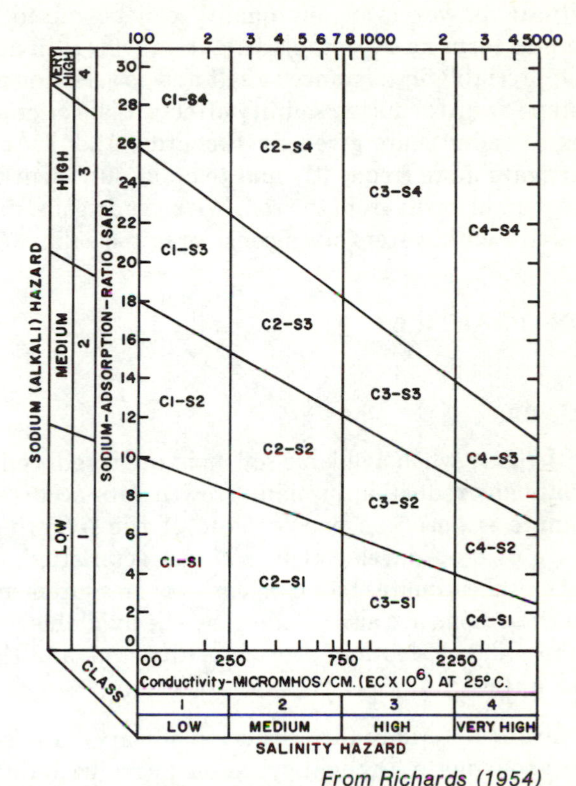

From Richards (1954)

FIG. 8.1. DIAGRAM FOR THE CLASSIFICATION OF IRRIGATION WATERS

Because nearly all irrigation waters contain sufficient salt to damage crop plants if allowed to accumulate, irrigation requires careful attention for salt control. The first preventive condition is good drainage, thus management of low quality water is easier on sandy soils. A second preventive condition is application of some excess irrigation water periodically to leach excess salts below the crop root zone and preferably from the soil. Excess sodium is more difficult to manage because exchangeable cations on clay are in equilibrium with the cations in the soil solution. If irrigation water is low in salts a high sodium adsorption ratio in irrigation water can be reduced by adding powdered gypsum to the water to increase the Ca^{++} content. Such a practice becomes impractical, however, as the conductivity of the soil solution increases, particularly above about 750 micromhos per centimeter. General management practices for saline and sodic soils have been discussed by Thorne and Peterson (1954).

In some situations waters of low quality can be mixed with higher quality waters to form an acceptable irrigation water. In addition, crop plants vary appreciably in tolerance to salt and so selection of crops may be another step toward reducing salinity effects. Lists of crops according to tolerances to salinity are given by Richards (1954). Another alternative is to irrigate more frequently and to apply sufficient excess water to leach salt accumulations from the soil. With sandy soils, drip irrigation has been effective with waters often considered too saline for irrigation.

IRRIGATION PRACTICES

When to Irrigate

It is time to irrigate when available soil moisture is reduced to the point at which significant reduction in plant growth rate occurs. In practice, such an estimate is based on one or more of the following: plant appearance, stage of crop development, soil characteristics, soil moisture measurements, and estimates based on climatic measurements. The periods between irrigations are also affected by the available water holding capacity of the soil, the depth of rooting of the crop, and the quality of the irrigation water.

Plant Appearance.—Many farmers use plant appearance as an indication of when to irrigate. Drought symptoms vary with different crops and these may change with stage of growth. Leaves of many crops such as beans, cotton, and groundnuts become dark green as soil moisture stress becomes severe. Maize and sorghum leaves curl and change their angle or orientation as moisture stress increases. In other crops leaf wilting becomes pronounced.

Critical Stages in Crop Development.—Where irrigation water is limited there may be opportunities for selecting the time of application in relation to stage of crop development. Such critical periods for water application include those in which water stress may greatly reduce yields. Much of the available data has been reviewed by Yaron, Danfors, and Vaadia (1973) and Doorenbos and Pruitt (1977). Some of the data of these authors are summarized in Table 8.2. Irrigation during the periods indicated seems to benefit crops most if water stress at other periods does not reach critical permanent wilting levels.

TABLE 8.2. CRITICAL SOIL WATER STRESS PERIODS FOR DIFFERENT CROPS

Crop	Critical Periods
Cereal Crops	
Wheat	Booting, flowering, and grain formation early growth
Barley	Early booting, tillering, soft dough
Oats	Beginning of ear emergence to flowering
Maize	Flowering and early seed development prior to tasseling, grain filling period
Sorghum	Tillering to booting, heading, flowering, and grain development
Pearl Millet	Heading and flowering
Legume Crops	
Alfalfa	Immediately after cutting for hay, start of flowering for seed production
Peas	Flowering and pod formation
Beans	Flowering and pod setting, earlier, ripening period
Groundnuts	Flowering and seed development, germination to flowering
Broadbeans	Flowering
Snap Beans	Flowering and fruit development
Oil Crops	
Castor Beans	Requires good moisture during full growing period
Sunflower	Flowering and seeding
Soybeans	Flowering and fruiting, period of maximum vegetative growth
Miscellaneous Crops	
Sugar Beets	3 to 4 weeks after emergence
Sugar Cane	Period of maximum vegetative growth
Tobacco	Knee high to blossoming
Tomatoes	Flowering and period of rapid fruit development
Watermelon	Flowering to harvest
Broccoli, Cabbage	During head formation and enlargement
Tree Fruits	
Apricots	Flowering and bud development
Cherries	Period of rapid fruit growth
Citrus	Flowering and fruit setting
Peaches	Period of rapid fruit growth
Mangoes	Flowering, some water stress before flowering considered desirable
Other	
Bananas	Requires good moisture throughout growth and fruit development periods, particularly during periods of hot weather
Coffee	Flowering and fruit development, some water stress before flowering generally held desirable
Olives	Just before flowering and during fruit enlargement

Source: Doorenbos and Pruitt (1977).

Soil Characteristics.—Various schemes such as that of Diebold (1953) have been developed for estimating the relative proportion of available moisture in soils based on such physical properties as cohesiveness and plasticity in relation to texture. Such observations are helpful as guides to soil moisture status and irrigation need, but are often replaced by more direct and accurate procedures.

Soil Moisture Measurements.—Field moisture measurements are being used increasingly as a guide to when to irrigate. The procedures involved and their application have been reviewed in Chapter 4 and in more detail by Taylor, Evans, and Kemper (1961).

Soil moisture measurements involve the use of scientific equipment that is seldom available on farms and there must also be knowledgeable technicians to install the equipment and to supervise its operation. In the U.S. it is moderately common for such measurements to be handled by consulting companies. Irrigation is commonly scheduled when the soil matrix potential reaches a certain value such as 0.5 bar at a depth of 25 cm, or when a certain proportion, such as 50 or 60%, of the available soil moisture in the major root zone has been depleted.

Climatic Data.—Various formulas have been devised for estimating potential evapotranspiration losses of water by different crops.

The major formulas employed and procedures for their use have been reviewed by Doorenbos and Pruitt (1977) and discussed in Chapter 4 of this book. If the available moisture capacity of the soil, depth of rooting of the crop, and the quantity of potential evapotranspiration of the crop are known, irrigation can be initiated when soil moisture depletion has reached a predetermined figure. Thus, if the soil has an available moisture capacity of 150 mm per meter, if the depth of rooting is 0.6 m, and if irrigation is planned when half of the available moisture in the soil root zone has been depleted, then irrigation should be scheduled when estimates indicate 45 mm of soil moisture has been depleted.

Rate and Quantity of Water to Apply

The quantity of water to apply per irrigation should approximate the quantity needed to restore the soil moisture content to field moisture capacity throughout the depth of crop rooting. Allowance must also be made for water application efficiency. An example of a procedure for estimating specifications for water application is the following:

(1) Estimate the total depth, d, of water required to bring the required depth of soil to field capacity. This procedure has been described in the previous section.

(2) Estimate the infiltration rate, I, of the soil. This establishes the time of water application to have a given depth of water enter the soil.

(3) Estimate the time, t, the water should be applied. If the depth, d, required to increase the soil moisture to field capacity and the infiltration rate of the soil are known, the time the water should be applied is given by:

$$t = d/I$$

Thus, if d is computed to be 45 mm and I = 2 cm/hr, t = 2.25 hr.

In some cases, particularly with sprinkle irrigation, water entry into the soil may be limited by the rate of water application, R, rather than by infiltration rate. In such instances the time would be estimated from $t = d/R$, in which R can be expressed in a convenient unit such as cm or mm depth per hour.

(4) Estimate or measure the size of the irrigation stream, q, in liter/sec or m^3/hr. Land area is ordinarily expressed in hectares.

(5) The values secured may be used to calculate the area covered, using the following equation:

$$a = qt/d$$

Thus, if the stream size (q) is 10 liter/sec (36 m^3/hr), the area irrigated in 2.25 hr is:

$$a = \frac{36\ m^3/hr \times 2.25\ hr}{0.045\ m} = 1800\ m^2, \text{ or } 0.18\ ha$$

(Doorenbos and Pruitt 1977).

The calculations assume a uniform depth of water being applied over the area irrigated and no deep percolation losses from surface runoff. Such precise irrigation conditions can seldom be met under farm situations.

Water Application Efficiency.—An ideal irrigation is assumed to be one which supplies enough water to wet the soil uniformly to field moisture capacity throughout the root zone. The degree to which such an application is attained is defined in terms of water application efficiency, or the ratio of water that is stored in the root zone of the soil to the amount of water delivered to the farm, or in which Ws represents the quantity of

$$Ea = 100\ Ws/Wf$$

water stored in the rooting zone of the soil, and Wf represents the quantity of water delivered to the area irrigated. Ea is the water application efficiency, expressed in percentage. Water application efficiency on farms is often low, frequently in the range of 35 to 60%.

In the example cited above, a stream of 10 liter/sec was found to supply 45 mm depth of water to 0.18 ha of land in 2.25 hours, assuming 100% efficiency. If the water application efficiency is 50%, the water would need to be applied 5 hours to the same area to complete the irrigation, instead of 2.25 hours.

Where water supplies are limited and the costs of water are relatively high, improved facilities and practices are justified to apply water more efficiently. Levelling land to zero slope or to a small uniform grade is necessary for uniform water application by surface methods. Also, catching runoff water in a reservoir and reusing it saves water. With basins, borders, or furrows a quick coverage of the soil is needed to secure uniform penetration over a field. Shifting to well designed sprinkler irrigation systems or even to trickle irrigation are further steps that usually improve water application efficiency. With good facilities and careful management, water application efficiencies of 70 to 85% are attainable.

The allowable time between irrigations and the quantity of water that would be applied per irrigation for a given crop depend on differences in the water holding capacities of soils. These vary in large measure with texture. Some approximate values for a range of differences in soils of temperate regions are shown in Table 8.3. Tropical soils (oxisols, utisols) usually have much lower water holding capacities.

TABLE 8.3. RELATION BETWEEN SOIL WATER TENSION IN BARS (ATMOSPHERES) AND AVAILABLE SOIL WATER IN mm/m SOIL DEPTH

	Soil Water Tension in Bars			
	0.2	0.5	2.5	15
Soil Texture	Available Soil Water in mm/m			
Heavy Clay	180	150	80	0
Silty Clay	190	170	100	0
Loam	200	150	70	0
Silt Loam	250	190	50	0
Silty Clay Loam	160	120	70	0
Fine Textured Soils	200	150	70	0
Sandy Clay Loam	140	110	60	0
Sandy Loam	130	80	30	0
Loamy Fine Sand	140	110	50	0
Medium Textured Soils	140	100	50	0
Medium Fine Sand	60	30	20	0
Coarse Textured Soils	60	30	20	0

Source: Doorenbos and Pruitt (1977).

Also to be considered are differences in the rooting depths of various crop plants. Rooting depth is usually greater in sandy soils, but some average values of rooting depths and commonly accepted values for the depletion of available moisture for these crops as reported by Doorenbos and Pruitt (1977) are given in Table 8.4.

TABLE 8.4. GENERALIZED DATA ON ROOTING DEPTH OF FULL GROWN CROPS, FRACTION OF AVAILABLE SOIL WATER, AND READILY AVAILABLE FOR DIFFERENT SOIL TYPES (mm/m SOIL DEPTH) WHEN ET CROP IS 5 TO 6 mm/DAY

Crop	Rooting Depth m	Fraction of Available Soil Water[1]	Readily Available Soil Water (mm/m) Fine	Medium	Coarse
Alfalfa	1.0–2.0	0.55	110	75	35
Bananas	0.5–0.9	0.35	70	50	20
Beans	0.5–0.7	0.45	90	65	30
Carrots	0.5–1.0	0.35	70	50	20
Citrus	1.2–1.5	0.5	100	70	30
Clover	0.6–0.9	0.35	70	50	20
Cotton	1.0–1.7	0.65	130	90	40
Deciduous Orchards	1.0–2.0	0.5	100	70	30
Grains, Small	0.9–1.5	0.6	120	80	40
Grains, Winter	1.5–2.0	0.6	120	80	40
Grass	0.5–1.5	0.5	100	70	30
Groundnuts	0.5–1.0	0.4	80	55	25
Maize	1.0–1.7	0.6	120	80	40
Melons	1.0–1.5	0.35	70	50	25
Palm Trees	0.7–1.1	0.65	130	90	40
Peppers	0.5–1.0	0.25	50	35	15
Potatoes	0.4–0.6	0.25	50	30	15
Safflower	1.0–2.0	0.6	120	80	40
Sorghum	1.0–2.0	0.55	110	75	35
Soybeans	0.6–1.3	0.5	100	75	35
Sugar Beet	0.7–1.2	0.5	100	70	30
Sugar Cane	1.2–2.0	0.65	130	90	40
Sunflower	0.8–1.5	0.45	90	60	30
Sweet Potatoes	1.0–1.5	0.65	130	90	40
Tomatoes	0.7–1.5	0.4	180	60	25
Vegetables	0.3–0.6	0.2	40	30	15
Wheat	1.0–1.5	0.55	105	70	35
Wheat (ripening)		0.9	180	130	55
Total Available Soil Water			200	140	60

Source: Doorenbos and Pruitt (1977).
[1]When ET crop is 3 mm or smaller increase values by some 30%; when ET crop is 8 mm/day or more reduce values by some 30%, assuming non-saline conditions (EC_e 2 mmhos/cm).
[2]Higher values than those shown apply during ripening.

Methods of Irrigation

Most of the methods of irrigation are well understood by farmers who are experienced in irrigation. The methods have also been described in numerous publications such as Israelsen and Hansen (1962); Thorne and Peterson (1954); Hagan, Haise, and Edminster (1967). Rather than repeating these extensive descriptions, Table 8.5 summarizes the principal irrigation methods, conditions favoring their use, and the major limitations and advantages of each.

TABLE 8.5. ADAPTATIONS, LIMITATIONS, AND ADVANTAGES OF VARIOUS METHODS OF IRRIGATION

Method	Adaptation	Limitations	Advantages
Furrow	All row crops	Moderate labor requirements	Uniform water application
	All irrigable soils	Engineering design essential for high efficiencies	High water application efficiency
	Slopes up to 5% with row crops and up to 15% for contour furrows in orchards	Some runoff usually necessary for uniform water application	Good control of irrigation water
		Erosion hazard on steep slopes from rainfall	Control equipment available at low cost, such as spiles, gates and siphons
Border Strip	All irrigable soils	Extensive land grading required	High water application efficiency possible
	Close growing crops	Engineering design necessary for high efficiency	Efficient in use of labor
	Slopes up to 3% for grains and forage crops	Relatively large flows required	Applicable on all soil types
	Slopes up to 7% for pastures	Shallow soils cannot be economically graded	Low maintenance costs
		Dikes hinder cultivation and harvesting	Positive control over irrigation water
Checks or Level	All irrigable soils	Extensive land grading often required	Good control of irrigation water
	Orchards and close growing crops	Large flows required	High water application efficiency
	Slopes up to 2.5% or more when benched or terraced	Initial cost relatively high	Uniform water application and leaching
		Dikes hinder equipment operations	Low maintenance costs
		Maintenance problems on steep slopes	Erosion control from irrigation and rainfall
		May affect yields on crops sensitive to inundation	Large streams can be utilized

TABLE 8.5 (CONT.)

Sprinkler	All irrigable soils All crops	High capital costs Labor costs may be high Leaf damage to citrus and some other crops with low quality water May encourage some foliar diseases	Can apply relatively uniform and accurate quantities of water Adapted to uneven topographies Can be used for fertilizer distribution, temperature control, and frost protection
Portable	(Main lines and laterals may include pump)	With band portable labor costs are high With self-propelled equipment and maintenance costs are high	Equipment costs relatively low Labor costs low
Semi-portable	(Stationary pumps and main lines, laterals portable)	Equipment and installation costs high	Reduced labor
Self-propelled	All but steep slopes	Equipment maintenance costs high	Labor costs low
Stationary		Equipment and installation costs very high, usually too high for for most field crops	Labor costs very low Equipment in place for irrigation at any time
Subirrigation	Relatively level soils with slowly permeable substrations, height of ground water can be regulated Often operated in conjunction with drainage system Soil highly permeable Area to be irrigated must be large	Ideal conditions occur in limited areas Adjoining land owners must follow similar practices Needs high quality water and soil low in salts Drainage and leaching practices must be intensive Choice of crops may be limited to narrow range of rooting characteristics In arid regions one annual surface irrigation may be needed to leach soils	Effective on coarse textured soils with dense subhorizons Labor requirements low for well designed systems Special tillage for conveying surface water eliminated Evaporative loss of water from soil surface is low Crop response generally good

TABLE 8.5 (CONT.)

Trickle	Relatively permeable soils High value perennial field crops, berries, or tree crops Low quality irrigation water	High installation costs Systems interfere with cultivation, must be removed for harvest of field and vegetable crops Clogging of tricklers, necessitate filtering system With saline waters soil leaching is necessary between crops	Efficient use of water, ET may be reduced by one-third Effective with waters of low quality Fertilizers may be applied in water

Source: Adapted from Hagan, Haise, and Edminster (1967).

The statements in Table 8.5 were prepared for productive farms and generally adequate supplies of water. In developing countries, some adjustments in concepts often may be necessary. Methods that use more labor and do not require costly facilities may be preferred. Also, with small farms all operations are on a smaller scale. Basins are smaller and are often hand levelled. Furrows and beds are often hand made and confined in small and relatively level basins. A common type of hand made furrow and ridge is shown in Figs. 8.2. and 8.3.

FIG. 8.2. HAND CONSTRUCTED LEVEL FURROWS AND RIDGES FOR IRRIGATING SMALL AREAS OF LAND. NEAR DAMASCUS, SYRIA, 1977

Systems

Four major types of systems are briefly described. However, it should be recognized that equipment for sprinkling is changing rapidly.

Movable Pipe.—Sprinkler pipes with attached sprinkler nozzles come in lengths of 20 to 40 ft. They can be moved by hand from one irrigation set to the next. These systems are the lowest in capital costs of sprinkler systems but labor requirements are relatively high. Two laborers are often needed to move pipe efficiently and to reduce the length of time the system is inoperative. The pump can be stationary and furnish water under pressure through a main line, or the pump can be moved with the pipe.

Mechanical-Move Sprinkler Laterals.—These systems are moved between operating locations by engine, tractor, or handpower and are mounted on wheels or skids. One example is the drag type of aluminum pipe with sprinkler heads hooked onto a tractor and pulled to a new location. Travelling guns or "hose-dragger" are propelled by waterpower or by an auxilliary engine. They travel in alleyways in the field and distribute water through a large-diameter nozzle. The side-roll wheelmove systems are mounted on wheels of sufficient height to clear the growing crop. The wheels are turned by electric or hydraulic drivers and the pipes extend between the wheel hubs. The lines move automatically across the field and sprinkler nozzles are mounted between the wheels at designated intervals.

FIG. 8.3. CONSTRUCTING BORDERS FOR BASIN IRRIGATION, NEAR DAMASCUS, SYRIA, 1977

Center Pivot Systems.—These are one distinctive modification of mechanical-move systems. The wheel-mounted sprinkler line attaches to a pump or a fixed outlet and the sprinkler line rotates in a circle around this water source. Wheels are moved by water-pressure driven motors or by electric or hydraulic motors. Such systems typically irrigate a circular field of 40 to 60 ha. A large system at Kufra in the Libyan-Sahara Desert irrigates about 100 ha per installation during the cooler winter months and this is reduced to about 60 ha in the dry hot summer season. Equipment costs are rather high. A system to irrigate 60 ha in Nebraska in 1976 was reported to cost $27,000 for the well, pump, and motor, and $33,000 for the center pivot system (Sheffield 1977).

Stationary or Solid-Set Sprinkler Systems.—In these systems pipe is buried to form a permanent grid network with risers and sprinkler nozzles at prescribed distances. Such systems are becoming common for high-value crops that require frequent irrigations, such as some fruit and vegetable crops. In one modification wide-spacing (87 m) and high pressures (100 psi or 7.5 bars) of water were used. Portable nozzles or "guns" are mounted on the risers. Labor costs with stationary and center pivot systems are appreciably less than with hand-moved pipes.

FIG. 8.4. LARGE CENTER PIVOT SPRINKLER AT KUFRA IN THE LIBYAN–SAHARA DESERT IN LIBYA, 1975. IRRIGATES 100 ha IN COOL SEASON AND ABOUT 60 ha IN THE HOT SUMMERS

Drip or Trickle Irrigation

This system, developed in Israel, is being installed at an increasingly rapid rate for fruit and other high value crops because of savings in water and labor and because of improved yields with water of low quality. It is estimated that about 70,000 ha were being irrigated by this method in the U.S. in 1975 and that this would be increased to over 200,000 ha by 1980. The emitters are spaced to provide water to up to 60% of the root zone.

Drip systems are usually chosen for very sandy soils, but reports from the San Joaquin Valley of California indicate that drip systems have helped solve water penetration problems in slowly permeable clay and silty clay soils. It has also decreased soil crusting.

Comparable yields can be obtained with most of the methods of irrigation listed. However, in order for surface methods of irrigation to reach a high level of irrigation efficiency, the land will usually have to be carefully levelled to small uniform slopes, and facilities will be required to reuse the runoff water and to apply water uniformly over the land. Where slopes exceed 2 or 3%, contour planting and irrigating or the use of low application-rate sprinkler or trickle irrigation methods are usually advisable to avoid erosion and to secure a high water application efficiency.

DRAINAGE AND SALT BALANCE

Where large areas of land are under irrigation usually part of the area becomes waterlogged and must be drained. Water requiring drainage ordinarily originates from excess water applied in irrigations. Additional water may come from seepage from ditches, canals, and reservoirs or from nearby areas with excess irrigation or precipitation.

Where ground water accumulates and leaching of salts is restricted, problems of soil salt occur. Problems of poor drainage are so general that it is estimated that about 122 million hectares, or over half of the world's irrigated land, have developed drainage and salinity problems. As more arid and semi-arid lands are brought under irrigation these problems will increase.

The solution to these extensive problems includes the shaping of the land surface and the installation of drains and the maintenance of low water tables so crops can root effectively. The solution also involves removal of excess soluble salts by leaching and proper irrigation water management to maintain a favorable salt balance in the soil root zone.

Where water tables occur, the level may rise during irrigation or with substantial amounts of rain. Even a temporary rise in water table may greatly depress plant growth and crop yields by reducing air supply for the roots. Where the water table is high in the spring or during rainy seasons and drops rapidly as the season advances, plants often have shallow root systems that are unable to develop fast enough to prevent drought effects. Drainage, therefore, allows crops to develop and maintain good root systems.

Drainage Systems

Installation of drainage systems is expensive and the design of an effective operating system requires intensive studies of soil permeability, of levels of water tables throughout the year, of sources of excess water, and of the nature of any salt problem. Such studies indicate the most effective type of drain, location of drains, the depth and spacing of drains, the quantities of water to be removed, and associated water management or other practices needed for effective operation of the system.

For irrigated lands the principal drainage systems are open field drains, covered or tile drains, wells, and pumps. The selection, design, and operation of drainage systems have been discussed in detail by Luthin (1957) and others.

Because of varying and often distant sources of water, the need for drainage outlets, and management problems, drainage systems usually extend beyond farm boundaries and their installation and effective operation require the cooperation of many farmers. The intensity of the problem can usually be reduced by adopting more efficient irrigation methods. Many irrigators apply twice as much water as is required by crops; attempts should be made to redesign irrigation systems and practices toward attaining a water application efficiency of 70 to 85%.

Salt Balance

Salt-affected soils have excessive concentrations of salt, or more than 15% exchangeable sodium, or both. In irrigated soils most salts enter as solutes in irrigation or seepage waters. One requirement of an irrigation system, therefore, is to achieve a salt balance in which the quantity of salt entering the soil for a given area is equalled by the quantity of salts leaving the area through drainage waters or runoff. This salt balance concept often needs a correction for the precipitation of such salts as $CaCO_3$ and $MgCO_3$ in the soil.

Leaching Requirement.—Leaching requirement (LR) is defined as the fraction of the irrigation water that must be leached through the root zone to keep the salinity of the soil below a specified value (Richards 1954). The general equation is:

$$LR = Dd/Di = ECi/ECd$$

in which LR is the leaching requirement, Dd is the depth of drainage water that passes beyond the bottom of the root zone, Di is the depth of irrigation water applied, ECi is the electrical conductivity of the irrigation water, and ECd is the electrical conductivity of the drainage water. This equation takes into account only the salt contained in the irrigation water and salt removed from the soil in the drainage water. The quantities of other salts added in fertilizers and amendments and the salts precipitated in soil and those removed by the crop are usually relatively small compared with those in the irrigation water. In some cases, however, corrections for other salt sources or losses may be needed. The equation is also based on a steady state condition in which water is being added and drained from the soil with the indicated concentrations of salt. Actually, irrigation and drainage flows are intermittent and the first flushes of drainage water with leaching contain more salts than the equation would indicate. In some instances, particularly with low quality water, the equation may overestimate the drainage water required. The equation does, however, serve as a guide to water management in relation to the control of salt.

Where rainfall is sufficient to significantly affect the water supply, corrections must be made for the dilution effect of the additional water. The depth of irrigation water required to maintain a specified level of salinity in the soil in relation to the evapotranspiration or consumptive use of the crop can be estimated as follows:

$$Di = Dc(1/1 - LR)$$

where Dc is the equivalent depth of water representing evapotranspiration losses. A similar relationship gives the depth of irrigation water needed in relation to the salt situation.

$$Di = \frac{ECd}{ECd - ECi} Dc$$

In such estimates the permissible level of salt (ECd) in the drainage water must be established. Tables of salt tolerance of crops based on salt levels that would result in a 50% reduction in yield have been developed by the U.S. Salinity Laboratory (Richards 1954). Adoption of such values for maximum concentration in drainage water would involve a potential optimum crop yield of 90 to 95% and would be impractical for all but high quality waters.

IRRIGATION AND CROP PRODUCTION

Irrigation in the arid tropics can increase production from a small yield or complete failure of one crop to high yields of three crops per year. Furthermore, a broad selection of crops is made available. Of near equal importance is the opportunity that irrigation gives for timing planting and other operations to benefit most from seasonal temperature or natural precipitation conditions or for favorable marketing periods.

Irrigation can also increase cropping intensity in the arid and semi-arid warm temperate zones. Under these conditions irrigation can supplement normal rainfall to allow small grains or other appropriate winter crops in addition to the customary crops during the summer months.

Irrigation is increasing in humid zones because of the insurance it provides against drought. High value specialty crops such as tobacco, vegetable, and fruit crops have been irrigated for many years. Such crops as rice and sugar cane require a good supply of moisture during their growth period with somewhat less as the plants mature. Such crops are greatly benefited by irrigation in all but the most favorable rainfall conditions. Increasing acreages of maize and soybeans are being irrigated in humid zones.

Maximizing total yields under irrigation is often associated with increasing the number of crops produced per year. This, in turn, requires careful selection of crops and varieties. Generally early maturing crops are preferred, thus 3 short-season rice crops usually have the potential of producing more than 2 long-season crops. Intercropping and relay cropping are also commonly practiced in developing countries and are favored by available irrigation water. In Southeast Asia irrigated rice is often intercropped with maize, soybeans, or groundnuts being planted on raised beds. Such cropping systems are described in a workshop report produced by the Asian Development Bank (Anon. 1973).

Integration of Irrigation with Other Management Practices

Irrigation research has emphasized that the most profitable irrigation regime can be determined when such other practices as fertilizer application and plant population are also adjusted. This principle is illustrated by results on sweet corn obtained by Peterson and Ballard (1953) and shown in Table 8.6.

TABLE 8.6. EFFECT OF NITROGEN AND MOISTURE ON THE YIELD OF SWEET CORN

	Kg Nitrogen per Ha			
Soil Moisture Condition	0	56	112	224
	Yield in Tons			
Wet	7.44	10.11	12.89	18.38
Medium	6.84	10.16	14.69	12.78
Low	6.88	11.43	13.14	12.69

Source: Peterson and Ballard (1953).

The "wet" plots were irrigated when the 15 or 30 cm depth of soil reached a water tension of 0.3 bar. The "medium" moisture plots were irrigated when about half of the available moisture in the root zone was utilized. The "dry" plots were irrigated when the soil moisture in the root zone approached the wilting percentage. The "dry" plots corresponded closely in irrigation practice and yield obtained to that of the local farmers. The soil was Nibley clay loam, the second year out of alfalfa.

In the data of Table 8.6 we see that the low moisture regime was adequate for maximum yields with the unfertilized soil. But as the available nitrogen was supplemented by 112 kg of fertilizer nitrogen per hectare, the medium moisture condition increased yields beyond the low moisture. Highest yields were attained only by combining the highest moisture regime with the greatest application of fertilizer nitrogen. It is evident that there was no benefit in fertilizing for 18 t of ear corn when the irrigation practice would support only 13 t.

The pyramiding of yields by combining best practices is significant for the improvement of yields under irrigation agriculture. Such results point to the futility of designing irrigation practices as an independent exercise. On the basis of the early irrigation studies in which soil fertility was ignored, the conclusion would have been reached that the low moisture regime was adequate for maximum yields.

REFERENCES

ANON. 1973. Regional workshop on irrigation water management. Asian Development Bank, Manila, Philippines.

DIEBOLD, C.H. 1953. Simple soil tests tell when to irrigate. Crops Soils *5*(6) 14-15.

DOORENBOS, J., and PRUITT, W.O. 1977. Crop water requirements. Irrig. Drainage Pap. *24.* FAO, U.N., Rome.

HAGAN, R.M., HAISE, H.R., and EDMINSTER, T.W. 1967. Irrigation of Agricultural Lands. American Society of Agronomy, Madison, Wisconsin.

HELLER, J., and BRESLER, E. 1973. Trickle Irrigation. *In* Arid Zone Irrigation. B. Yaron, E. Danfors and Y. Vaadia (Editors). Springer-Verlag, New York.

ISRAELSEN, O.W., and HANSEN, V.E. 1962. Irrigation Principles and Practices, 3rd Edition. John Wiley and Sons, New York.

LUTHIN, J.R. 1957. Drainage of Agricultural Lands. Monograph VII. American Society of Agronomy, Madison, Wisconsin.

PETERSON, H.B., and BALLARD, J.C. 1953. Effect of fertilizer and moisture on the growth and yield of sweet corn. Utah Agric. Exp. Stn. Bull. *360.*

PRES. SCI. ADVIS. COMM. 1967. The World Food Problem, Volume II. The White House, Washington, D.C.

RICHARDS, L.A. 1954. Diagnosis and improvement of saline and alkali soils. U.S. Dep. Agric. Handb. *60.*

SHEFFIELD, L.F. 1977. The economics of irrigation. Irrig. J. *27*(1)18-22.

TAYLOR, S.A., EVANS, D.D., and KEMPER, W.D. 1961. Evaluating soil water. Utah Agric. Exp. Stn. Bull. *426.*

THORNE, D.W., and PETERSON, H.B. 1954. Irrigated Soils, 2nd Edition. McGraw-Hill Book Co., New York.

YARON, B., DANFORS, E., and VAADIA, Y. 1973. Arid Zone Irrigation. Springer-Verlag, New York.

9

Crop Characteristics and Cropping Systems

D. Wynne Thorne

Our important crops are the result of human ingenuity in selecting and improving plants found in nature. Only a few hundred plant species have become important cultivated crops (Harlan 1975, 1976). During the past 100 years there has been an increasing selection from, and dependence on, a relatively few of these plant species. The principal crops now contributing directly to human food are shown in Table 9.1. Each of the 32 crops shown now contributes one million or more tons to the annual total of world food. Wheat, rice, maize, and potatoes contribute more tonnage to the total than the other 28 crops combined.

In addition to the crops listed in Table 9.1, there are many minor crops of fruits, vegetables, and condiments. Also, some food crops are important in restricted areas such as teff (*Eragrostis abyssinica*) used as a cereal in Ethiopia, and quinoa (*Chenopodium quinoa*), grown in the Andes region of Bolivia, Ecuador, and Peru. Also, a large group of forage crops are important in farming systems and contribute much to human food through the indirect route of feed for animals.

Initially many of our important crops had limited areas of adaptation. The potato was grown in the Andean highlands and was only slowly adapted in recent centuries to conditions in other parts of the world. Rice was once a major crop only in the tropics of Asia, but is now grown world-wide and has been adapted to shorter seasons and to cooler regions. Only 200 years ago the tomato was used mostly as an ornamental and was widely considered as poisonous.

TABLE 9.1. SOME OF THE MOST IMPORTANT CROPS OF THE WORLD, AREA PLANTED, AND TOTAL PRODUCTION

Average for years 1973, 1974, and 1975

Crop	Area 1000 ha	Production 1000 t
Wheat	224,111	363,945
Rice, Paddy	137,395	329,358
Maize	112,346	311,030
Potato	20,937	303,024
Barley	89,109	165,157
Sweet Potato	14,705	134,395
Cassava	11,319	102,624
Soybeans	45,264	63,308
Grapes	10,014	62,334
Sugar Cane, Sugar	11,677	57,850
Oats	32,250	54,374
Sorghum	43,269	52,800
Millet	67,927	45,370
Banana	2,576	34,978
Cotton, Seed Cotton	32,861	38,181
Tomato	1,642	33,564
Sugar Beet, Sugar	8,198	32,850
Oranges	29,338	
Rye	15,970	28,485
Groundnuts, in Shell	19,084	17,868
Beans, Dry	24,915	12,769
Sunflower	9,643	12,209
Peas, Dry	9,456	10,939
Buckwheat	2,482	6,218
Broadbeans	5,378	6,112
Chickpeas	9,974	6,008
Tobacco	4,148	4,831
Coffee	8,551	4,193
Oil Palm, Oil		2,537
Sesame	5,633	1,937
Pigeon Peas	2,792	1,858
Lentils	1,785	1,092

Source: Anon. (1976); Harlan (1976).

Through plant breeding and selection our present cultivated crops have been systematically modified in many important growth and quality characteristics and in areas of adaptation. Also, with the development and extensive use of such technologies as mechanized tillage and harvesting machinery, fertilizers, lime, irrigation, drainage, and pesticides to control weeds, insects, and diseases, the characteristics sought in crop plants have changed. To some degree the characteristics considered most important vary in different parts of the world. In general, however, crops are selected for high yielding capabilities, desired quality traits, resistance to insects and diseases, and for dependable economic returns. The primary goal of high production is different from the goals of agricultural systems based heavily on self-sufficiency and security. In the older security-oriented systems, crops were commonly grown in mixtures. Crops and production practices were selected to provide security in food supply with protection against total disaster, but such systems would generally not attain high levels of productivity. Fertilizers and pesticides were lacking and diversity was used to ensure at least a minimum yield. In many developing areas farmers are still concerned more with subsistence than with producing for markets. Whatever the purpose of production, the preferences of people affect the quality ratings of color, texture, and flavor.

CROP CHARACTERISTICS IN RELATION TO CROPPING SYSTEMS

A primary goal in crop production systems is to produce high yields of desired quality products at minimum costs, with costs being measured per unit of product. In developed countries labor has become one of the major costs, and it is becoming increasingly scarce in many countries. In consequence, labor-saving equipment, inputs of pesticides, and the conservation and supplemental application of irrigation water have become important aspects of farming systems and affect the selection of crops. High yields have become necessary to minimize the cost per unit of production and crops are increasingly selected that will have the potential for making full use of available yield-promoting inputs.

In many developing countries modern inputs of improved machinery, fertilizers, pesticides, and crop varieties are often not readily available, or the costs may be excessive for the operator of a small farm. In such instances compromises in the selection of crop types and of technologies are frequently necessary.

In developed countries major attention is given to the yield of the primary marketable portion of the crop. A major objective in selecting crop varieties and production practices is to attain a high harvest index (the proportion of the total dry weight of a crop that is harvested). In cereal crops the harvested portion is grain and a harvest index of 40% or more is usually sought. A high harvest index indicates an efficient crop in that a high proportion of the total photosynthesis is used. However, in developing countries, where livestock depend on straw or stover for feed, a lower harvest index of grain may be acceptable or even desired by the farmer if this substantially increases the livestock feed supply.

There are a number of characteristics of crop plants that are important in crop selection for any given situation and in determining the farming practices that should be used. Characteristics are summarized in Table 9.2 for a selected group of extensively grown crops.

TABLE 9.2. SOME GENERAL CHARACTERISTICS AND GROUPINGS OF CROPS

Crop	Photoperiod for Flowering	Growing Period Days to Maturity	Growth Habit	Soil & Root Depth	Harvested Part	Climatic Zone Rainfed
Wheat	LD/DN LD	100–130 (SC) 180–240 (WC)	D D A	SIL CL <50	Seed	III A B IV A B V B
Barley	LD/DN LD	100–130 (SC) 180–240 (WC)	D D A	SIL CL <50	Seed	III A B IV A B V B
Oats	LD/DN	110–130 (SC) 210–270 (WC)	D D A	SIL CL <50	Seed	III A B IV A B V B
Maize	DN	80–100 (EC) 110–130 (MEC)	D A	SIL C <50	Seed	III A IV A V A B
Pearl Millet	DN SD	70–100 (EC) 120–240	D D A	SIL SCL <75	Seed	IV A B V A B C
Finger Millet	DN	70–90	D A	SIL SIC L <75	Seed	V A B
Rye	LD/DN	110–130 (SC)	D	SL SLL <50	Seed	IV A V A B III A B
Rice	DN SD	110–130 (EC) 140–180 (LC)	D A	SIL CL <50	Seed	IV A V A B
Sugar Cane	SD/DN	270–365	D P	SIL C <50	Stem	V A B
Sugar Beets	LD	160–240	D B	L SICL <75	Root	III A I V A

Potato	V	190–110 (EC) 120–140 (MEC) 150–180 (LC)	ID A	L SICL <75	Tuber	III A IV A V B (H)
Sweet Potato	SD	150–210	ID	SIL CL <75	Tuber	V A B
Cassava	SD	180–270 (EC) 270–165 (LC)	ID P	SIL CL <100	Tuber	V A
Soybeans	DN SD	190–130 (EC) 140–180 (LC)	D ID	SIL–CL <75	Seed	III A IV A V A B
Beans, Dry			A			
Broadbeans		190–120	A			
Pigeon Pea			P			
Chick Pea	SD/DN	190–120 (EC) 130–180 (KC)	ID A	<50	Seed	IV A B V B
Cowpea	DN	190–120 (EC) 140–180 (LC)	D ID A	<75	Seed	IV A V A B
Sesame	SD/LD	100–130 (UBC, EC) 140–180 (BRC, LC)	ID	L – C <75	Seed	IV A V A B
Rape	LD/DN	100–110	D A		Seed	III A IV A
Tomato	DN	110–130 (UTC) 120–140	D ID	L–CL	Fruit	III A IV A V A B
Sorghum	DN DN/SD SD	90–110 (EC) 120–130 (MEC) 140–240 (LC)	D D D	SIL–C <50	Seed	III A B IV A B V B

TABLE 9.2 (CONT.)

Crop	Photoperiod for Flowering	Growing Period Days to Maturity	Growth Habit	Soil & Root Depth	Harvested Part	Climatic Zone Rainfed
Groundnut	DN	90–110 (SBC)	ID	SL–CL <75	Seed	IV A VAB
Cotton	SD/DN	130–140 (OLC) 160–180 (NLC)	ID	L–CL <100	Seed Cotton	IV A V A B

EXPLANATION OF SYMBOLS

Plant Characteristics Symbols

BRC—Branched cultivar
D—Determinate
DN—Day neutral
EC—Early cultivar
ID—Indeterminate
LC—Late cultivar
LD—Long day
MEC—Medium cultivar
SC—Short day
UBC—Unbranched cultivar
V—Variable
WC—Winter cultivar

Soil Texture Symbols

L—Loam
SL—Sandy loams
SIL—Silt loams
CL—Clay loams
SCL—Sandy clay loams
SICL—Silty clay loams
SIC—Silty clay
SC—Sandy clay
C—Clay

Climatic Zones Symbols

IIIA—Humid cool temperate zone
IIIB—Arid cool temperate zone
IVA—Humid warm zone
IVB—Arid warm temperate zone
VA—Humid tropical zone
VB—Semi-arid tropical zone
VC—Arid tropical zone

Source: Sarrae and Kowal (1977).

Growing Period

The number of days required between date of plant emergence and maturity is important both in determining the climatic zone in which the crop variety can be grown and also in fitting a particular cultivar into a multiple cropping system. Recent breeding programs have produced a series of varieties of the major crops, such as wheat, rice, maize, and sorghum that have broad differences in the time required from seedling emergence to maturity. In consequence, the required growing period must be interpreted broadly.

The potential yields of annual crops are not necessarily related to length of growing period. Each variety must be selected on the basis of growth characteristics as verified in field trials under selected conditions of climate and soil. Temperature conditions are important in determining days to maturity of crops, particularly those that are day-neutral. The effect of temperature on plant development is commonly expressed in terms of heat units required to bring a crop to the flowering stage. *Heat units* are usually measured in terms of the accumulated product of the number of days above a certain minimum temperature required for normal growth of a particular crop times degrees of temperature above this minimum. If the minimum temperature for normal development of the crop is 8°C, then a day with a 20°C temperature would add 12 degree days (heat units) to the total. In some instances more precise estimates of plant development are obtained by using the hours accumulated above a determined minimum of temperature rather than days. Varieties within a crop may vary substantially in heat unit requirements for development.

Photoperiodism

Many plants are sensitive to the length of day for initiation of certain stages of development such as flowering, tillering, or dormancy. For many plants the length of night (darkness) rather than length of daylight is critical. In such cases short-day plants are those that require prolonged darkness daily to induce flowering, and long-day plants initiate flowering when nights are relatively short. Even short light exposures at night may hasten flowering of long-day plants and prevent flowering of short-day plants.

A number of plants are day-neutral, and carry out sequences of development without regard to daylength. Breeding and selection programs have resulted in cultivars of many major crops that are largely day-neutral. Day-neutrality is a quality generally sought because it adapts a crop variety to a relatively large range of latitude.

Daylength is about 12 hours throughout the year in tropical zones; in other parts of the world daylength is 12 hours only at the vernal and autumnal equinoxes. The relative degrees of short- and long-daylengths begin from this 12-hour value. Warm or cool temperatures may in part compensate for daylength with some plants. When crop varieties are planted outside of their zone of daylength adaptation they are usually too early or too late in stages of development for best performance in relation to season. In some plants the sequence in change in daylength may be important in inducing changes in development. Increasing daylength may help initiate flowering, while in the fall the advent of shorter days may promote fruiting, maturity, or dormancy.

The growth habits of crop plants are important in determining production and management practices. Dwarf varieties are generally preferred over tall counterparts because of their upright growth habit, greater ease of harvesting by machine, reduced likelihood of lodging, earlier fruiting, and often a higher harvest index. Bush type varieties are preferred over vine types because they possess many relatively dwarfed branches which bear fruit uniformly. Bush or determinant plants complete most of their vegetative growth before shifting to reproductive growth. Indeterminate plants carry on both vegetative and reproductive types of growth simultaneously.

Root Systems

There are two different types of root systems common to crop plants: fibrous and tap roots. Fibrous roots permeate the soil rather intensively and tend to hold the soil particles together. As a result, the grasses promote good soil structure and help protect soils against erosion. Tap rooted crops include those with roots commonly harvested for food or feed, such as sugar beets, mangels, carrots, and turnips. Tap rooted plants tend to be relatively deep rooted, as in the case of alfalfa and trees. Alfalfa plants, for example, have been found to have roots extending as deep as nine meters. However, not all tap rooted plants root to greater depths than do fibrous rooted crops.

With most crop plants the major volume of roots occurs in the upper 30 cm of soil. Root hairs developing behind the root tips provide the major surface for water and nutrient absorption. Root hairs are very fine and are highly branched. Studies of root hairs in grasses have recorded several million per plant. Raper and Barber (1970) studied two varieties of soybeans. One had 83% more root surface in the upper 80 cm of soil than the other. The one with the greater root surface also had 37% of the root surface below 30 cm compared with 28% in the other variety.

Table 9.2 also gives common depths of intensive rooting of a number of crops. This is affected by soil moisture, texture, compaction, and aeration, and the supply of available plant nutrients (Pearson 1974; Rappaport and Sachs 1976).

EFFECTS OF CROPS ON SUCCEEDING CROPS

When the same crop or different crops are grown in sequence on the same area of land, situations may develop that benefit or harm succeeding crops.

Buildup of Harmful Insects or Disease Organisms

This is a well established relationship that causes great concern and frequent losses to farmers. When nematodes develop in sugar beet fields a long rotation with a series of nematode-resistant crops such as cereals, alfalfa, potatoes, maize, or beans becomes the most practical remedy. The land must also be kept free from susceptible crops such as mangels or turnips, or of susceptible weeds. Other crops have similar harmful organisms that live over in the soil and are best controlled by rotation with nonsusceptible crops. In India, rotations with nonsusceptible crops are recommended for barley root rot, the brown spot and head smut of maize, bacterial blight of cowpeas, and root rot of groundnuts.

Depletion of Essential Nutrients

An 11,000 kg per hectare maize crop may remove as much as 250 kg of nitrogen, 50 kg of P_2O_5, and 225 kg of K_2O per hectare. Sorghum may remove even more nitrogen and comparable amounts of phosphorus and potassium. Other crops also draw heavily on the major nutrients. The solution to the effects of such nutrient depletion is to match a fertilizer program to each crop or to sequences of crops. For example, wheat, barley, or oats following well fertilized maize are often adequately supplied with residual phosphorus and potash. Soil nutrient depletion is now well understood and is considered less of a reason for rotating crops than formerly.

Moisture Depletion

Under rainfed conditions some plants use up so much soil moisture that provision must be made to replenish moisture for ensuing crops. Sorghum, for example, has a dense root system that can extract moisture to low levels in the root zone. In multiple and relay cropping, careful attention to moisture status of soil must be given before planting the succeeding crop. Also, in interplanted crops, differences in rooting depths and in peaks of moisture demand among companion crops allow more effective use of moisture supplies.

Beneficial Effects

The use of legumes in crop sequences and the associated beneficial effects of increased nitrogen supply on subsequent crops have been extensively noted. Under favorable conditions alfalfa may fix as much as or more than 200 kg of nitrogen per hectare per year, and after a few years of hay production, particularly if a good growth is plowed under, the soil may be enriched by several hundred kg of nitrogen per hectare. Annual legumes such as soybeans, dry beans, and chickpeas usually do not obtain all of the nitrogen required from the nitrogen fixation process. Most annual legume crops also leave only limited quantities of residue with the soil. Such crops, therefore, leave only modest benefits for succeeding crops, but are not as demanding of nitrogen as most of the nonlegume crops.

Green manure crops are of questionable economic benefit if a harvested crop is displaced. In some situations, such as in Australia and Mediterranean countries where the use of hard-seeded and self-seeding medicago and clover cultivars are established to alternate with wheat, a successful forage and nitrogen source helps the wheat and also furnishes forage for livestock. Such a system is well established in Australia and is being adopted in several Mediterranean climate countries. It provides reduced erosion of soil and reduced needs for nitrogen fertilizer for wheat.

Allelochemics

Allelochemics is a term applied to chemicals released from plants that affect the growth, health, or population biology of another plant or of lower forms of life in the soil, but excluding products that may be used as food by the second species. Such chemicals may be released as root exudates, they may be washed from leaves, or released or produced in the decomposition of residues from some plants. These chemicals may be absorbed on clay or organic matter particles in soil and have shown prolonged effects on roots and microorganisms. This little studied area has been investigated more in the field of general ecology than as a factor in different combinations and systems of cropping (Whittaker and Feeny 1971).

There are several instances where such products are important to crop interrelations and succession. In Europe oats were found to release scopoletin and related unsaturated lactones that inhibited the germination of oats, peas, and some other plant seeds (Borner 1960). Flax also excretes a self-toxic phenolic compound that injures young flax seedlings. Also, peaches excrete amygdalin and apples, phlorozin, that inhibit the growth of another similar tree planted in the same location (Tukey 1969). Rain water washes a chemical, juglone, from the leaves of walnuts that is injurious to many plants growing beneath. It has also been noted that decomposing wheat, barley, and rye straw may release fenulic acid that inhibits many germinating seeds.

Also with important implications for intercropping is the fact that as many plants mature, large quantities of potassium, amino acids, and other materials may be leached from the leaves by prolonged rains. Associated or succeeding plants are in position to make use of such leached nutrients.

CROPPING SYSTEMS

Crop production systems include both cropping systems and the associated crop production practices and technologies used to achieve the harvested yield of crops. Cropping systems may consist of the monoculture of one continuous crop such as maize or wheat. It may also consist of formal sequences of crops repeated in an orderly pattern to constitute a rotation. Cropping also includes flexible arrangements of crops that may change from year to year, and also intensive successions of crops within single years or even within seasons. Cropping systems vary greatly with differences in soil and climate and with economic and social systems under which farming is practiced.

Several interactions among crop plants are involved in developing or choosing a cropping system:

(1) Cropping systems should be devised to provide growing crops with high photosynthetic capabilities for as great a portion of the year as is practical. In intercropped or mixed crops, plant height, the shape and angle of leaves, and the rate of growth and time period required to reach maturity are important characteristics that determine the compatability of photosynthesis in plant communities.

(2) A major goal should be to maximize annual crop production or net economic gains per unit area of land. Thus, two short-season crops may provide greater total yields than one long-season crop. On the other hand, the one long-season crop may in some instances provide greater economic gains than the two crops. Decisions as to crop intensities must be based on the best available evidence for each particular combination of conditions.

(3) In order to promote sustained high yields and profits, cropping systems should be designed to maintain soil organic matter and tilth; to reduce the incidence of weeds, insects, and diseases; to help keep plant nutrients in balance; to conserve water; and to minimize soil erosion.

(4) Roots take up water, nutrients, and oxygen from the soil. To utilize

these resources effectively, roots should form an active and extensive network throughout the soil. Good crop combinations have compatible root systems that permeate the soil to a depth of 25 to 30 cm with some roots extending much deeper. One crop may root deeper than another to good advantage.

Crop Rotations or Monoculture

The relative merits of a crop rotation consisting either of an orderly or a flexible sequence of crops, or of a monoculture of one continuous crop must be assessed by specific climate, soil, and economic situations. The relative merits of the two systems are adapted below, from Tisdale and Nelson (1975):

Rotations

(1) Deep-rooted legumes may be grown periodically over all fields.
(2) There is more continuous vegetative cover with less erosion and water loss.
(3) Tilth of soil may be superior.
(4) Crops vary in feeding range of roots and nutrient requirements; deep-rooted versus shallow-rooted, strong feeder versus weak feeder, and nitrogen-fixer versus nonlegume.
(5) Weed and insect control are favored.
(6) Disease control is favored. Changing the crop residues fosters competition among soil organisms and may help reduce the pathogens.
(7) Broader distribution of labor and diversification of income are affected.

Continuous Cropping or Monoculture

(1) Profits may be greater.
(2) A soil may be especially adapted to one crop; for example, maize, rice, or forage.
(3) The climate may favor one crop; for example, maize is more suited than oats in the Corn Belt.
(4) Machinery and building costs are often lower.
(5) The grower may prefer a single crop and become a specialist. Few people can become well enough informed to do an expert job of growing a large number of crops and also produce livestock.

Balanced crop rotations typically include a legume or sod crop to improve the nitrogen and organic matter and tilth of the soil, a cultivated crop to aid in weed control and also to serve as one of the major cash income crops, and a close-grown cereal crop. Crop sequences in rotations may be arranged in descending order of lime or fertilizer nutrient requirements to obtain maximum benefits from soil treatments. In some soils deep-rooted crops are advantageously alternated with shallow-rooted ones.

Multiple Cropping

The production and harvesting of two or more crops from the same area of land in one year is most practical in warm temperate subtropical and tropical climatic zones (Zones IV and V). With a series of short season cultivars such as 110-day wheat, 100-day potatoes, and 120-day rice, 3 crops are possible where water supply is adequate and the growing season extends through all or most of the year.

Under all but the high rainfall zones, multiple cropping must be carefully timed to have adequate moisture for crop growth and sufficient dry weather for harvest and seedbed preparation. In the rice growing regions of Asia two crops of rice per year are often grown. In Taiwan production has been intensified to include 2 crops of transplanted rice with 2 intervening short-season vegetable crops. Under rainfed conditions in the semi-arid tropics of India it is possible to produce two crops in place of the one traditional crop (see Chapter 17).

Soil preparation and planting of a second crop is hastened, where appropriate technology is available, by no-tillage or limited-tillage procedures in which plowing or disking are omitted, a herbicide treatment is applied to control weeds, and a disk is used to cut the crop residues preceding the seed drill. These operations can be carried on in one operation with appropriate equipment.

Long-season crops such as cotton are difficult to use in multiple cropping unless irrigation water or timely precipitation is available for the second crop. An extensive discussion of multiple cropping is given by Papendick *et al.* (1976).

Relay Cropping

Relay cropping is the interplanting of a second crop before the harvest of a maturing crop. It has been practiced under irrigated agriculture and under favorable rainfall situations in humid climates. With the development of high-yielding short-season varieties there has been renewed interest in relay cropping. Relay planting, one to three weeks before harvest, offers an opportunity for reducing the risk of failure in establishing a second crop. Successful relay planting requires relay crops that are not sensitive to shading in the seedling and early development stages. It also requires procedures for controlling weeds and stubble of the first crop that would compete with the relay crop. Timing of the planting of the relay crop and of the harvest of the first crop are essential for the success of such a cropping system. The careful use of herbicides and limited-tillage systems may fit well into some relay cropping programs.

Examples of successful relay cropping systems include sorghum planted in early monsoon stages in the semi-arid tropics and interplanted with pigeon peas. A special example is the intercropping of annual crops between trees. Cassava, soybeans, peanuts, and maize are often grown between rows of coconut, oil palm, and young rubber trees.

Intercropping

Intercropping involves two or more crops growing simultaneously in rows in a definite pattern. A related system is that of mixed croppings practiced in some areas in traditional agriculture in which two or more crops are planted in nonsystematic mixtures. In event of a drought or other disaster, such as insect infestations, the chances for success with mixed or intercrop systems may be increased because one crop might survive and give a harvest.

Evidence has accumulated that total yields per unit area of land can be increased by growing alternate or pairs of alternate rows of two different crops in place of solid plantings of either crop alone. The results of intercropping experiments at the International Crops Research Institute for the Semi-Arid Tropics (ICRISAT) at Hyderabad, India are shown in Table 9.3. Greater benefits from intercrop components exploit the environmental supplies of growth factors in different ways. Gains of 10 to 40% in yields from intercropping have been reported (Trenbath 1974).

A crop using sunshine most of the growing season is needed to provide a canopy. A fast-growing crop, maturing early, may fit well with one that grows slowly. Figure 9.1 shows the percentage of light interception by pigeon peas and sorghum planted separately and as intercrops at ICRISAT, Hyderabad, India in 1977.

Ratoon Cropping

Some crops such as sugar cane have the capacity to send up new shoots from the base after a crop is harvested. In sugar cane the growing of one or more ratoon crops after harvesting the original crop is standard production practice. Some cultivars of pearl millet, sorghum, rice, pineapple, ramie, and cotton also have ratooning capabilities. Current experiments with pearl millet, rice, and sorghum are still in the trial stages and involve tests for the production of both forage and grain.

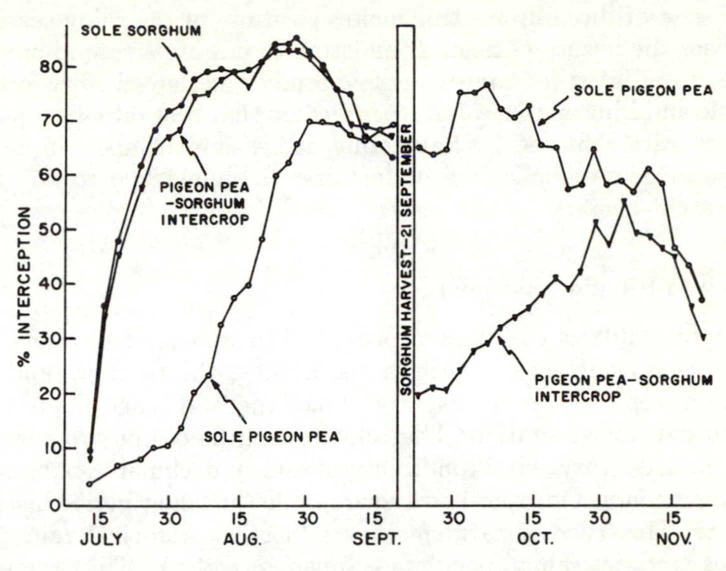

Courtesy of R.W. Willey, ICRISAT, India

FIG. 9.1. PERCENTAGE LIGHT INTERCEPTION BY PIGEON PEAS AND SORGHUM AS SOLE CROPS AND AS INTERCROPS, HYDERABAD, INDIA

TABLE 9.3. YIELDS OF FOUR INTERCROP COMBINATIONS BY METHOD OF INTERCROPPING WITH PIGEONPEA ON BLACK SOIL, HYDERABAD, 1973

	Method of Cropping			
Intercrop	Solid q/ha	Alternate Rows q/ha	Paired Rows q/ha	LSD (05) q/ha
Mungbean–Pigeonpea				
Mungbean	1.7	1.8	2.7	0.4
Pigeonpea	3.8	4.4	6.8	1.1
Soybean–Pigeonpea				
Soybean	3.9	4.8	7.6	2.1
Pigeonpea	3.9	4.9	6.5	1.2
Finger Millet–Pigeonpea				
Finger Millet	15.4	17.3	26.1	1.9
Pigeonpea	3.9	4.9	8.2	1.1
Sunflower–Pigeonpea				
Sunflower	10.1	11.9	23.1	1.9
Pigeonpea	2.5	3.0	4.7	1.6

Source: Anon. (1974).
Harvest dates: Mungbean: Aug. 27. Soybean: Oct. 10. Sunflower: Oct. 19. Finger Millet: Oct. 20. Pigeonpea: Nov. 30.
Fertilizer treatment: All crops received 205 kg/ha of 28–28–0 at planting or a total of 57 kg of nitrogen and 25 kg of phosphorus.

Under semi-arid conditions this makes good use of the rainy season. It minimizes the hazard of needing moisture to initiate a second crop and reduces total labor for the two crops. Studies in Hawaii with sorghum (Escaldo and Plucknett 1977) have indicated that best ratoon crops with sorghum were achieved by harvesting at a height of about 8 cm and applying a heavy application of fertilizer to supply 200 to 250 kg of nitrogen per hectare.

Shifting Cultivation Systems

The term "shifting cultivation" is applied to agricultural systems that involve clearing forest or brush lands, burning the residues, and then growing a crop for a few years, after which the land is allowed to revert back to natural vegetation. The relative lengths of the cropping and fallow periods vary with conditions of soil and climate, but usually cropping is conducted from 1 to 4 years while the fallow period lasts 4 to 20 years. This type of farming occurs mostly in tropical rain forest climates and in thinly populated savanna regions. This system is discussed in more detail in Chapter 19.

PLANNING CROPPING SYSTEMS

Successful farming systems are designed to make near optimum use of climate, soil, supplemental water, and other technological inputs to attain maximum yields and qualities of crops per unit area of land on a year-round basis. In consequence, the planning of a cropping system involves first an assessment of climate, soil, and water resources.

Temperature

The length of growing season limits the potential for producing one or more crops. Except for a few short-season crops, a growing season of 180 to 200 or more days is commonly required to produce more than one major crop. In areas with mild winters a hardy legume or cereal grain crop can be grown during the winter and harvested in time to produce a second crop during the regular growing season. Much of the warm-temperate climatic zone meets such requirements.

The average daily temperature and the range of temperature between night and day also limit the suitability of specific crops. Potatoes, for example, grow best under cool weather conditions, when the mean monthly temperature is below 21°C. Maize does better with warmer mean temperature ranging between 22° and 26°C. Sorghum stands temperatures in excess of 30°C better than most crops.

Moisture

Under rainfed conditions the occurrence of drought is critical to crop growth. Dates of seedbed preparation, planting, and harvesting of crops need careful planning to take advantage of high probabilities of favorable moisture conditions. Such strategies are described under the various climatic zones.

Crops vary in ability to withstand drought. In warm rainfed regions upland rice would typically occupy the higher rainfall areas. But much of the land producing upland rice gives low yields because of inadequate or poorly distributed rainfall. Many such areas could do much better with maize. As the zone of rainfall suitable for maize becomes marginal, sorghum takes over as a major crop. In the arid zones with only short periods of precipitation, sorghum is displaced by pearl millet. Such systems of crop adaptability have been identified by long experience of farmers, but often there is reluctance to adopt the most suitable crop because of tradition or personal preferences.

Irrigation increases the available alternatives for variations in crop selection and in cropping patterns. Even small quantities of water harvested and stored in a farm pond can be of strategic value during a drought period or in establishing a second crop near the end of the monsoon season.

Soils

Cropping systems are at times planned to conform with special soil properties. We have seen that some crops are more tolerant of acid soils than others and that following lime application, sequences of crops can be arranged in order of increasing tolerance to acidity or aluminum. Salinity and excess lime in soils also necessitate a careful choice of crops to secure good production.

If such special conditions as acidity, alkalinity, or salinity are not dominant, the relationships of soil properties to crop adaptation are largely concerned with the water relationships of the soil. Soils that hold good quantities of available water in the root zone help crops avoid or endure periods of drought.

Cropping systems should also be planned to reduce such hazards as erosion. In general, close sown crops, crops with fibrous root systems, and the use of appropriate tillage and other management practices can be adapted to such situations.

The combination of crops grown in any season should be selected to:

(1) Minimize peaks in labor requirements.
(2) Minimize peak periods for use of power machinery.

(3) Minimize demands for highly specialized or little-used expensive equipment.

(4) Provide a reliable source of income at a moderately uniform level.

(5) Minimize buildup of insects or plant diseases.

(6) Protect soils against deterioration—erosion, waterlogging, salinity.

(7) Where irrigation is practiced, peaks in water requirement should be minimized.

The foregoing guidelines can be helpful, but they are not a substitute for field testing varieties, cropping systems, and production practices under local conditions. Most government agricultural ministries, agricultural universities, and many international and bilateral assistance programs are making such tests. The local verification of crops and technology is essential to wise decision making.

REFERENCES

ADAMS, J.E. 1974. Residual effects of crop rotation on water intake, soil loss, and sorghum yield. Agron. J. *66*, 299-304.

ANON. 1974. Annual report 1973-74. International Crops Research Institute for the Semi-Arid Tropics, Hyderabad, India.

ANON. 1976. Production Yearbook 1975. FAO, U.N., Rome.

BEETS, W.C. 1977. Multiple cropping. World Crops *29*(1) 25-27.

BORNER, H. 1960. Liberation of organic substances from higher plants and their role in the soil sickness problem. Bot. Rev. *26*, 393-424.

DONALD, G.M., and HAMBLIN, J. 1976. The biological and harvest index of cereals as agronomic and plant breeding criteria. *In* Advances in Agronomy, Vol. 28. N.C. Brady (Editor). Academic Press, New York.

ESCALDO, R.G., and PLUCKNETT, D.L. 1977. Ratoon cropping of sorghum. Agron. J. *69*, 341-346.

HARLAN, J.R. 1975. Crops and Man. Crop Science Society, Madison, Wisconsin.

HARLAN, J.R. 1976. The plants and animals that nourish man. Sci. Am. *235*(3) 89-97.

HUBER, D.M., WARREN, H.L., NELSON, D.W., and TSAI, C.Y. 1977. Nitrification inhibitors—new tools for food production. BioScience *27*, 523-529.

LITZENBERGER, S.C. 1974. Guide for Field Crops in the Tropics and Subtropics. Technical Assistance Bureau, U.S. Agency for International Development, Washington, D.C.

MARTIN, J.H., LEONARD, W.H., and STAMP, D.L. 1976. Principles of Field Crop Production, 3rd Edition. Macmillan, New York.

MILLER, E.C. 1938. Plant Physiology, 2nd Edition. McGraw-Hill Book Co., New York.

NELSON, L.R., GALLAHER, R.N., BRUCE, R.R., and HOLMES, M.R. 1977. Production of corn and grain sorghum in double cropping systems. Agron. J. *69*, 41-45.

NEWMAN, J.E. 1971. Measuring corn maturity with heat units. Crops Soils *23*(8) 11-14.

PAPENDICK, R.I., SANCHEZ, P.H., and TRIPLETT, G.B. 1976. Multiple Cropping. Spec. Publ. *27*. American Society of Agronomy, Madison, Wisconsin.

PEARSON, R.W. 1974. Significance of rooting patterns to crop production and some problems of root research. *In* The Plant Root and Its Environment. E.W. Carson (Editor). University Press of Virginia, Charlottesville.

PLUCKNETT, D.L., EVENSON, J.P., and SANFORD, W.G. 1970. Ratoon cropping. *In* Advances in Agronomy, Vol. 22. N.C. Brady (Editor). Academic Press, New York.

RAPER, C.D., JR., and BARBER, S.A. 1970. Rooting systems of soybeans. I. Differences in root morphology among varieties. Agron. J. *62*, 581-584.

RAPPAPORT, L., and SACHS, R.M. 1976. Physiology of Cultivated Plants. Rappaport and Sachs, Davis, California.

RUTHENBERG, H. 1976. Farming Systems in the Tropics, 2nd Edition. Clarendon Press, Oxford.

SANFORD, J.O., MYHRE, D.L., and MERWINE, H.C. 1973. Double cropping systems involving no-tillage and conventional tillage. Agron. J. *65*, 978-982.

SARRAE, S., and KOWAL, J. 1977. Provisional grouping of main crops in respect of their eco-physiological responses and characteristics with draft soil and land suitability criterion. *In* FAO Soils Bulletin on Land Evaluation Standards. U.N., Rome.

TISDALE, S.L., and NELSON, W.L. 1975. Soil Fertility and Fertilizers, 3rd Edition. Macmillan, New York.

TRENBATH, B.R. 1974. Biomass productivity of mixtures. *In* Advances in Agronomy, Vol. 26. N.C. Brady (Editor). Academic Press, New York.

TROUGHTON, A., and WHITTINGTON, W.J. 1969. The significance of genetic variation in root systems. *In* Root Growth. W.J. Whittington (Editor). Plenum Press, New York.

TUKEY, H.B. 1969. Implications of allelopathy in agricultural plant science. Bot. Rev. *35*, 1-6.

WHITTAKER, R.H., and FEENY, P.E. 1971. Allelochemics: chemical interactions between species. Science *171*, 757-770.

10

Planning Farms for Improved Production

D. Wynne Thorne

Farm planning is making decisions about a farm. It involves bringing together best uses of farm resources and applicable production technologies into an operations program that will make a productive and prosperous farm unit.

Each farm is a unique entity with combinations of climate, soils, water, crops, livestock, equipment, and other facilities. A farm plan should express the interests, resources, and capabilities of the farm family or operating unit. Just as there are great diversities in farm situations, there are many differences in the goals, backgrounds, and capabilities of the farmer, and in the kinds, purposes, and details of farm plans.

Planning may have a prime purpose of conserving the soil and related water, plant, and animal resources. It may aim primarily at maximizing farm income. It may be limited to producing desired products. In addition to traditional farm production, it may examine many alternatives such as recreation, wildlife, and fish production. At the best planning level, all of these facets and alternatives are involved.

Farm plans may also be oriented toward helping the farm family attain social as well as economic goals. Too often, however, planning is fragmented into expediencies with often unrelated or even conflicting decisions. One area of a farm is drained; another is brought under irrigation; a field is planted to a fruit orchard; or a special item of equipment is purchased, possibly for a crop that may soon thereafter be taken out of the cropping system. Frequently plans are abandoned for lack of resources, or because of an unforeseen event such as drought, fluctuations in commodity prices, or changes in farmer interests. In all cases, though, farm planning must be accepted as a dynamic and continuing process, subject to frequent reviews and changes. Flexibility is necessary.

With changes in new technologies, the ever increasing demands for food, and the rising expectations of farmers and their families, there is increasing public interest in integrated farm plans to help farmers achieve goals through systematically examining resources and alternate courses of action. In most developed and in many developing countries farm planning has become a major factor in improving agriculture. The Soil Conservation Service reports that by 1973 about 1.5 million farmers and ranchers in the U.S. had developed written conservation plans for their units.

BACKGROUND FOR FARM PLANNING

Since the intrinsic nature of farm planning is making decisions about a farm, the key planner and final decision maker must be the farmer or the responsible farm operator. Agronomists, economists, engineers, and other specialists can advise and help with evaluating resources and in identifying alternatives and preparing detailed plans. But if the farmer is not the key planner and decision maker he is unlikely to feel committed to carrying out the plan and the entire planning program becomes a futile exercise.

Well developed farm planning requires the contribution of several specialists. Farm planning begins with one or a series of meetings with the farmer. The farmer provides information about the farm, its history, the crops grown, yields, livestock, equipment, disposition of farm products, and problems encountered. But most important of all, the farmer establishes objectives and goals to be attained through a plan of action. In the beginning stages these goals may be limited or general. He may look forward to attaining certain yields or a level of income. However, the general type of farm should be agreed on and some concepts should emerge on the type and degrees of change that would be acceptable.

The practicality of specialists meeting with individual farmers and working with them on an individual basis in developing farm plans varies from country to country and with the size of the farm. Where farms are predominantly small, perhaps less than 10 ha, farm planning is almost necessarily approached on a group basis with the farmer following general directions of specialists in helping assemble information, but with a public agency providing maps of soil and of contour lines. The specialists also present some of the most feasible alternatives for the major types of farms. With some help the farmer can usually identify his farm on an aerial photo, or in its absence the farmer can draw an approximate scale map of his farm, outlining the various fields and showing principal improvements, roads, buildings, and other features.

This farm can then be located on the soil and contour maps. The detailed work of drawing up plans and preparing a report of agreed-on improvements and practices can be a cooperative undertaking by the farmer and the technical adviser.

The steps in developing and implementing the plan are outlined below.

EVALUATING FARM RESOURCES

The Farm Map

Acquiring an accurate and detailed map of the farm is a primary first step. Aerial photographs are the most useful background for a base map for detailed farm planning. A clear aerial photograph at a scale of about 1:5000 is needed for showing details. For small farms a scale of 1:2500 is more desirable. Aerial photographs are helpful because they show land use, field boundaries, farm roads, buildings, and other features important in planning. If aerial photographs are not available a map will need to be constructed with the major identifiable features carefully located.

The initial map should show all present features, including field boundaries, fences, permanent irrigation and drainage channels, roads, buildings, and land use at the time the map is made. If information is available, drains should also be sketched in.

One of the first features to be added to the map are topographic lines, usually at intervals of about 25 cm for surface irrigated lands, but somewhat wider intervals can be used for rainfed lands. These lines will be important in laying out water management programs, realigning field boundaries, and in helping to identify many problems and in devising solutions to them.

Climate and Hydrologic Data

Climatic data should be obtained from weather stations nearest the farm. The data should be for as long a period as possible. Data for temperature, precipitation, killing frosts, and lengths of growing season should be obtained. The amount of precipitation should be for weekly intervals and for individual storms, if available. Departures from averages are also needed. The probabilities of periods of drought or of substantial amounts of rain within restricted periods of time throughout the year are needed to plan cropping systems and on-farm water management. Often climatic data have been analyzed previously for the probability of such occurrences.

If planning is for a new area where farm experiences have not been related to published data, allowance must be made for differences in climate between the farm and the nearest weather station with regard to elevation, topography, and vegetation. Frequently, local climatic records are of short duration so that judgments are necessary as to likelihood of substantial deviations from available records.

Soils

A detailed map of the soils on the farm is needed for reliable planning of most farm management programs. Some agencies doing extensive farm planning do not map the details of soil families and series. All recognize, however, the need for mapping principal differences in topography, texture of the topsoil, depth of surface soil and permeability, water holding capacity, and adequacy of aeration in the subsoil. Each soil mapping unit should include only soils that can be advantageously managed with the same practices and which will produce about the same yields under uniform management. Special features such as major differences in slope, soil depth, degree of erosion, water tables, salt, pH, stones, hardpan, poor structure, and likelihood of surface flooding must be carefully observed and affected areas outlined on the map. The major principles involved are given in more detail in Chapter 3.

Soils data and related information on topography, water supplies, and climate are used to prepare an interpretive map showing land use capability classes. There are several different classification systems, but many of them correspond rather closely to the land-use capability classes being used by the USDA Soil Conservation Service (see Chapter 3).

Water Resources for Irrigation

When planning is for an irrigated farm, the source, quality, quantity, dependability, and cost of water are important. Ideally it is desirable to have irrigation water available on demand so crops can be irrigated when needed. This is possible where water is drawn from adequate and well managed storage and distribution systems, where the farm has its own temporary storage facilities, or with farms obtaining water from wells or other controlled sources.

In the absence of water on demand, information should be assembled on the schedule and quantities of water likely to be available throughout the growing season. Water diverted from natural streams without storage is likely to vary in quantity during the season. Water rights laws are complex and vary among countries and even between states and regions within countries. It is necessary, therefore, that as a basis for farm planning the details of rights to water and estimates of supplies during the various parts of the growing season be estimated as accurately as possible.

Examination of the hydrology of a farm should also include identification of possible reservoir sites and estimates of the water requirements of various crops throughout the growing season. This, combined with information about the water holding capacity of the soils and of average or probable amounts of precipitation, forms the basis for estimates of water that can be harvested and stored in reservoirs, of additional irrigation water required beyond available supplies for good crop yields, or of the need for surface or ground water drainage.

FARM PLANNING ACTIVITIES

Information Needed

Types of information generally needed as a basis for developing a farm plan are itemized below. Where applicable, data should be compiled by fields.

(1) Farmer and farm family—Information obtained should include: (a) the number in family, ages, education; (b) the total days of labor per year available for the farm; (c) the experience in farming, types of crops, livestock; (d) the type of farm, crops, and livestock preferred and those the farm and farmer are best suited to handle; (e) the food and income needed from the farm; and (f) the family interests and goals.

(2) Climate—This includes: (a) the length of growing season; (b) the temperature, rainfall, humidity, sunshine data, periods of drought; and (c) prevailing winds and intensities.

(3) Map of farm—This preferably should be on an aerial photograph and should show: (a) the outline of fields, size, and shape; (b) the location of buildings, farm roads; and (c) the location of drainage and irrigation channels.

(4) Topographic map—This should show the contour intervals of about 25 cm for somewhat level land, or wider intervals for more sloping lands.

(5) Map and information about soils on the farm—These should indicate: (a) boundaries of soil units for the farm; (b) descriptions of profiles of principal soil units; and (c) special features such as rock outcrops, water tables, poorly drained areas, salt problems, erosion, and general nutrient levels of soils.

(6) Water resources—Information compiled should include: (a) water requirements of crops through the growing season; (b) estimates of

adequacy of precipitation and available supplemental water to meet demands; (c) for irrigated farms, data should include source, quality, and quantity of irrigation water, type of water right, cost of water, frequency and duration of water turns; (d) irrigation facilities; (e) drains; (f) areas needing draining should be identified for all farms; and (g) sites and potentials for water harvesting and storage are also often needed.

(7) The farmstead—This should be plotted out on a large scale showing: (a) locations of buildings and uses; (b) arrangement of facilities, access routes for farm; and (c) access to public highways.

(8) Livestock—An inventory should show: (a) types, breed, quality, numbers, and ages; (b) productivity records and potentials; and (c) feed required.

(9) Miscellaneous facilities—These should be itemized, including: (a) farm machinery, type, size, and condition; and (b) fences—their location, construction, and needs for improvement.

(10) Crop yields—Data should be compiled on: (a) crops grown in recent years, areas, yields, qualities; and (b) reasons for reduced yields.

(11) Management practices—The various farm practices carried out in preparing seedbeds, planting, growing, harvesting, storing, and marketing crops should be described.

(12) Economics—This should include: (a) costs of inputs, labor utilized; (b) detailed returns from various crops and livestock; (c) estimates of net returns from various crops; and (d) returns for labor and capital.

After information gathering and preliminary analysis have been completed in accord with the above outline, the next step is to determine how farm design, management, and productivity can be improved.

Identification of Limiting Factors

Examination of soil and land resources and their management should indicate adequacy of past farm practices for high crop yields and potentials for gains through more efficient use of resources. Data on crops grown and yields obtained, when compared with experimental results under comparable soil and climatic conditions, or when compared with yields obtained by the best farmers of the area, should also indicate potentials for increasing yields and qualities of crops. Some of the common weaknesses in farm production programs and limitations of factors for high crop yields include:

(1) Timing of farm practices—planting, weed control, insect control, water application, harvest.

(2) Soil fertility—inadequate nutrients supplied in relation to crop needs; consideration is needed of fertilizer applications, including types, amounts applied, method of application, and soil organic matter maintenance.

(3) On-farm water management practices—lack of quick removal of excess water, not maintaining good soil infiltration conditions, seriousness of periods of drought, and inadequate alternatives for alleviating them.

(4) Losses from pests and inadequate practices for their control—insects, diseases, birds, rodents; pre-harvest and post-harvest losses.

(5) Weed infestations and their control—tillage, hand weeding, crop rotations, applications of herbicides.

(6) Harvest—delay in timing, losses in harvest, inadequate equipment; improvements needed in care, storage, and marketing of products.

Having identified important factors that would have been limiting in attaining high crop yields, the next logical step is to devise changes. Usually some order of priority is needed. Also a strategy is needed regarding which factors can be most readily improved and what the costs and benefits will be to the farmer.

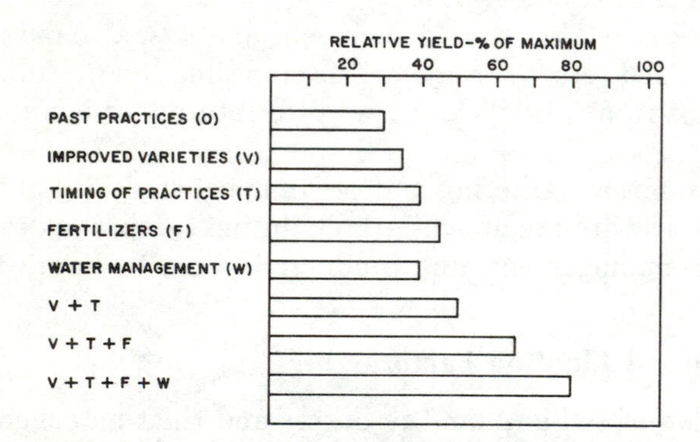

FIG. 10.1 RELATIVE YIELD OF A CROP LIMITED BY CROP VARIETY, POOR TIMING OF PRACTICES, LOW SOIL FERTILITY, AND POOR WATER MANAGEMENT AS INFLUENCED BY IMPROVEMENT OF THESE FACTORS, INDIVIDUALLY AND IN COMBINATIONS

Some of the principles involved in securing better yields through improved practices are illustrated in Fig. 10.1. In the assumed situation four factors limiting crop yields are identified and practices are initiated for their improvement. First, local unimproved crop varieties are replaced by well adapted improved varieties. Second, planting is made earlier to take advantage of improved moisture conditions and to have better crop environmental seasonal conditions. Third, a balanced fertilizer program is adopted to apply nutrients in accord with soil supplies and crop needs. Fourth, crops are planted on a contour grade to remove water after heavy rains and broad furrows are made between rows to increase the entrance of water into the soil and to fill the soil to field capacity. As a result of these innovations there were some gains in yield from each of the new practices, but when the practices were combined, the total benefits obtained totaled a yield increase of 267%.

Type of Farm

To assure success, a farm must be operated as a unit. The major farm activities should contribute to desired goals. An important consideration in achieving unified operations is agreeing on the type of farm and the major products to be obtained.

Choice of type of farm depends on the interests and aptitudes of the farmer as well as on the character of the farm resources and the market for the products. There are, of course, physical and economic factors that often make one type of farming profitable and another unprofitable on a given farm. These factors should be emphasized in working out a farm plan with the farmer, but the farmer must make the final decision as to type of farming system adopted. Some types of farms and general considerations for planning of a successful operation are outlined in Table 10.1.

Adjustments Between Crops and Livestock

On farms with a significant livestock component the balancing of crops to provide needed feed becomes important. The needs for feed vary with types and numbers of livestock. A farm's capacity to produce feed depends in turn on the size of the farm, its productive capacity, the growing season, the crops grown, and production practices used. Appropriate government agencies, extension services, and agricultural experiment stations publish bulletins on feeding livestock that can be used as a basis for estimates. Table 10.2 presents data on feed equivalents for various forage crops and other livestock feeds as produced in temperate climates. The daily feed requirement for a "standard" chopped air dry forage is about 3 kg per 100 kg live weight for sheep and cattle.

There are several more precise methods for evaluating livestock feeds and feed requirements. Such details can be found in Crampton and Harris (1969) or other books on animal nutrition.

TABLE 10.1. SOME TYPES OF FARMS AND FACTORS FAVORING SUCCESS

	Type of Farm			
Factors	Field Crops	Vegetable, Fruits or Special Crops	Dairy	Livestock Production
Size of Farm	Variable, small to large	Variable, small to medium	Medium to large	Large
Climate	Limits kinds and varieties of crops	A restriction for many specific crops	Adapted to cool climate	Adapted to wide range in climate
Soil	Classes I to III	Class I, but many crops have special requirements	Classes I to III Needs good moisture capacity	All classes Classes I to III for harvested forage
Water	Good water supply, can adapt to some arid conditions	Needs good water supply	Needs good water supply	Can be adapted to wide range
Labor Requirements	Varied	Generally high	High	Varied
Specialized Labor	Medium	High for some crops	High, veterinary service	High, veterinary service
Capital Investment Machinery, Buildings	Varied	Generally high	High	Varied
Fertilizer Requirement	High, especially nitrogen	High and varied	Medium, low nitrogen	Medium, low nitrogen
Pest Control	Varied nitrogen	Varied, high for some crops	Medium, low nitrogen	Medium, low nitrogen
Use of Crop Rotations	Varied, often lacking	Varied	Common	Common

TABLE 10.2. CONVERTING VARIOUS LIVESTOCK FEEDS TO AN ALFALFA HAY EQUIVALENT

(Data Given in Tons, Air Dry, to be Equivalent to One Ton of Alfalfa Hay)

Kind of Feed	T.D.N. Basis	Average Use Base
Alfalfa	1.0	1.0
Timothy	1.06	1.5
Wild Hay		2.5
Maize Fodder, Dry	0.88	4.0
Oat Hay	1.11	2.0
Wheat Hay	0.95	2.0
Barley Hay	0.94	2.0
Oat Straw	1.13	4.0
Wheat Straw	1.40	4.0
Barley Straw	1.21	4.0
Maize Silage	2.9	2.5
Sugar Beets	3.68	2.5
Beet Pulp, Wet	6.97	5.0
Beet Molasses	0.88	0.75
Potatoes	3.02	3.0
Wheat Bran	0.85	0.75
Cottonseed Cake, Cold Pressed	0.73	0.33
Wheat	0.63	0.63
Maize	0.63	0.63
Oats	0.73	0.73
Barley	0.65	0.65

Cropping Systems

Based on the many considerations outlined in Chapter 9 and elsewhere, one or more tentative cropping systems should be drawn up for each field and for the farm as a whole. Major income crops should receive special consideration. A first consideration is the relative suitability of the various possible crops to the climate of the area, soil conditions on the farm, the equipment available, and the labor and background experiences of the farm family or available workers.

The inputs required for each crop in terms of farm machinery, power, labor, and such materials as seeds, fertilizer, and pesticides should be detailed. According to weekly or other convenient intervals such an analysis should show the labor required on the farm for the production of each crop, machinery use and costs, and the amount and cost of fertilizers and pesticides. Allowances should also be made for interest on investments, credit costs, and the amount of family labor used. Finally, the expected yields and estimated total and net income should be computed. Hilton (1968) and other books on farm management show formats of useful tables and give suggestions for making detailed inputs and costs and returns analysis.

Such an analysis provides an estimate of the profitability of each crop and of the total cropping system, and also helps to identify special problem areas. Among these are the labor requirements by weekly intervals. If the combinations of crops result in excessive labor requirements during one period and inadequate needs for labor at other periods, some adjustment in crops or practices may help reduce the peaks to manageable levels.

The analysis will also indicate capital requirements for the crop season. Costs of such materials as fertilizers, new seeds, and pesticides and of new or additional equipment must be met by use of available resources or through credit. The alternatives and costs of meeting these expenses should be anticipated.

With irrigated farms, an analysis of water requirements in relation to supply during the course of the growing season is also important. Adjustments in crops, in dates of planting, or other alternatives can be used to minimize excessive peaks in water demand that cannot be reasonably taken care of with available supplies. Similarly, the impact of foreseen water deficits on crop alternatives and potential yields can be estimated. Thus, in temperate zones in which the water supply declines in mid or late summer, a larger proportion of such early-maturing cereal crops as winter wheat or barley can be grown, and the area of such late-season water users as cotton or maize can be held within potential water supplies. Several such alternatives can be found for nearly every farm situation.

Special Considerations

Since each farm is a unique situation there are likely opportunities for special facilities or practices. Among these are drainage of wetlands, planting crops on the steeper sloping lands on the contour, and developing grass waterways to carry away excess water. Stone removal may provide for improved tillage or other mechanized practices. Changing the directions of rows and of tillage practices and leaving the surface rough or with crop residues partially on the soil surface may reduce wind damage to crops and both wind and water erosion of soils.

For irrigated farms there are usually several alternatives to help improve uniformity and efficiency of water application. Investments in water control or water-application facilities may decrease the labor needed for irrigation and increase capability for applying water more uniformly and accurately. Among such alternatives are water control structures and devices to control and equalize the quantity of water entering each furrow, land levelling to ensure more uniform applications, shifting to one or more of several types of sprinkler equipment, or even the use of drip irrigation for special situations. Helpful suggestions for improving irrigation practices will be found in Chapters 8 and 15.

THE FARM PLAN AND REPORT

When the major decisions have been made regarding the layout, land use, cropping system, major production practices, and the many other details that are involved in an effective farm plan, these must be recorded in written form.

The Farm Layout

The farm layout of fields, land use, roads, permanent irrigation channels, and open drains must be considered from the viewpoint of efficiency of operation, accessibility to fields, irrigation distribution systems, and methods of water application (for irrigated farms), surface and underground drainage and waterways, storage ponds feasibility and locations, contour planting, and other erosion control practices, soil characteristics, and general attractiveness in appearance. The farm layout of fields, the farmstead land use, and special practices should be indicated on the farm map and explained in the accompanying report.

The Farmstead

Usually planning is done for farms already in operation and the farmstead is established and difficult to move. In principle, however, the farmstead should be located as a natural control center for the farm. Long farm roads to fields use up farm land, are inefficient, and may be expensive to maintain. If possible, the farmstead should be near a highway or a good public road, have a good quality domestic water supply, and, for irrigated farms, be near the point of access to irrigation water. Suggestions for planning homesteads on U.S. farms are given by Betts (1931).

Field Arrangements

In general, farms should be divided into fields of approximately equal size. Where somewhat systematic rotations are followed, there should be one field for each year of a rotation system, but this general rule needs frequent modification. Two small fields may be used for one large field, and land capability classes may indicate that substantial areas may be best kept in permanent pasture or transferred to woodland.

Field planning, particularly for mechanized farms, should take advantage of the principle that a rectangular field is the most efficient shape, with the field being about twice as long as wide for the most efficient use of farm machinery. Some studies have found that even with small to medium sized tractors and equipment, triangular fields require about 20% more machine time than rectangular fields. Also, fields of 2 ha in size require 10% more machine time than fields 6 ha or larger in size.

For irrigated farms, fields should be arranged for easy access to and convenient management of water. With surface irrigation systems, a field is usually longer than one run of water, so one or two temporary cross ditches or portable tubing arrangements are used. With sprinkler irrigation, modest variations in topography can be handled. Convenient lengths or widths of fields vary with the types of systems used. Long fields are convenient for self-propelled, wheel mounted systems. Round fields are used with central pivot systems, with special arrangements for irrigating the spaces between circles where land is valuable. Obviously, field shapes on irrigated farms are a compromise among land classes, irrigation method, and the equipment used.

Drains

Drains are an important aspect of the farm layout. Tile drains and schemes for surface drainage should be located and installed before other land improvements are initiated. Tile drains do not disrupt the farm layout. Open drains should be adjusted wherever possible to farm boundaries or to borders of permanent fields.

Conservation and Water Management

Where there is seasonal excess water and where water runoff or wind cause erosion, the problems require special farming practices and facilities.

Contour planting and furrowing are effective if carefully laid out on a uniform grade. Adjustments in slope and the outlet at the end of the row can help regulate water absorption into the soil, and can serve to divert or provide surface drainage of excess water to a grass or otherwise protected waterway. Small reservoirs to store water coming down waterways have been found useful for supplemental irrigation in India, Australia, and elsewhere.

For wind erosion careful selection of crops, ridging of soil, and planting with rows at right angles to the direction of the prevailing winds are common arrangements for minimizing damage to growing crops. In some areas and with high value crops, wind break plantings are used. These permanent plantings are necessarily restricted to spaces between fields, and must be properly oriented to be effective for the prevailing direction of winds.

Examples of various farm maps used in the planning process and in final reports, together with brief conservation reports, can be found in two U.S. Soil Conservation Service Reports (Anon. 1964, 1973).

The Written Report

A written report should accompany the detailed map of the farm plan. The report should be a record of decisions made by the farmer in consultation with the specialists and should contain recommendations for putting decisions into practice. The report may be brief as shown in the Soil Conservation Service 1973 report. The written report should describe the soils and major classes of land and their management needs and suitabilities for various crops and practices. Recommendations should be given for irrigation structures, land levelling operations, drainage systems, contour farming systems, and other new developments. A proposed cropping system should be outlined with recommendations for production practices, including a schedule of operations, fertilizer treatments, pest control practices, etc., for each field. Irrigation practices should be outlined with recommended rates, quantities, and frequencies of application.

Important in the farm plan is a schedule for development. Since it is seldom feasible to redesign all the farm at one time, the report should outline the work to be done each year. This schedule should fit in with the cropping plan. For example, it is convenient to level land during non-cropping seasons.

Farm planning involves application of the principles of all phases of agricultural science. It demands the best in technical skill along with insights and imagination. Each farm should be planned from the broadest possible viewpoint. The work of soil scientists, engineers, crop scientists, livestock specialists, landscape specialists, and economists should be combined with due regard for the interests and abilities of the farmer. The making and following of farm plans will lead to productive, efficient, and attractive farms.

REFERENCES

ANON. 1964. What is a ranch conservation plan? U.S. Dep. Agric., Soil Conserv. Serv. *PA-637.*

ANON. 1973. What is a farm conservation plan? U.S. Dep. Agric., Soil Conserv. Serv. *PA-69.*

ANON. 1976. A framework for land evaluation. FAO Soils Bull. *32.* U.N., Rome.

BERNARD, C.S., and NIX, J.S. 1973. Farm Planning and Control. Cambridge University Press, London.

BETTS, M.C. 1931. Planning the farmstead. U.S. Dep. Agric., Farmers' Bull. *1132.*

CASE, H.C.M., JOHNSTON, P.E., and BUDDEMEIER, W.D. 1960. Principles of Farm Management, 2nd Edition. J.P. Lippincott, Chicago.

CRAMPTON, E.W., and HARRIS, L.E. 1969. Applied Animal Nutrition, 2nd Edition. W.H. Freeman and Co., San Francisco.

DAWSON, J.A., and DOORNKAMP, J.C. 1973. Evaluating the Human Environment. St. Martins Press, New York.

HILTON, N. 1968. Farm planning in the early stages of development. FAO Agric. Serv. Bull. *1.* U.N., Rome.

KLINGEBIEL, A.A., and MONTGOMERY, P.H. 1961. Land-capability classification. U.S. Dep. Agric. Handb. *210.*

THORNE, D.W., and PETERSON, H.B. 1954. Irrigated Soils, 2nd Edition. McGraw-Hill Book Co., New York.

UPTON, M. 1973. Farm Management in Africa. Oxford University Press, London.

11

Crop Production Systems in Humid Cool Temperate Zones

Marlowe D. Thorne

GEOGRAPHICAL AREAS

This zone is characterized by a summer growing season of at least 150 days and with mild to cold winters. Precipitation varies from about 800 mm to about 1500 mm per year, with a greater proportion coming during the warm season than during the winter period. Annual precipitation exceeds annual evapotranspiration, but during the mid-summer period of peak water use, evapotranspiration may exceed precipitation for a number of months in succession. Most of the zone is between 35° and 55° latitude.

The world distribution of this zone is shown in Fig. 11.1. It includes most of Europe, with the exception of Spain, Portugal, and parts of Italy and the Balkan countries. Much of western USSR and an arm of land between 50° and 55°N extending east to Omsk, and eastern USSR south of 50° are in this zone. Also included are northern Japan, part of Korea, and much of northern China lying between 30° and 50°N. Small areas in Australia, New Zealand, Tasmania, and Chile are also in the zone, as indicated in Fig. 11.1.

In North America, the zone is bounded on the west by about the 98th meridian and on the south by the 37th parallel. It extends into the lower part of the Canadian provinces east of the 98th meridian and includes a thumb-shaped area extending westward from Winnepeg across Saskatchewan to near Edmonton in Alberta. This zone includes the Corn Belt of the U.S., one of the world's most favorable areas of climate and soils for feed grain and oilseed production. Precipitation increases from west to east and temperature increases from north to south.

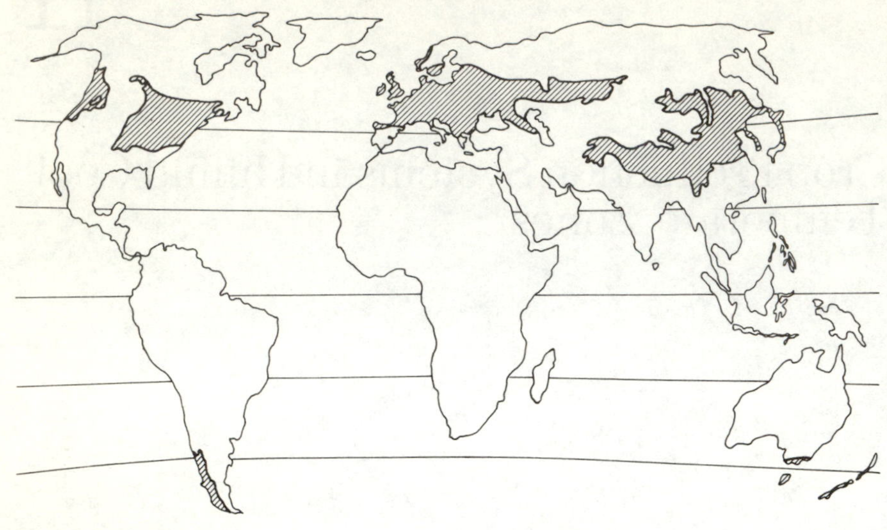

Adapted from Troll and Paffen (1965)

FIG. 11.1. MAP OF HUMID COOL TEMPERATE ZONE

SOILS OF THE ZONE

Soils of the zone are generally highly productive. They include Mollisols and Alfisols primarily, with some Spodosols, Ultisols, Entisols, and Histosols. Organic matter contents and nutrient supplying capabilities are generally high. When they are cleared of native vegetation and placed under intensive cultivation, they may produce quite satisfactory yields without fertilization for many years. The soils are usually mildly to highly acidic and liming is generally required. Moisture holding capacities are high and rainfall is such that moisture conditions are favorable for summer crops. Excessive water may be a problem, particularly in the spring and fall. Surface and/or subsurface drainage practices may be needed. With adequate drainage, these soils may be tilled quite easily in the optimum moisture range and will retain a structure favorable to root development.

CURRENT CROP PRODUCTION SYSTEMS

In the U.S., the general cropping pattern in this zone includes maize, soybeans, and forage crops. A few decades ago rotations including maize, a legume crop, and a small grain were practiced in most of the area. The legume provided nitrogen for the maize and was used as feed for the livestock which were raised for meat, dairy products, and farm power.

The small grain made a good "nurse-crop" to get the legume established and it was used as livestock feed or as a cash crop. Maize was fed to livestock—primarily beef cattle, dairy cattle, and hogs—or was sold as a cash grain crop.

The cropping pattern in the U.S. Corn Belt has changed appreciably in the past few decades. Synthetic nitrogen has become readily obtainable and soybeans have developed as a major crop. While over 70% of the milk and 85% of hog production in the U.S. come from this zone, their production is increasingly more concentrated in specialized operations with closer confinement of the animals. There are fewer producers, each having large numbers of animals. It is common to find farms in the area with no beef or dairy cattle or hogs on the farm. It is now unusual to find all three kinds of animals on the same farm.

In the Corn Belt, most farmers rotate maize with soybeans, but the crops may not appear in any one field in alternate years. Some land is found to be more productive for maize and some for soybeans. Thus a field may be left in maize or in soybeans for many years in succession. There is good evidence to indicate that maize following soybeans generally has higher yields than continuous maize even on the most productive soils in the area (Welch 1977).

Nearly half of the world's maize is now produced in the U.S. and the five Corn Belt states (Illinois, Indiana, Iowa, Missouri, and Ohio) normally produce about half of the country's maize. U.S. produces about 65% of the world's soybeans with the Corn Belt accounting for over half of the U.S. crop.

Even though this zone is not the most important one for wheat production in the U.S., wheat is one of the most important of the secondary crops grown. Much of the area is suitable for production of winter wheat but winters are usually too long and too cold to permit extensive fall and winter grazing of wheat as is done in zones closer to the equator. Some of the winter wheat area is suitable for double-cropping, but harvest of wheat is so late that a sufficiently long season is not available for the second crop in much of the zone.

Three crops may be produced satisfactorily in two years in some parts of the zone but two crops each year is generally not possible. In the cooler parts of the zone survival of fall planted wheat is not well assured and production of spring planted varieties is more common. Barley and oats were once grown extensively in this zone in U.S. but areas of these crops have decreased as horse power has been replaced by tractor power.

Summer rainfall and temperatures in this zone are conducive to high yields of forage crops. The conversion of these crops to hay may be difficult, however, because of untimely rains. Conversion to silage avoids this problem generally but the product is not easily transported long distances for marketing. Thus forage crops not consumed on the farm are marketed primarily as hay. The states in the humid cool temperate zone thus produce an appreciable portion of the forage crops in the U.S., but are not the most important ones in producing hay for sale. In other parts of the humid cool temperate zone, forage crops are relatively more important than they are in the U.S.

Improvement in mechanization of forage harvesting and handling has helped keep forage crops competitive even where labor rates are high. The crops can be handled as green chopped material or as hay with small human labor expenditure. Such equipment requires a sizeable capital investment, of course, which only large-scale production can justify. Timely harvesting helps to ensure satisfactory quality of the final product.

Crop production in the humid cool temperate zone in North America currently is dependent upon high inputs. Specially bred crop cultivars are planted which have a high degree of disease and insect resistance and which produce high yields under optimum conditions. High plant populations are utilized and chemical weed and insect control practices are followed.

In the parts of Europe and Asia lying in this zone a much higher proportion of the land is devoted to wheat, oats, barley, rye, sorghums, and millet than in the U.S. Only in France, Italy, and the People's Republic of China does maize constitute as much as one-fourth of the total grain production. Nearly three-fourths of the cultivated cropland in the U.K. is devoted to small grains, principally wheat and barley. Livestock products may account for a larger part of gross farm revenues than in this zone in North America. In Sweden, 40% of the agricultural land is in grass and a like amount in cereals (Duckham and Masefield 1970). Potatoes and sugar beets are important crops in many of these countries. Yields per hectare in the European countries are among the highest in the world.

Farms in the part of Europe lying in the humid cool temperate zone are normally smaller than in North America. In 1960, the average number of hectares per farm worker was 6.3 in the Netherlands, 11.1 in Denmark, 40 in U.K., and 120 in U.S. (Duckham and Masefield 1970). In Europe a higher proportion of the land is devoted to forages and crop rotation is followed more regularly.

In China, multiple cropping and intercropping reach very high intensities, with as many as 12 crops being harvested in a single season, in extreme situations. Wheat is second only to rice both in area sown and in total production. Maize is grown as the primary crop in northern China and as a component of a multiple cropping scheme in the south. Maize may be grown in combination with wheat, soybeans, or other crops in the north. Soybeans are grown throughout the country. In the north they are an important field crop and in the rest of the country they are grown in small fields, gardens, ditch banks, and other waste lands. Water control practices, including drainage, irrigation, and land leveling, are practiced on 40% of the cultivated area. An efficient system for utilization of human and animal wastes and of crop residues is followed (Sprague 1975).

Some of the most productive part of the Soviet Union, including the Ukraine, lies in the humid cool temperate zone. However, much of the zone in the U.S.S.R. is north of the limits of the zone in North America. Because of climatic and topographical limitations, agriculture in the U.S.S.R. is limited to about one-quarter of the total land area. About 10% of the land area is in arable farming with 2% of the land area in hay and 14% in pasture (Symons 1972). The U.S.S.R. produces about 75% as much total grain as the U.S. It produces nearly twice as much wheat as U.S. in some years, but less than 10% as much maize. Per hectare yields of grain crops in U.S.S.R. are close to World average yields (USDA 1977). In contrast to the situation in U.S., temperature and rainfall belts in western U.S.S.R. run more nearly parallel to each other and the part of the zone with highest mean annual temperature has the lowest annual rainfall. The Ukraine is thus more subject to severe droughts than the Corn Belt in U.S.

Rotations are generally followed in the U.S.S.R. and are characterized as "field systems" such as an 8-field rotation or a 6-field rotation. The rotational practices for the fields vary because crops are left on the land for unequal periods of time. Crops grown include maize (for grain or for fodder), legumes, wheat (planted in either spring or fall), sugar beets, oil crops (such as soybeans, sunflowers, and rape), potatoes, flax, vegetables, and other crops. About 120,000 ha are devoted to grain crops, with about half in wheat (USDA 1977). There has been a significant increase in area devoted to fodder crops in the past few decades. Numbers of cattle, pigs, and sheep have also increased appreciably during this period.

Agricultural production in this zone of the U.S.S.R. is mechanized. Soil erosion is a continuing problem here as elsewhere. In the more intensively cultivated areas, as high as 20% of the agricultural land is estimated to be moderately or severely affected by water erosion. Also large areas of sandy soils and high river terraces are particularly vulnerable (Symons 1972).

CONSTRAINTS TO HIGH YIELDS

Probably the most common and most serious restraint to high yields in this zone is excessive moisture. Unusually heavy and/or prolonged rains in the spring make it difficult to achieve timely planting. The optimum period for planting is earlier in the spring in the parts of the zone nearest the equator and becomes increasingly later as latitude increases. Planting after this optimum period may result in a serious yield decrease of corn or soybeans. While topography in this zone is relatively level, most farms contain depressions in the surface configuration which tend to fill with water during heavy rains and may retain ponded water for several days. Tillage and planting must be delayed in those areas, of course.

Drought may also be a constraint to high yields in the humid cool temperate zone. Although the overall moisture balance is quite favorable, almost every year has one or more periods when crops suffer from moisture stress. Periodic droughts of several months duration may be encountered when yields are reduced significantly. Yield reductions are most severe on sandy soils, of course, because of their low moisture storage capacities.

Soil erosion can be a serious constraint to continued high yields in this zone. It is generally known that soybeans leave the soil more susceptible to erosion than does maize. Siemens and Oschwald (1976) used a rainfall simulator to apply intense "storms" to plots receiving different tillage treatments. They found that a higher percentage of the applied water ran off with all tillage treatments after soybeans than after corn. With fall plowing, soil loss was 2 to 3 times greater after soybeans than after corn. Even though 125 mm of applied water caused the loss of over 24 metric tons of soil per hectare following fall plowing, loss was reduced to 6.7 metric tons per hectare with fall chisel plow treatment and 4.5 metric tons per hectare with zero tillage.

POTENTIAL IMPROVEMENTS IN CROP PRODUCTION SYSTEMS

Cropping Systems

Possible improvements in cropping systems for the humid, cool temperate zone must be directed towards increasing the portion of the year in which crops may be grown as well as increasing the output per unit of land during the major production season. Double-cropping of wheat and soybeans is one system which extends the productive season. Winter wheat can be grown in much of the region, but in the colder latitudes the wheat harvest extends so far into the warm season that the period left for the soybean crop may be too short for satisfactory production. The seeding of the new crop before the harvest of the previous crop may offer decided advantages for double-cropping in the colder latitudes. This is difficult to accomplish on a field scale basis without damage to the standing, nearly mature crop. Seeding via airplane may offer much potential. The success of double-cropping anywhere in the zone is highly dependent upon adequate moisture in the surface layer of soil because the seed may not get completely covered with soil, particularly if the surface soil is dry and hard. Irrigation can provide this moisture and continuing expansion of both double-cropping and irrigation can be expected.

Development of better yielding, shorter-season soybeans and earlier-maturing wheat will also facilitate double-cropping in the colder parts of the zone. Cultivars of both crops with these characteristics are available, but yields generally are lower than with their longer season, later-maturing counterparts. Perhaps crops other than wheat and/or soybeans may be found which will produce a greater annual value of product in a multiple cropping sequence. Various cropping schemes have been tried with some success, but none has thus far seemed to offer such advantage that it has been widely adopted.

Tillage Practices

As mentioned in Chapter 6, reduced tillage or conservation tillage programs offer many advantages. The soil has a cover of a growing crop or of a crop residue to protect it over a greater part of the year, thus reducing erosion from wind and water. Bone (1978) reports that conventional tillage for soybeans requires 3 times as much labor and equipment operating hours and 8 times as much fuel as a no-tillage system requires. Conventional tillage involved plowing plus secondary operations prior to planting.

Specialized planting equipment is required for reduced tillage systems. Generally a sharp disk (sometimes fluted) is required ahead of the planting shoe in order to cut through crop residues. Accurate regulation of depth of planting must be achieved and there must be means to vary the depth rather easily. The seed must be covered and the soil firmed around the seed even though considerable variation in soil moisture content will be encountered. The increasing popularity of reduced tillage systems has come only after considerable progress has been made in planting equipment development.

Adequate chemical weed control is essential to the success of reduced tillage systems. There is little opportunity to control weeds by cultivation because of the crop residues on the surface and the cloddiness of the soil. Zero tillage systems may depend on the killing of the previous crop by chemicals to avoid its competition with the new crop. Crop residues may provide shelter for insects and accentuate the need for adequate insect control practices, as well as to make them more difficult to achieve.

Soil and Water Management and Conservation

As other management factors have been more intensively applied in this zone, the importance of water management and conservation has increased. Shaping of the land surface to facilitate runoff at a rate moderate enough to prevent serious erosion can do much to alleviate the problem of excessive moisture. Spreading the water more uniformly over the surface ensures more uniform infiltration and maximizes the depth of wetting from the rain, as well as prevents ponding and the resultant problems with field operations in wet areas.

Drainage.—Subsurface drainage practices must be followed on many of the soils in this zone. Development of some of the areas which are now most productive was delayed until drainage tile was readily available and its proper installation and use understood. Plastic drain tubing has been developed which may provide as good drainage as clay tile and be cheaper and easier to install. Investigations are underway to determine if claypan soils can be drained economically by small diameter plastic tubing installed at shallow depth and with close spacing.

A study in Illinois has indicated that the number of days in which field operations can be performed on a Flanagan silt loam soil during the month of May might be as low as 5 with poor drainage and can be increased to 20 with adequate surface and subsurface drainage practices (Elliott *et al.* 1975). Many of the fine textured soils have inadequate drainage and are slow to warm in the spring. Thus, moldboard plowing in the fall continues to be a preferred practice in the U.S. Corn Belt, particularly following a corn crop, even though it results in loss of more soil and more water than occurs with some of the reduced tillage operations.

The development and use of larger horsepower tractors, particularly those with 4-wheel drive, now allows most farms in the U.S. Corn Belt to get their entire corn or soybean acreage planted in a few days. Thus the length of the period of favorable weather needed to plant or to harvest is much shorter and the probability of getting the field operations done at the optimum time is increased. There is concern that increased soil compaction may be occurring with the use of these large machines, particularly when soil moisture is not optimum.

Progress is being made in controlling water erosion in this area while simultaneously increasing yields, as pointed out by Aldrich *et al.* (1976). They reported that the total amount of sediment deposited in 4 river reservoirs in central Illinois was only half as much from 1945 to 1966 as during the 20-yr period preceding 1945. The overall yield, in terms of grain equivalent, increased 2½ times during this total period on the land in the watersheds of these reservoirs.

Irrigation.—Irrigation is increasing in the humid areas as well as in the arid and semi-arid. In U.S. there have not been large federally funded projects to develop irrigation water in the humid areas as there have been in the arid. Thus irrigation development in the humid zone has been primarily in areas where ground water is available in adequate quantity. There is some diversion of water from streams, primarily by individual farmers. Large-scale development of surface water resources in the humid area can be undertaken when demand for increased production requires it. Since river valleys are not so deep as they are in arid areas, the type and size of individual impoundments may be quite different.

Even though irrigation equipment is expensive (Chapter 15), costs of such equipment have risen less rapidly than land prices in U.S. during the past decade. Thus it may be more advantageous to invest in irrigation equipment to increase and stabilize production rather than to invest in more land. Other costs of production have also risen to a level which makes a crop failure, or even a partial failure, a much more serious problem to the farmer and the nation.

Maize and soybean yields as high as 12.5 and 4 t per hectare, respectively, have been obtained on sandy soils in Illinois with irrigation and with high levels of fertilization. Unirrigated yields were about half as much. Currently yield increases of about 2.5 t maize per hectare and 1 t soybeans per hectare are needed to pay annual costs of irrigation. On claypan soils in central Illinois, maize yield increases as high as 6 t per hectare have been obtained by irrigation when drainage and fertilization were adequate and mid-summer rainfall unusually low.

Soil Fertility Maintenance

Soil testing procedures have been developed which give accurate predictions of the amounts of phosphate, potash, lime, and some secondary nutrients needed to maintain fertility and to secure satisfactory yield levels on most soils in this zone (see Chapter 5). A knowledge of the amounts of nutrients contained in various crops is also helpful in planning a fertilization program (Table 11.1). Soybeans and alfalfa, being legumes, get a high proportion of their nitrogen from the symbiotic nitrogen-fixing organisms in their root nodules. In all other cases the nutrient must be supplied by the soil or by fertilizer additions.

Lime is needed on most soils in this zone which have been cropped for a number of years. Since precipitation normally exceeds evapotranspiration for many months of each year, there is a net downward flow of water through the soil profile with consequent leaching of cations. The principal crops of the zone have optimum growth at pH levels between

TABLE 11.1. APPROXIMATE AMOUNTS OF NUTRIENTS CONTAINED IN VARIOUS CROPS

Crop	Yield t/ha	Nutrient Content (kg/ha)				
		N	P_2O_5	K_2O	Ca	Mg
Corn Grain	11.3	191	77	54	5	18
Stover	9.0	79	34	215	39	38
Soybean Grain	4.0	280	54	9	19	19
Straw		95	18	65	11	11
Wheat Grain	5.4	157	52	29	2	13
Straw	6.7	45	10	148	4	13
Alfalfa Hay	17.9	504	90	538	224	45
Orchard Grass Hay	13.5	336	113	404	56	28

Source: Barber (1977).

6.0 and 7.0 and a good management program will maintain pH within those limits. Soil pH and lime requirement tests are easy to run and control of acidity is not difficult or expensive.

Maize has been grown continuously for over 100 years on the Morrow Plots at the University of Illinois. Yields in plots where best known technology is applied average about four times the yields where no fertilizer has been applied. Maize in a maize-oats-clover rotation outyielded continuous maize only if additional nitrogen in the form of inorganic nitrogen fertilizer or manure was added to the rotation maize. Maize with only the nitrogen provided by the legume in the rotation yielded 63 to 70% of that in continuous maize with adequate fertilization. Another comparison available since 1967 shows that maize in a 2-year maize-soybean rotation (the most common cropping system now followed in the Corn Belt) yielded 15% more than continuous maize, when high fertility practices were followed in both plots (Welch *et al.* 1976).

The nitrogen provided by decomposing soil organic matter and plant residues is thus not sufficient for high level yields of nonlegumes. The quantity of supplemental nitrogen to apply is determined most satisfactorily by yield trials. Such trials indicate that the average economic nitrogen rate for maize production in Illinois varies from 2.2 to 2.4 kg of nitrogen per 100 kg of maize produced, when nitrogen is applied in the spring (Graffis *et al.* 1976).

Pest Control Practices

Intensification of management requires the adoption of good pest control practices. This includes weed, insect, and disease control. Weed

Courtesy of University of Illinois

FIG. 11.2. THE MORROW PLOTS AT THE UNIVERSITY OF ILLINOIS WHERE CONTINUOUS MAIZE AND CROP ROTATIONS HAVE BEEN COMPARED FOR OVER 100 YR

control is largely by application of chemical herbicides although rotation of crops, timely planting and harvesting, proper row spacing, and other management practices may be most helpful. Knowledge of the weed species which are to be controlled must be coupled with knowledge of the tolerance of crop and weeds to specific chemicals in order to plan a good weed control program. Some herbicides are not effective in dry soils, others may be lost by excessive leaching if heavy rains follow application. Both concentration of herbicide and total amount applied may have to be controlled within specified limits. Applying too little herbicide can result in poor weed control. Applying too much can damage the crop or leave such a high residual amount that succeeding crops will be damaged.

The use of resistant varieties is always a good precaution against diseases and insects. However, varieties may not be available which have adequate resistance as well as high yield capability and other desirable qualities. Rotation of crops may be important for pest control. Changes of planting date may give significant advantage to the crop over the invading pest.

Control of pests by chemical and/or biological means may be essential for high yields. Many insects can be controlled by predators which can be introduced. The more common means of insect or disease control is by application of chemicals which control the undesirable organisms. Proper dosage to apply and method and timing of application are highly important for adequate control with minimum hazard.

Harvesting Practices

Soybeans and maize are both harvested primarily by self-propelled combines in this zone. The maize combine commonly shells the kernels from the cobs in the field and deposits the cobs on the ground. Some farmers have reported that net marketable yields increased 10% with the adoption of the picker-sheller combine. The combine is fitted with a different kind of head for harvesting soybeans and small grains. In spite of improvements which have been made, further improvements are needed. In years when high amounts of rainfall come after soybean harvesting, germination of beans which have been lost in the harvesting may produce enough seedlings to completely shade the soil.

Improvements in combine heads for soybeans appear to be on the horizon. The most promising involves combine heads which channel individual rows into the combine similar to the action in combining maize.

The ability to harvest the crop in a short period of time is most important in this zone where rainfall may delay harvesting operations. Wide, low-pressure tires increase the combine's ability to move through the field when the soil is wet. However, the grain may be at too high a moisture content to allow it to be handled in the combine satisfactorily with minimum damage. Drying of maize grain by blowing heated air through it following harvest has become a rather common practice in the zone. It allows harvesting to proceed at an earlier date, but it requires expensive equipment and energy costs may be high.

REFERENCES

ALDRICH, S.R., SCOTT, W.O., and LENG, E.R. 1976. Modern Corn Production, 2nd Edition. A and L Publications, Champaign, Illinois.

BARBER, S.A. 1977. Plant nutrient needs. *In* Our Land and Its Care. Fertilizer Institute, Washington, D.C.

BONE, S.W. 1978. Reduced tillage systems for soybean production. Soybean News *29*(2).

DUCKHAM, A.N., and MASEFIELD, G.B. 1970. Farming Systems of the World. Praeger Publishers, New York.

ELLIOTT, R.L., LEMBKE, W.D., and HUNT, D.R. 1975. Tile drainage as a means of increasing tillage days. Ill. Res. *17*, 3-4.

GRAFFIS, D.W. *et al.* 1976. Agronomy Handbook. Univ. of Ill. Ext. Circ. *1129.* Urbana.

SCOTT, W.O., and ALDRICH, S.R. 1970. Modern Soybean Production. S & A Publications, Champaign, Illinois.

SIEMENS, J.C., and OSCHWALD, W.R. 1976. Corn-soybean tillage systems: erosion control, effects on crop production, costs. ASAE Pap. *76-2552.* St. Joseph, Michigan.

SPRAGUE, G.F. 1975. Agriculture in China. Science *188*, 549-555.

SYMONS, L. 1972. Russian Agriculture. John Wiley and Sons, New York.

TROLL, C., and PAFFEN, KH. 1965. Seasonal climates of the earth. *In* World Maps of Climatology, 2nd Edition. H.E. Landsberg *et al.* (Editors). Springer-Verlag, New York.

USDA. 1977. U.S. Dep. Agric., Foreign Agric. Serv. Circ. *F.G. 5-77.*

WELCH, L.F. 1977. Soybeans good for corn. Soybean News *28*, 3-4.

WELCH, L.F., MELSTED, S.W., and OLDHAM, M.G. 1976. Lessons from the Morrow Plots. Ill. Res. *18*, 3-4.

12

Crop Production Systems in Arid and Semi-Arid Cool Temperate Zones

Hayden Ferguson and James Krall

This zone is generally characterized by long cold winters, cool or cold soils, short growing seasons, high summer insolation, high daytime summer temperatures, potential evapotranspiration rates that exceed precipitation for much of the growing season, low relative humidity, cool nights, a summer precipitation pattern, and persistent winds that often become high, especially in the fall and spring. The world distribution of this zone is shown in Fig. 12.1.

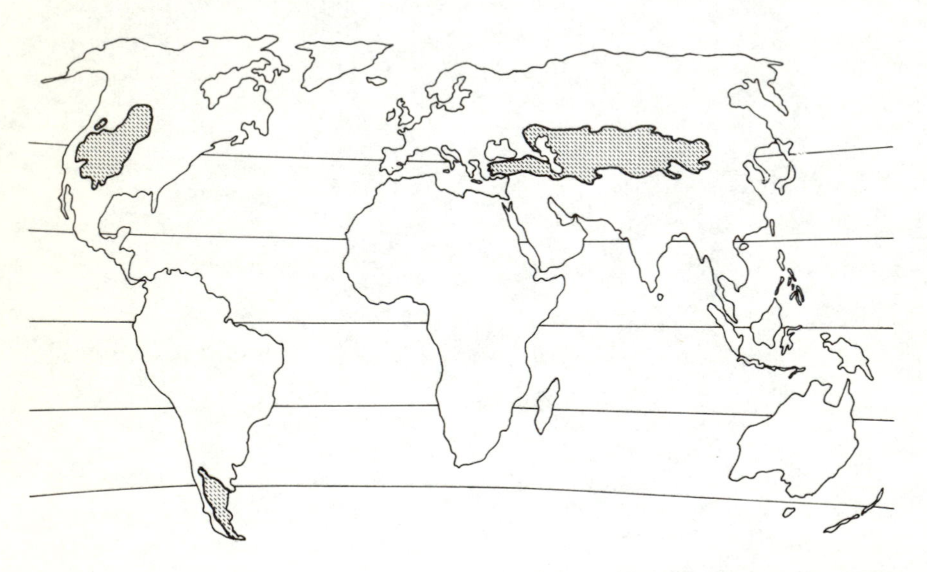

Adapted from Troll and Paffen (1965)

FIG. 12.1. MAP OF ARID AND SEMI-ARID COOL TEMPERATE ZONES

Soils of the zone are relatively unweathered, are generally dominated by 2:1 layer clays, have pH values of 7 or greater because of the dominance of calcium, magnesium, and sodium on the exchange complex, often contain significant soluble salts, have thin A horizons, and are low in organic matter.

Although the soils are generally productive, the climate creates an environment too harsh for the production of many staple crops. Without irrigation, severe water stress occurs on almost a yearly basis and drought is a common occurrence. Damaging late spring and early fall frosts are common. Crops are often damaged by wind or by soil particles carried by wind, especially in the seedling stage. Unless soils are protected, wind erosion is almost a yearly occurrence that becomes severe during droughts. Water erosion from violent storms or melting snow is common.

With these limitations it is not surprising that efforts to turn these areas into "bread baskets" were, until recent times, largely a failure. The technology of humid Europe and North America could not be successfully transplanted into these areas, as early settlers learned in uncounted disheartening experiences.

MANAGEMENT UNDER DRYLAND CONDITIONS

The foundation for stable, successful dry land crop production in these areas was established by research conducted between about 1910 to 1935. Droughts prior to and, especially, during the 1930s forced adaptation of this research and encouraged innovations by farmers in order to survive. Tillage and seeding machines currently in use are largely based on farmer innovations. A great surge in terms of crop production stability in these areas occurred after about 1945 with the availability of large tractors and chemicals such as weedicides and fertilizers. Continuing inputs of research and industrial technology and farmer innovation have, at present, created crop production systems that have made, once common, complete crop failures very rare, in spite of the fact that at least some areas have suffered from severe droughts in the last two decades. The newness of this technology, at least on a worldwide basis, can perhaps be emphasized by the fact that Goseen and Bieroff were 1972 recipients of the Lenin prize, one of the highest honors available to Soviet citizens, for the development of a system of stable crop production in the central Siberian Plains. Their work was begun in about 1961 and the system derived is nearly identical to that used in many areas of the North American plains.

Crops and Crop Growth

Climate limits the crop species. The two major factors are the short growing season and the almost continuous water shortage. The dominant crops are small grains—spring wheat, barley, oats, and in some areas winter wheat; oil seed crops—safflower, rape seed, flax, sunflowers, and mustard; forages—alfalfa, sainfoin, and cool season grasses. Varietal selection is highly important to circumvent disease and pest problems and to take advantage of changing technology.

With adapted varieties, available water is most often the limiting factor in production. About 10 cm of available water are required to produce any seed of the crops listed and about 25 cm to produce satisfactory yields. Because of the stage of crop development-water requirement interactions, either a bumper crop or a near failure might occur with 25 cm of available water. High plant water stress at approximately the bloom stage, caused by either a lack of water or high atmospheric demand—hot dry winds—will almost always cut yields drastically regardless of the water available at other periods. On the other hand, precipitation or low atmospheric water demand at the right time can result in high yields with considerably less than 25 cm of water available.

The relationship between yield and available water, both that stored in the soil and growing season precipitation, has been studied extensively for some major crops in the zone. This information is used to predict yields, to help in determining crop rotations, and in determining fertilizer requirements.

Table 12.1 presents equations describing the average yield of spring wheat as a function of available water for four large political subdivisions. These equations indicate that the production efficiency of

TABLE 12.1. YIELD OF SPRING WHEAT AS A FUNCTION OF SEASONAL WATER SUPPLY FOR FOUR LARGE AREAS

Location	Regression Equation
South Dakota	y[1] = 1.31SW[2] + 2.47R[3] – 12.99
North Dakota	y = 2.33SW + 2.33R – 4.15
Montana	y = 2.21SW + 2.28R – 19.13
Saskatchewan	y = 2.59SW + 2.41R – 8.45

Source: Adapted from Bauer (1972).
[1]Bushels/acre; 1 bu/a = 67.2 kg/ha.
[2]Inches of available soil water in the root zone; 1 in. = 2.54 cm.
[3]Inches of precipitation during growing season.

stored soil water increases in the zone from south to north. Also, in North America, north of about the 45th parallel (north of South Dakota) the efficiency of stored soil water and precipitation is about the same; south of that line precipitation is more efficient than stored soil water. Stored soil water is highly dependent upon texture and with coarse-textured soils stored water becomes less important in determining yields. One of the direct effects of proper fertilization is that the increased root growth, both extension and root density, makes more soil water available to crops.

Table 12.2 gives estimated yields of spring wheat based on available moisture stored in the soil and rainfall during the growing season under good soil fertility. These data are based on recent experimental data obtained in Montana and North Dakota. Similar predictive data are available for winter wheat and barley.

TABLE 12.2. ESTIMATED SPRING WHEAT YIELDS BASED ON STORED SOIL WATER AND SEASONAL RAINFALL

Growing Season Precipitation cm	Available Soil Water at Seeding in 120 cm of Soil							
	2.5 cm	5.0 cm	7.5 cm	10.0 cm	12.5 cm	15.0 cm	17.5 cm	20.0 cm
	Yield in Tons/Hectare							
0	With less than 5 cm of seasonal ppt yields will be extremely erratic to nonexistent							
5.0								
7.5	0.0	0.0	0.34	0.67	1.00	1.34	1.68	2.02
10.0	0.0	0.34	0.67	1.00	1.34	1.68	2.02	2.35
12.5	0.34	0.67	1.00	1.34	1.68	2.02	2.35	2.69
15.0	0.67	1.00	1.34	1.68	2.02	2.35	2.69	3.02
17.5	1.00	1.34	1.68	2.02	2.35	2.69	3.02	3.36
20.0	1.34	1.68	2.02	2.35	2.69	3.02	3.36	3.70
22.5	1.68	2.02	2.35	2.69	3.02	3.36	3.70	4.03
25.0	2.02	2.35	2.69	3.02	3.36	3.70	4.03	4.37
27.5	2.35	2.69	3.02	3.36	3.70	4.03	4.37	
30.0	2.69	3.02	3.36	3.70	4.03	4.37	Dryland	
32.5	3.02	3.36	3.70	4.03	4.37	Yields seldom exceed 4.3 t/ha		

Water Conservation

In this zone the yearly potential evapotranspiration, PET, exceeds the precipitation (ppt); more importantly, the average daily PET at the peak of the evaporation season generally far exceeds the average daily precipitation at this same time. This is of major importance in terms of crop production because some of the most critical stages of crop growth, in terms of seed production, from boot stage through kernel filling, generally coincide with the period of peak daily PET. In view of this difference between PET and ppt, soil storage of precipitation occurring before the growing season is essential for production.

Late summer-early fall ppt, usually occurring as rain, is generally a small percentage of the total annual ppt but the dry soil tends to absorb this ppt readily. PET is still high during this period so a considerable quantity of this ppt, perhaps up to 70%, is lost either through direct evaporation or transpiration by post harvest weeds. The quantity lost by evaporation is highly dependent upon ppt patterns; nearly all ppt occurring in light (<0.5 cm) intermittent storms is generally lost. Conversely, precipitation from substantial storms may be largely conserved.

From 20 to 30% of the annual precipitation of the zone is as snow with a precipitation range from 250 to 500 mm; this amounts to from 5 to 15 cm of precipitation that can, if conserved, be used for crop production. Snow, in this zone, is often accompanied and always influenced by high winds. Thus, unless protected in some way, the snow is swept from the fields, piled up in some downwind coulee, and makes almost no contribution to crop production. Frequently with unprotected soils, over 90% of the winter precipitation is not stored in the soil.

Stubble is an effective means of conserving snow. Soil storage efficiencies of winter precipitation of from 37% (at 50°N) to 99% (at 40°N) due to standing grain stubble have been reported. Stubble height, water content of the surface soil prior to snowfall, and soil freezing affect soil moisture storage. Stubble height controls the depth of snow retained against the wind. In one test in Montana the depth of wet soil was measured in the spring under 3 treatments of fall tillage—no standing stubble, stubble 20 cm high, stubble 40 cm high. The depth of wet snow was over 20 cm, 30 cm, and >75 cm, respectively. The winter winds removed most of the snow from stubble only 20 cm high; whereas the 40 cm stubble remained nearly full of snow. A deep layer of snow is an effective insulator that significantly reduces the intensity of soil freezing. Wet frozen soil is relatively impermeable. As the water content of the soil increases before freezing in the fall, conservation of winter precipitation decreases. Also, deeper snow packs tend to start melting somewhat earlier in the spring and a larger absolute quantity of water may run off, but more is generally retained in the soil.

Intermittently spaced barriers have proven to be effective means of conserving water in snow. Relatively short, stiff stemmed plants such as tall wheatgrass, sunflowers, and mustard have been used as barriers. Perennial tall wheatgrass barriers have been studied extensively by Black and Siddoway (1976) in the northern Montana plains. The grass, which attains a height of about 1.2 m, is planted in single or double rows approximately perpendicular to the prevailing winds at interval spacings of 10 to 15 m. This combination of height-barrier spacing has resulted in soil storage efficiencies of winter precipitation of about 60%, compared to about 30% outside the barriers. Erosion from the spring melting of snow in the barrier system has not been a problem, probably because of the relatively even snow distribution.

Temporary barriers, 2 to 3 rows of mustard forming barriers 10 to 12 m apart, are used rather extensively in Russia for winter snow management. Mustard, seeded in July on fallow land, grows rapidly and has a very stiff straw that stands well over winter.

Mechanically formed snow ridges are another temporary barrier that are apparently used extensively in Russia and have been briefly evaluated in North America. The Soviets have a specially designed snow-ridging machine which they use in the central Siberian plains, where the snow reaches depths of 30 to 40 cm between the ridges. Yield increases due to snow ridging have been reported from both Russia and Canada in areas with substantial snowfall.

Tree and shrub shelter belts are less successful in efficient snow retention and distribution because they give uneven snow accumulation and take up an excessive amount of land. Work in Canada has shown significant increases in grain yield in the areas of the snow accumulation but overall increases in yield were slight.

Level bench terraces have been shown to significantly increase the soil water at seeding time in North America, especially when they are associated with snow barriers. The level bench retains the melted snow water and allows greater infiltration. The author has observed deep (approximately 1 m) contour ditches filled with straw in Russia that were reputed to be effective means of collecting winter and spring runoff. Supposedly, the straw prevents the bottom of the ditch from freezing and, thus, water entering the ditch can easily enter the soil.

Conservation of summer precipitation in fallowed land is largely dependent on the condition of the soil surface, the intensity of storms, and weed control. Soils of this zone are generally low in organic matter and, thus, tend to puddle easily. Infiltration is generally enhanced by tillage practices that tend to maintain a high concentration of stubble on the soil surface. The stubble protects the soil surface from the mechanical forces of falling and running water and thus tends to maintain macropores through which much of the infiltration occurs. In general, the deeper water is moved into the profile, the longer it is retained against evaporation.

Normal fallow periods, from harvest of one crop to seeding of the next crop, are about 14 months if winter grain is seeded and 21 months if a spring crop is planted. Even under the best management practices, most of the water retained by the soil is accumulated during the first fall and winter period. Long-term experiments show that of the total water retained by the soil, approximately 75% is stored during the first fall and winter (9 months), 16% is stored during the following spring and summer (5 months), and 9% during the second fall and winter (7 months). The storage efficiency of the precipitation that occurs during these same periods is about 72%, 8%, and 9%, respectively. The relatively high soil water content that results from the first period storage reduces infiltration during the second and third periods.

Intermittent barriers can be beneficial in conservation of soil water during the summer. Soils protected by barriers retain water, especially near the surface, for a longer period allowing the slow process of unsaturated flow to move the water deeper in the soil where it is retained longer against evaporation.

Summer fallow has been a significant factor in stabilizing yields in this zone. The approximately 2.5 cm of additional soil water accumulated during the summer and second winter is an important factor. However, the accumulation of nitrates in the soil during the summer is probably the major benefit of summer fallow. Summer fallow also gives better weed control during the cropping season and distributes the work load on the large farms of the zone. Because of these factors the alternate year crop-fallow system became almost universally accepted in North America. In Russia, a policy of 2 to 3 years crop, 1 year fallow is generally followed.

It is now apparent, however, that in some areas summer fallow can have a major detrimental effect. Conservation and downward movement of moisture causes "saline seeps" over a vast area of the states of North Dakota, South Dakota, and Montana, and the Canadian provinces of Alberta and Saskatchewan. The soils of this area are underlain at depths from a few centimeters to perhaps 100 m by vast marine shale formations that vary from about 500 to 800 m in thickness with the landscape paralleling the shale surface. The lower part of the soil profiles over the shale contains large quantities of soluble salts, largely sodium and magnesium sulfates. It is not uncommon to encounter electrical conductivities of from 8 to 12 mmhos per centimeter at .8 to 1.2 m in depth.

Prior to farming, this area supported a broad spectrum of plants with variable rooting depths and growth habits. Thus, water entering the profile was largely utilized by the plants before it could escape the root zone. Farming and especially summer fallow resulted in relatively long time periods with no water removal from the soil by plants. Water stored in the soil often moves below the root zone resulting in the buildup of a shallow water table with a very high concentration of soluble salts on top of the marine shale. The water tends to move laterally over the shale and when it approaches within about 1 to 1.2 m of the surface it tends to

FIG. 12.2. AERIAL VIEW OF ALTERNATE CROP FALLOW PRACTICED IN THE ZONE

The strips are perpendicular to the prevailing winds and are from 120 to 150 m in width. This photo shows the saline seep across a portion of the strips and these seeps make strip cropping difficult.

move up to the surface and evaporate. This results in a thick deposit of salts on the soil surface in many downslope areas. Such a saline seep (a deposit of salt on the surface) tends to grow larger as long as water is supplied to the water table.

In areas conducive to the development of saline seeps, a modification of the standard crop-fallow system will have to be developed. This will probably require more intensive cropping and more diverse rotations. In order for annual cropping to be about as successful as the crop-fallow system, greater efficiency of storage of winter precipitation, modification of weed control systems, especially grass-type weeds and changes in machinery, especially drills, will be required.

Erosion

Both wind and water erosion are common in the zone; wind erosion is the more spectacular and probably does the most short-run damage but a substantial amount of the total erosion is from water. Soil and management factors that influence erosion are slope, surface texture, surface roughness, and soil protection.

The most serious wind erosion occurs during the strong winds of late fall and early spring on tilled lands when the soil surface is often dry and before planted crops are large enough to protect the soil. Large clods on the soil surface are significant deterrents to wind erosion because they modify the wind velocity near the surface and protect small erodable particles. Unfortunately, the processes of weathering and tillage for weed control often break clods down to erodable sized aggregates.

The two most successful methods of wind erosion control in the zone are stubble mulch tillage and strip cropping with alternate fallow-cropped strips. Stubble mulch tillage is generally an effective means of wind erosion control on silts or finer textured soils because these soils are not so erodable and because they generally produce enough plant growth so that adequate plant residues can be maintained on the surface during the critical periods. Approximately 1500 kg per hectare of plant residues are required after tillage to control wind erosion on these soils. Erosion is more difficult to control on sandy loam or loamy sand soils. These soils tend to be highly erodable because of their dominant particle size and because stable clods are almost impossible to maintain. Moreover, because of their lower water-holding capacity, dry matter production is generally less. Residue covers of approximately 2000 kg per hectare are required to protect these soils.

In areas of especially high winds or highly erodable soils, stubble mulch tillage is generally combined with alternate crop-fallow strips to control wind erosion. With this system, a standing barrier of stubble is alternated with erodable strips during periods of high erosion potential. The stubble controls erosion by reducing wind velocity in the covered strip and downwind from the strip for a distance of about ten times the stubble height. Soil that moves on the erodable strip is trapped in the stubble. The desirable strip width is largely a function of surface soil texture; with less erodable silty clays and fines, strips may vary from 50 to 150 m in width. Strips on more erodable soils should be 30 m or less and very erodable soils, loamy sands, may require strips as narrow as 8 to 9 m in width.

Intermittent barriers can also effectively control wind erosion. Permanent barriers can be formed with rows of tall grass, shrubs, or trees; temporary barriers can be formed with almost any rapid-growing, relatively tall plant such as mustard, sunflowers, or sorghum. In general, barriers will reduce lea side wind velocities by 40 to 50% at a distance of 8 to 12 times the barrier height (H). For barriers of 90 to 120 cm height, this spacing generally provides adequate field protection even when the wind direction deviates significantly from perpendicular to the barriers. With trees the spacing must be considerably less than 8 to 12 H in order to provide field protection from winds not perpendicular to the barriers. It appears doubtful that tree barriers are a generally promising field scale means of erosion control because: (1) if properly spaced to provide acceptable erosion control they occupy a large amount of valuable field space, (2) they utilize soil water and, thus, reduce yields for a significant distance from the rows, (3) they often cause excessively large snow banks and uneven distribution of snow, and (4) they provide a harbor for weeds.

However, some farmers are successfully using tree barriers. Also, tree barriers are common in Russia, but the spacing is generally too great to be an effective means of wind erosion control.

Permanent grass barriers appear to have real promise for controlling wind and improving snow distribution. Grass barrier establishment on about 7% of a total field area can provide permanent wind erosion control in conjunction with intensive between-barrier soil cultivation. Moreover, established permanent grass barriers retain their effectiveness during periods of drought when crop residue production is too scant to provide protection by normal summer fallow methods.

During drought periods it is sometimes necessary to resort to short-term emergency tillage for controlling wind erosion. The formation of rough, cloddy surfaces by chiseling, disking, or plowing can provide protection for non-sandy soils. Sandy soils should be protected by a permanent plant cover or properly spaced permanent barriers.

Water from storms or melting snow running across unprotected land can result in significant sheet erosion. Control measures for water erosion are similar to those for wind erosion. Merely farming across slope perpendicular to the direction of water flow will reduce runoff by 20 to 25% on slopes up to about 12%. Rough, cloddy surfaces, especially if protected by plant residues, will significantly reduce runoff and erosion. Excessive tillage accelerates water erosion because it smooths the surface, generally decreases surface storage and infiltration, and destroys protective residues. In areas with steep slopes and high probability of intense storms or rapid melt of deep snow pack, contour sod strips should be used to reduce the erosion potential.

Fertilization

Soils of this zone are generally unleached and have pH values in the range of 7 to as high as 10; values above about 8.5 to 8.7 generally indicate a dominance of sodium with a resultant poor structure and productivity. The productive soils generally have a pH range of 7 to 8.7. Most essential elements are relatively soluble in this pH range and, thus if present in the soil, are available to plants. Exceptions, such as manganese, are usually required in such small amounts that deficiencies are not common. However, it is a vast zone and many unique fertilizer problems that defy generalities exist; for instance, in the Palouse area of eastern Washington the addition of very small quantities of molybdenum is almost essential for legume production. Such problems must be solved "on site," usually through field experiments.

Phosphorus is often deficient in the zone, partly because the soils are often calcareous and the calcium-phosphates have low solubilities. Soil tests for available phosphorus are generally very reliable and should be used. Most phosphorus fertilizers can be applied with the seed without damage; exceptions are newer materials such as urea ammonium polyphosphate (28-28-0) and urea ammonium phosphate (24-42-0). Rates of these materials that provide more than about 20 kg per hectare of nitrogen with the seed can damage stands. In general, phosphorus is utilized by grain plants more efficiently if placed near or with the seed. Broadcasting generally requires 2 to 3 times more phosphorus than banding for the same response.

The small grains and oil seed crops grown in the zone require approximately .05 kg per hectare of NO_3-N to produce 1 kg per hectare of seed yield. Nitrogen fertilization should be based upon the quantity of soil nitrate in the root zone at the beginning of the season, expected nitrate mineralization during the growing season, available soil water at the start of the growing season, and expected growing season precipitation. Soil tests are used to determine the nitrate in the root zone. The available soil nitrogen at the beginning of the growing season is highly dependent upon management practices; after fallow, from 40 to 90 kg per hectare of NO_3-N might be found in the root zone of a soil ($\sim$ 1.2 m) and 15 to 30 kg per hectare might be expected in a soil that had been continuously cropped. Growing season mineralization must be estimated. In Montana an average of from 30 to 40 kg per hectare of NO_3-N becomes available during the growing season in fields that have had most of the straw residue returned to the land.

With "optimum" available water, i.e., 30 cm, a yield of winter wheat of 2600 kg per hectare might be expected. This yield would require about 130 kg per hectare of nitrogen and the quantity of fertilizer nitrogen needed can be obtained by subtracting available soil nitrogen plus estimated growing season mineralization from 130. With less than "optimum" available water both the expected yield and the amount of fertilizer nitrogen required would be less. Thus, some estimate of expected yield based on available water is essential. The available stored soil water at the beginning of the growing season can be measured or can be estimated from a measurement of depth of wet soil using data such as that in Table 8.3. Growing season precipitation can only be estimated, based on the long-term average for a particular area. Given these two values, an estimated yield can be obtained from data such as that in

Table 12.2. All of these emphasize the importance of stored soil water in determining yield and required nitrogen fertilization.

It is generally advantageous to apply some nitrogen fertilizer with the seed usually in some combination with phosphorus, even if the budget approach described above indicates little or no nitrogen fertilizer requirement. However, rates greater than about 20 kg per hectare should be broadcast (dry material) or injected (liquid or gas) in the late fall or early spring.

Significant quantities of nitrogen can be added to soils by including legumes in the rotation. In addition to nitrogen additions perennial legumes are attractive in parts of the zone because their long season growth habit and deep root systems are beneficial in saline seep control. However, they dry the soil deeply and it is generally advisable to fallow for one season after removing the legume.

Soils of this zone generally contain a high level of potassium. However, crops that make significant growth during cold, early spring periods often respond to potassium fertilization even on soils that contain as much as 1000 kg per hectare of extractable potassium in the surface 15 cm. Recent work in the Northern Plains has shown significant economical yield response to potassium fertilization over 50% of the time with winter wheat, over 40% of the time with barley, and over 25% of the time with spring wheat on soils that test high in potassium. This phenomenon cannot be a matter of the presence of potassium but must be associated with the influence of temperature on the release of potassium, the diffusion to roots, or the uptake by plants.

Soils that test low or medium in potassium should generally be fertilized by broadcasting 25 to 60 kg per hectare of K_2O in the fall or spring. On soils that test high in potassium, it is advisable to fertilize a limited area and carefully check the yield. This should be made for 2 or 3 years before a final decision is made because weather is so important in potassium response.

Weeds

Weeds are a major factor determining both the production costs and the productivity of dryland agriculture of the zone. Weed control is an absolute necessity for efficient production but it is difficult and expensive.

Among the more important weeds are pigweed, Russian thistle, Canada thistle, field bindweed, perennial sowthistle, wild buckwheat, wild oats, pigeon grass, downy brome, and volunteer small grains. The difference in growth habits and physiological responses within even this small group defies almost any combination of control methods.

Summer fallow is one of the most effective means of weed control. The fallow period breaks the reproduction cycle of many weeds and reduces the population. However, viable seeds of almost every species will remain in the soil without germinating during the fallow period and will then germinate during the crop growth season. Warm season plants such as wild oats and pigeon grass can effectively evade tillage by not germinating during the fallow year nor until relatively late during the crop year, thus escaping fallow and early spring tillage.

Many weeds, especially broad leaf types, can be controlled in the crop by chemicals. Others, especially perennials, are only set back or have only the top killed. Timing of chemical application is important; for instance, wild buckwheat can be killed easily with 2,4-D if sprayed at the right growth stage but spraying at that stage will often damage small grains. Canada thistle is best killed by spraying at the bud or blossom stage but by then the crop has been badly damaged if the thistle population is high.

Grass-type weeds such as wild oats, downy brome, and pigeon grass are especially difficult to control in small grains. Selective herbicides are available for wild oats but the results are often mixed.

Weeds can be best controlled when crops are diversified. For instance a sequence of winter wheat-spring grain will allow better weed control than a monoculture of either, and'winter wheat-fallow-spring crop is still better. The fallow period will allow for control of many of the perennials and warm season plants and the late fall and/or early spring tillage before seeding the spring crop will help control the winter annuals. The inclusion of a broad leaf crop in the cropping system will further improve weed control because chemicals effective against grasses can be used with the broad leaf crops and those effective against broad leafs can be used with the small grains.

Weed infestations, including volunteer crop plants, that occur after harvest can be a serious problem especially if saline seep or other factors dictate annual cropping. Tillage for control of these weeds may reduce the stubble so that it is ineffective in retaining snow and controlling erosion.

Saline-Sodic Soils

Under nonirrigated conditions, reclamation of saline and sodic soils is difficult and slow. Because summer fallow generally increases the movement of water through the profile, saline soils are often improved by this practice. The process can be accelerated by maintaining a thick straw mulch on the soil; this decreases evaporation and runoff and results in greater water movement through the profile.

Deep plowing can be effective in reclaiming sodic soils if the sodium is concentrated at or near the soil surface or if a more permeable strata can be brought to the surface. Addition of gypsum or some more soluble calcium salt can be beneficial. Such salts will, under some conditions, increase the infiltration characteristics of the soil and, thus, the rate at which sodium is removed.

Machinery

Available farm power has changed tremendously in 50 years. It has evolved from the use of animals, oxen and horses, to large 4-wheel drive diesel tractors. Power is available to speed up the operation on a larger scale. Seed beds for crops can be prepared rapidly. Large acreages can be planted on time so the crops can reach their most productive stage when rainfall is most abundant. Crops can be harvested as soon as they mature, thus greater production is obtained. Furthermore, due to the versatility of this power, lands that were previously considered unfit for crop culture have been brought under cultivation. Almost every operation from filling grain boxes on drills to moving the final product to markets is done by powered machinery.

Seed bed preparation equipment has had a dramatic evolution. Machinery designed for areas with greater rainfall, deeper soil, and generally wetter climates has been found inadequate for arid zones. The moldboard plow was soon discarded as it did not scour when the soil was moist, it turned under stubble and encouraged erosion, and the plow shear caught on glacial rocks. The moldboard was replaced by the rolling disk plow that rolled over rocks, turned under less stubble, and generally had lighter draft. Later models of these machines were hooked in gangs rather than singularly. They were commonly called the one-way plows, wheatland plows, and were mainly an innovation of the arid regions.

Both the moldboard and disk plow moved the soil laterally, creating depressions in the centers of the fields which resulted in gully erosion. The tandem disk that moves soil both directions has been adopted recently to avoid dead furrows. With the advent of stronger steel beams the tool bar cultivator or chisel plow came in use. This type of implement is the common tillage tool used throughout the zone. Compared to the plow it stirs the soil rather than turns it over and mixes the stubble with only the surface soil. For better weed control, a rotating rod attached to the rear gang of shanks is often used. Recent models have high clearance to avoid plugging. Such equipment is shown in Fig. 12.3.

Another innovation which proved practical for dryland small grain production was the deep furrow press drill. The seed metering mechanism remains the same but the openers are spaced in several ranks similar to the cultivators so the drill does not plug with residues. Instead

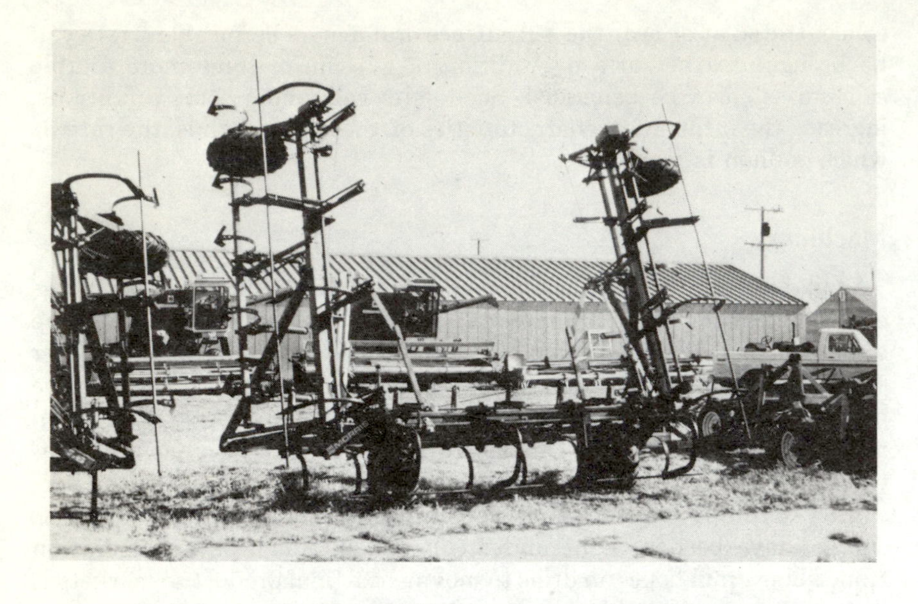

FIG. 12.3. TOOL-BAR CULTIVATOR WITH ROD-WEEDER ATTACHMENT COMMONLY USED FOR SUMMER FALLOWING

The wings are hydraulically lifted to facilitate moving from field to field. Width varies from 10 to 12 m.

of using single or double rotating disks, shovels, or listers, 10 to 20 cm wide and spaced 30 to 40 cm apart are used. The shovel opener accomplishes two objectives. First, it moves the surface dry soil aside so that the seed falls into moist soil. Second, the 4 to 8 cm deep furrow creates a more favorable microclimate that reduces frost damage or winter killing with winter wheat and other fall sown grain. It is estimated that this method of seeding has moved the winter wheat belt about 300 km north in some regions of this zone. Most furrow drills have a packer wheel, generally a pneumatic tire, behind each opener to firm the soil as it falls back over the seed. The standard width of these drills is 3 to 4 m; they are pulled in gangs of 4 to 5 drills.

The two most common types of equipment used for the application of herbicides are the ground sprayer and the airplane. Except for some specialized crops (oil seeds) most herbicides are applied post emergence. The boom width varies from 9 to 11 m and applicators are capable of spraying 20 ha per hour. The airplane is especially useful during wet soil conditions and for large areas and uneven land.

The greatest percentage of the cereal grains is harvested standing at maturity with a self-propelled combine cutting a swath 5 to 8 m wide.

One machine is capable of harvesting 20 to 30 ha per day. In areas with short growing seasons the grain is often swathed and left to dry until mature. It is then picked up and threshed with a pickup attachment on a regular self-propelled combine. Swathing has several advantages over harvesting standing grain. It can be accomplished 10 to 12 days earlier, thus reducing chances of storm damage. However, should wet weather occur after swathing the chances of sprout damage are increased compared to standing grain.

Cropping Systems

A fixed sequence of cropping, either crop-fallow or a 3-year rotation including 1 year of fallow, is the general pattern throughout the zone. The cereal grains do well in the harsh environment of the zone. Fallow is an insurance against drought in that the soil profile can generally be "filled" with water in a 14 to 21 month period, even during drought. Moreover, fallow aids in weed and disease control, results in soil nitrate accumulation, and makes for a good seedbed.

Conversely, the near monoculture of spring wheat-fallow-spring wheat has facilitated disease, weed, and insect problems. Excessive fallow has led to saline seep problems and has often resulted in inefficient use of precipitation. Fear of dry years is perhaps too dominant in determining cropping systems. During a recent 40-year period, drought occurred 16% of the time in the Northern Plains but wet years occurred 19% of the time. The rigid cropping system may have improved production during the dry years but it certainly has not made the best use of available water during wet years.

Wider adoption of flexible cropping systems is in order with root zone water storage as the decision controlling factor. If, using data similar to those of Table 12.1, the estimated yield based on measured stored soil water and average (expected) growing season precipitation is 65 to 75% of average fallow yields, a crop can be planted with a good chance of obtaining a satisfactory yield. If the estimated yield is less than 60% of average fallow yields, the land should be fallowed. Additional constraints of this system should be that land should not be fallowed 2 years in a row and that the same crop should not be planted 2 years in a row. A comparison of a flexible cropping system with three rigid systems is given in Table 12.3. In this example, 7.5 cm of stored soil water is taken as the crop or fallow break point. The alternate fallow or 3-year rotation would result in fallowing during years when, because of adequate stored soil water, the probability of growing a successful crop would be high. Applying the rule indicates that, with continuous cropping, failure would be probable in 2 of the 6 years. With the flexible systems, fallow would

TABLE 12.3. A COMPARISON OF FLEXIBLE AND RIGID CROPPING SYSTEMS

	Crop Year					
	1	2	3	4	5	6
	Soil Water at Planting (cm)					
Cropping System	9.0	11.5	4.5	5.5	12.0	13.0
Alternate Fallow	C[1]	F[2]	C	F	C	F
3-year Rotation	C	C	F	C	C	F
Continuous Crop	C	C	C	C	C	C
Flexible	C	C	F	C	C	C

[1]C = crop.
[2]F = fallow.

occur only in year 3. The rule indicates fallowing in year 4 also, but, because of erosion hazard, 2 successive years of fallow are excluded.

Probability of success of the proposed flexible system could be improved by doing a better job of snow conservation and by including broad leaf crops in the system for improved weed and disease control. Limited experience in the Northern Plains indicates that the flexible system is a practical improvement over rigid systems.

MANAGEMENT UNDER IRRIGATED CONDITIONS

On the average, the yield of all of the crops mentioned previously can be increased with proper irrigation. Also, crops that are less drought tolerant, such as sugar beets, potatoes, silage corn, beans, and some specialty crops, can be grown.

The greater diversity of crop rotations made possible by irrigation has both positive and negative attributes. In general, disease, insect, and weed control are facilitated because the diversity of crops breaks their life cycle. A wide range of weedicides can be used in rotations and the moldboard plow is an effective soil preparation-weed control tool on irrigated lands. Some of the crops, i.e., potatoes and sugar beets, are especially subject to insect damage but effective insecticides are available. The capital cost of crop diversity can be tremendous. A typical irrigated farm might grow potatoes, sugar beets, small grains, silage corn, and alfalfa; each of these has its own specialized equipment requirements. In recent years this high equipment cost has forced many farmers to revert to almost monoculture farming and, thus, not take advantage of the benefits of diversity.

The general principles and practices of irrigation farming in temperate zones are discussed in Chapter 15.

REFERENCES

ANON. University of Idaho agronomic guide. Idaho Coop. Ext. Serv. Moscow.

ANON. Management guide series. Montana Coop. Ext. Serv. Bozeman.

ANON. Growers guide for crops. Washington Coop. Ext. Serv. Washington State Univ., Pullman.

AASE, J.K., SIDDOWAY, F.H., and BLACK, A.L. 1976. Perennial grass barriers for wind erosion control, snow management, and crop production. Symposium: Shelterbelts on the Great Plains. Denver, Colorado, Apr. 20-22. Great Plains Agricultural Council Publ. *78*, 69-78. Lincoln, Nebraska.

BAUER, A. 1972. Effect of water supply and seasonal distribution on spring wheat fields. N. D. Agric. Exp. Stn. Bull. *490.*

BERMAN, J.W., HARTMAN, G.P., and BLACK, A.L. 1976. Safflower growing is better than summer fallowing. Mont. Farmer-Stockman *63*(13) 24-27.

BLACK, A.L., and SIDDOWAY, F.H. 1976. Dryland cropping sequences within a tall wheatgrass barrier system. J. Soil Water Conserv. *31*, 101-105.

BURROWS, W.C., REYNOLDS, R.E., STICKLER, F.C., and VAN RIPER, G.E. 1970. Proceedings, International Conference on Mechanized Dryland Farming. Deere and Co., Moline, Illinois, Aug. 11-13, 1969.

FERGUSON, H., BROWN, P.L., and MILLER, M.R. 1972. Saline seeps on non-irrigated lands of the northern plains. Great Plains Agric. Counc. Publ. (60)169-191. Lincoln, Nebraska.

FRANK, A.B., HARRIS, D.G., and WILLIS, W.O. 1976. Influence of windbreaks on crop performance and snow management in North Dakota. Symposium: Shelterbelts on the Great Plains. Denver, Colorado, Apr. 20-22. Great Plains Agricultural Council Publ. *78*, 41-48. Lincoln, Nebraska.

GREAT PLAINS AGRIC. COUNC. 1962. Stubble-mulch farming in the Great Plains. Proceedings of Workshop. Great Plains Agricultural Council, Lincoln, Nebraska, Feb. 8-9. Great Plains Agric. Council.

JOHNSON, W.C., and DAVIS, R.G. 1972. Research on stubble-mulch farming of winter wheat. U.S. Dep. Agric., Agric. Res. Rep. *16.*

SKOGLEY, E. 1976. Potassium soil test recommendations not always accurate. Crops Soils *29*(1) 15-16.

SMIKA, D.E., and WHITFIELD, C.J. 1966. Effect of standing wheat stubble on storage of winter precipitation. J. Soil Water Conserv. *21*, 138-141.

STAPLE, W.J., LEHANE, J.J., and WENHARDT, A. 1960. Conservation of soil moisture from fall and winter precipitation. Can. J. Soil Sci. *40*, 80-88.

TROLL, C., and PAFFEN, KH. 1965. Seasonal climates of the earth. *In* World Maps of Climatology, 2nd Edition. H.E. Landsberg *et al.*(Editors). Springer-Verlag, New York.

WILLIS, W.O., and HAAS, J. 1969. Water conservation over winter in the northern plains. J. Soil Water Conserv. *24*, 184-186.

WOODRUFF, N.P., LYLES, L., SIDDOWAY, F.H., and FRYREAR, D.W. 1972. How to control wind erosion. U.S. Dep. Agric. Inf. Bull. *354*.

13

Crop Production in Humid Warm Temperate Zones

Marlowe D. Thorne

GEOGRAPHICAL AREAS

This zone is characterized by humid summers of six months or more duration and mild winters. Precipitation varies from about 850 mm to 1800 mm per year and is generally ample for agricultural purposes. Precipitation is mostly well distributed throughout the year. Occasional droughts occur but widespread droughts are infrequent. The amount and distribution of precipitation depends on nearness to the ocean, direction of prevailing winds, and the topography (Nuttonson 1965). Most of the zone is between 25° and 35° latitude. In the U.S., most of this zone has greater than 8 cm of annual potential direct runoff and greater than 18 cm average annual potential percolation (USDA 1976).

The world distribution of the zone is shown in Fig. 13.1. Agricultural areas in southeastern U.S., in Argentina, South Africa, the southeastern coastal zone of Australia, northern New Zealand, southern Japan, and in China lie in this zone.

This presentation will be concerned primarily with the zone in North America where it is bounded on the north by the 37th parallel and on the west by about the 100th meridian. It includes the region commonly known as the U.S. South, extending from southern Virginia westward to eastern Oklahoma and south through east central Texas. It also extends into northeastern Mexico just slightly south of Monterey.

SOILS OF THE ZONE

The soils of this zone are primarily Ultisols and Alfisols, with some Entisols interspersed. They are intermediate in weathering between the soils of the cooler temperate zone and those in humid subtropical or tropical areas. While base saturation in the soils of this zone may be relatively low, the soils can be very productive under intensive management. Attention must be given to fertilization, liming, and protection from water and wind erosion. Even though the native fertility may be less than that of soils developed under somewhat lower mean annual temperatures, the longer growing season and somewhat better precipitation amounts and distribution contribute to high annual crop production in this zone.

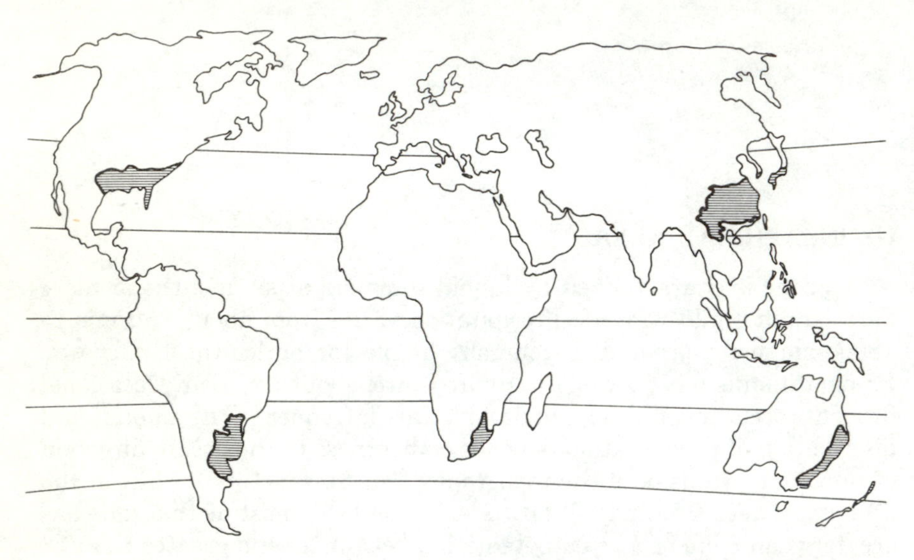

Adapted from Troll and Paffen (1965)

FIG. 13.1. MAP OF HUMID WARM TEMPERATE ZONE

CURRENT CROP PRODUCTION SYSTEMS

Most of the crops of major importance in North America are grown to some extent in this zone. Nearly all the rice, groundnuts, tobacco, and sugar cane and about two-thirds of the cotton crop grown in the 48 conterminous states are produced in this zone. Less than 25% of the zone's total land resources area is in cropland except for the Mississippi delta where as much as 75% may be in cropland.

China produces all of the major crops grown in the U.S. and nearly all are grown to some extent in the portion in the humid warm temperate zone. Rice is China's most important grain crop and it is the predominant crop in this zone. Soybeans, maize, and wheat are grown but this zone is not the major producing area for those crops. Soybean culture in south China is largely limited to small fields, gardens, ditch banks, and other "waste" lands. In northern China soybean production is on a field scale basis and it is an important crop. Multiple cropping is practiced extensively and three grain crops per year are commonly produced in this zone. A common cropping pattern is wheat or barley followed by indica rice, followed by photoperiod-sensitive japonica rice. A second pattern commonly used is barley, cotton, and rice. Vegetable crops are grown extensively and it is reported that as many as 12 crops may be harvested in a single season (Sprague 1975).

Cotton

Cotton was the leading crop in the U.S. Southeast from colonial times until fairly recently. It is still an important crop, but other crops are gaining in importance in the zone and the arid and semi-arid areas of southwestern U.S. are increasing their share of the nation's cotton production. Until the present century many areas of the South were planted to cotton continuously until declining soil fertility limited yields and sometimes resulted in cessation of cotton production on them. The ravages of the boll weevil also hastened the movement of much cotton production to areas further west where the pest infestation was less serious. In the past 10 years, the U.S. has produced about 25% of the world's cotton on about 15% of the cotton acreage (Chapman and Carter 1976).

Tobacco

Tobacco is a leading cash crop in this area of the U.S. because of the high dollar value per hectare, rather than a large total area in tobacco production. An individual grower seldom has more than 4 ha in production. It is estimated that tobacco requires over 700 worker-hours of labor to grow and harvest a hectare compared to only about 10 worker-hours per hectare for maize. There is strict governmental control over the amount of tobacco which can be marketed annually and land prices are tied closely to tobacco allotments in parts of the zone. Although U.S. exports of flue-cured tobacco have remained rather constant, the U.S. share of the world market has dropped from 60% in 1959 to about 33% at present. The U.S. growers produce about 25% of the world market of burley tobacco.

Tobacco has rather narrow limits of temperature and moisture for maximum production of high quality product. Sandy loams with their good drainage, relatively low moisture-holding capacities, and low nutrient supplying power produce plants with leaves desirably light in color and weak in aroma. Since the tobacco plant has a relatively shallow root system, it can be grown quite successfully on shallow soils.

Adequate soil moisture must be available so summer precipitation must be adequate or irrigation water must be added for top yields.

Rice

Rice is grown primarily in the Mississippi delta area of Arkansas, Tennessee, Mississippi, and Louisiana and in the gulf-coast areas of Texas and Louisiana. Since rice requires a constant and plentiful supply of water, it is usually grown on fine-textured soils with impeded drainage. It is generally grown two or more years in succession on the same area since it is necessary to level the field and build levees to retain the water. After a period in rice, the field levees are destroyed and alternative crops are grown for one or more years. Soybeans are increasingly being used in this kind of a rotation with rice. Soybeans generally do quite well on the soils to which rice is adapted and favorable soybean prices have encouraged their production.

Soybeans

About one-third of the U.S. soybean production comes from this zone. Some of the largest individual fields of soybeans in the U.S. are found in the Mississippi river delta region. A minimum frost-free period of 120 days and a mean summer temperature above 21°C are favorable to soybean production. Soybean plants are the most sensitive of all common crop plants to the length of day and are also sensitive to intensity of light. Varieties have been developed which are adapted to various latitudes. Most varieties give best full season production in a band usually no wider than 150 to 250 km from north to south (Scott and Aldrich 1970). Thus selection of a suitable variety is most important.

Forage Crops

The long growing season in the humid warm temperate zone makes forage production popular and generally profitable. Forage crops can provide direct feed for livestock over a high proportion of the year and can provide protection to the soil against erosion by water and wind. Grasses and legumes may be grown either separately or together. Some pasture mixtures contain as many as six species. Small grain grown as a grain crop may be grazed in the South for a high proportion of the winter. The cattle must be removed from the field in late winter or early spring to allow optimum development of the plants for grain production.

Bermudagrass is well adapted to this zone and marked improvement has been made in this crop through breeding and selection. Coastal bermudagrass, a recently developed variety, responds very favorably to high levels of management. Coastal bermudagrass fertilized with 224 kg of nitrogen per hectare has produced 726 kg beef per hectare per year (Anon. 1978A).

Other Crops

Groundnuts are an important crop in this zone, although U.S. production is only about 5% of the world total. Groundnuts generally require a frost-free period of about 200 days, although cultivars which mature in shorter periods have been developed. A mean annual temperature above 7°C is required, along with a mean mid-summer temperature of over 24°C. The crop is best adapted to light, well-drained loam soils and can tolerate acidic soils rather well. Groundnuts are usually grown in rotations with cotton, corn, tobacco, and pasture. Groundnuts, a legume crop, may leave the soil too high in available nitrogen for best production of a succeeding crop of tobacco (Chapman and Carter 1976).

This zone produces about 10% of the U.S. crop of maize grain and of maize silage. Some exceptionally high yields have been reported and some states are usually at or above the national average yield levels. Most states in the zone, however, normally have maize yields below the national average. Maize remains an important crop in some areas but for the zone as a whole, maize has lesser importance than many of the other crops mentioned.

Multiple Cropping

Because of the long growing season in the humid warm temperate zone, multiple cropping can be practiced with a higher degree of success than in the humid cool temperate zone. Such schemes may involve 2 crops in one year (double-cropping) or 3 crops in 2 years. Development of improved tillage practices, particularly no-tillage planting, improved weed control practices, and earlier maturing cultivars of soybeans and small grains have favored expansion of double-cropping in the zone (Lewis and Phillips 1976). It is estimated that 10% of all U.S. soybean hectarage planted each year is the second crop on that land, following winter wheat or other small grains (Anon. 1977).

CONSTRAINTS TO HIGH YIELDS

Pests

Because of its mild winters, higher mean annual temperature, and humidity, this zone has more serious problems with pests than do cooler zones. Insects, disease organisms, and weeds are better able to survive from one year to the next and control measures must be timely and effective to prevent serious yield reductions. Some areas with higher mean annual temperatures have a lengthy dry season each year which helps to keep pest populations in check. Litsinger and Moody (1976) have written an excellent paper on pest control practices which are applicable to this zone. They point out that the state of knowledge is not adequate to make good predictions of the consequences of new technology on pest problems. They warn that we may never have adequate predictions since pests can change to meet new stresses.

Soil Factors

Soils in this zone generally are more highly leached than in the humid cool temperate zone. They tend to be lower in organic matter with a more limited rooting zone and generally more rolling in topography. Thus soil and water management practices to provide nutrients as they are needed and to control loss of water, soil, and nutrients are highly essential for high levels of crop production. A fertility program based on soil testing is essential for satisfactory crop production, as pointed out in Chapter 5. In most of this zone if the ammonium form of nitrogen is applied in the fall for maize production the following spring the potential loss of nitrate is 50% or higher. This obviously increases the potential for nitrate pollution of water resources as well as increases the cost of production.

Most of the zone has average annual values of the rainfall-erosivity factor, R, equal to 250 or greater (see Chapter 6). Values in the U.S. Corn Belt are generally less than 200 (USDA 1976).

Drought can be a serious constraint to high crop yields in this zone, particularly on sandy or shallow soils. While improved fertilization and water management practices have done much to alleviate detrimental effects of drought, crop production is seriously limited in many years on shallow or sandy soils unless irrigation is practiced.

POTENTIAL IMPROVEMENTS IN CROP PRODUCTION SYSTEMS

Multiple Cropping

As earlier maturing cultivars of wheat and of soybeans which have yield potential equal to full season cultivars become available, multiple cropping will be practiced more extensively. Improvements in cultural practices such as no-tillage planting, pest control, and water management will also hasten the spread of multiple cropping in this zone.

While yields of soybeans grown as a double crop with a small grain are usually slightly less than yields of full-season soybeans, the total marketable crop yields and net income per hectare per year are generally higher with double-cropping (Lewis and Phillips 1976; Reicosky *et al.* 1977). Narrow row planting of the soybeans generally produces important increases in yield. Since late-planted soybeans will not produce exceptionally large plants, narrow row planting allows increased interception of sunlight, greater photosynthesis, and greater yields. Parks an- Livingston (1978) reported a 23% increase in soybean yields in 18-in. rows as compared to 36-in. rows. This was from 10 experiments over a 3-year period in Tennessee.

Total forage production per year may also be increased appreciably by multiple cropping. Small grain may be grown for silage, followed by maize or sorghum for silage. Or one of the crops may be grown for silage and one for grain. Lewis and Phillips (1976) report that triple-cropping was successfully accomplished in Georgia. This involved barley-maize (relay-intercropped in barley)-soybeans and barley-maize-snapbeans. Many other multiple cropping combinations are possible with resulting improvement in yields and income. Proper attention to fertilization and to pest control practices is vital to the success of such crop production systems.

Tillage Practices

No-tillage crop production may be the most revolutionary and effective soil conservation practice adopted in this century (Reicosky *et al.* 1977). This practice is especially well adapted to this zone where erosion can be a serious hazard and where no-tillage can help ensure successful multiple cropping.

Unger and Stewart (1976) have pointed out the advantages and disadvantages of no-tillage for double-cropping. Advantages include less delay in planting the second crop, higher soil moisture contents in the seed zone, slower drying by evaporation, lowering of maximum soil temperatures, reduced soil crusting, and improved soil structure.

Higher soil moisture content and reduced soil temperatures may also be detrimental, particularly early in the season. Poor soil-seed contact may be a problem in zero tillage systems, but improved planting equipment now available can avoid this problem in most cases. Pest control may be more difficult with no-tillage practices but adequate control practices are attainable. There are indications that repeated production of wheat with no-tillage in Texas has led to yield declines because of stubble buildup. This has apparently resulted in poor planting and weak plants after emergence. Limited tillage by undercutting or discing every second year seems to have alleviated the problem (Anon. 1978B).

Water Management

Crop acreage under irrigation in this zone has increased appreciably during the last decade. Even though irrigation with modern labor-saving equipment involves a relatively high per hectare investment, it can be justified, particularly on the sandy soils which occur in this zone. Land prices have generally increased faster than irrigation equipment prices and many farmers find it more advantageous to stabilize production at optimum levels on their present farms than to invest in additional land. While droughts occur less frequently in this zone than in the more arid zones, droughts can be disastrous to crop production and irrigation provides a way to ensure continued high production. It is anticipated that irrigation will increase in the years ahead.

Soil and crop management practices other than irrigation can also reduce detrimental effects of moisture deficiencies and excesses. Land shaping to improve surface runoff during periods of excess rainfall permits field operations to proceed at optimum times. Conservation tillage practices, particularly those utilizing crop residues for surface protection, can improve infiltration and recharge of soil moisture while also minimizing erosion from intense rainstorms. Contour planting and strip-cropping are practiced by many farmers, especially in the Piedmont and Upper Coastal Plains (Reicosky *et al.* 1977). Use of winter cover crops protects the soil against water and wind erosion and helps avoid water and air pollution. Small grains used for grazing and/or harvested for grain are replacing crops used strictly for soil protection.

If suitable sites are available, direct runoff from a farm can be reduced by constructing ponds to collect the runoff. This does little toward keeping the water on the cropland where it is needed, but such ponds can provide needed water for irrigation of the cropland. Using the stored water for irrigation increases the pond's effective storage of runoff water during the season also. Sediment collects in the pond and reduces its useful life, but contamination of surface water resources is reduced.

Pest Control

As previously indicated, we may never be able to predict the consequences of changing technology on pest problems. However, we have had excellent success in developing adequate control methods as new pest problems have evolved. The control methods include pesticides, resistant varieties, biocontrol, and cultural control. Current thinking is that no one of these methods can be used to the exclusion of the others. Since pests do not remain static, a control method successful in the short run may prove quite unsatisfactory over a period of years. Where several methods of pest control are used, crops continue to have protection; if one method fails, the others act as a buffer (Litsinger and Moody 1976).

Integrated pest management or integrated pest control are terms used to denote combinations of chemical, biological, cultural, or mechanical control techniques to eradicate pests or to keep their populations at or below accepted damage levels. Building of successful integrated pest management programs requires the efforts of teams of scientists of varied disciplines working with farmers for the desired goal. This subject is getting increasing attention in both developed and developing countries.

REFERENCES

ANON. 1977. Oats/corn double-crop "new" combinations for livestock farmer. Crops for Livestock *22* (1)3-5.

ANON. 1978A. Fertilize forages for profits. Better Crops Plant Food, Vol. XLI, Winter 1977-78, 23.

ANON. 1978B. Agronomists learn there are some drawbacks to no-till wheat. Successful Farming *76* (4) 26.

CHAPMAN, S.R., and CARTER, L.P. 1976. Crop Production Principles and Practices. W.H. Freeman and Co., San Francisco.

LEWIS, W.M., and PHILLIPS, J.A. 1976. Double cropping in the Eastern United States. *In* Multiple Cropping. R.I. Papendick, P.A. Sanchez and G.B. Triplett (Editors). Spec. Publ. *27.* American Society of Agronomy, Madison, Wisconsin.

LITSINGER, J.A., and MOODY, K. 1976. Integrated pest management in multiple cropping systems. *In* Multiple Cropping. R.I. Papendick, P.A. Sanchez and G.B. Triplett (Editors). Spec. Publ. *27.* American Society of Agronomy, Madison, Wisconsin.

NUTTONSON, M.Y. 1965. Global agroclimatic analogues for the southeastern Atlantic region of the United States and an outline of its physiography, climate and farm crops. American Institute of Crop Ecology, Washington, D.C.

PARKS, W.L., and LIVINGSTON, S. 1978. Soybeans, sunlight and yield. Better Crops Plant Food XLI, Winter 1977-78, 8-10.

REICOSKY, D.C. *et al.* 1977. Conservation tillage in the Southeast. J. Soil Water Conserv. *32*, 13-19.

SCOTT, W.O., and ALDRICH, S.R. 1970. Modern Soybean Production. S and A Publications, Champaign, Illinois.

SPRAGUE, G.E. 1975. Agriculture in China. Science *188*, 549-555.

TROLL, C., and PAFFEN, KH. 1965. Seasonal climates of the earth. *In* World Maps of Climatology, 2nd Edition. H.E. Landsberg *et al.* (Editors). Springer-Verlag, New York.

UNGER, P.W., and STEWART, B.A. 1976. Land preparation and seedling establishment practices in multiple cropping systems. *In* Multiple Cropping. R.I. Papendick, P.A. Sanchez and G.B. Triplett (Editors). Spec. Publ. *27.* American Society of Agronomy, Madison, Wisconsin.

USDA. 1976. Control of water pollution from cropland, Volume I. U.S. Dep. Agric. Rep. *ARS-H-5-1.*

14

Crop Production Systems in the Arid and Semi-Arid Warm Temperate and Mediterranean Zones

Peter A. Oram

THE GEOGRAPHY OF THE REGION

The region described here, corresponding to subzones IV, IV_2, IV_3, and IV_5 in Troll and Paffen's (1965) map of Seasonal Climates of the Earth is outlined in Fig. 14.1. These are broadly similar to the Cs, BSK, and BSh regions as defined by Köppen and Geiger (1936). The characteristic climatic pattern is one of hot dry summers and cooler winters, with

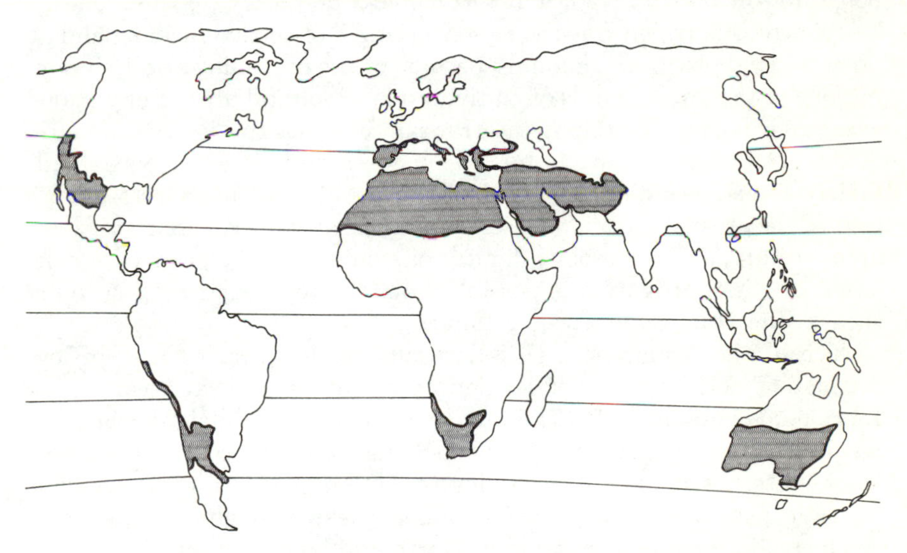

From Troll and Paffen (1965)

FIG. 14.1 MAP OF THE ARID AND SEMI-ARID WARM TEMPERATE AND MEDITERRANEAN ZONES

limited precipitation largely concentrated in the winter season. Such climates are found in all continents, lying between the Tropics and latitude 40°.

The largest single contiguous area of Mediterranean climates lies around and to the east of the Mediterranean Basin, stretching for some 5000 km from the Atlantic coast at longitude 10° W, to Afghanistan in the east, and lying between the Tropic of Cancer and latitude 44°. It merges into the cool temperate region to the north, the Sahelian semi-arid tropical climate to the south and the tropical monsoon area to the east.

In Australia and southern Africa the situation is a mirror image of that in North Africa—the more arid areas are found to the north; and the sea, bordered by areas of the "true" Mediterranean climate, lies to the south.

The Mediterranean areas of warm temperate climate in both North and South America lie along the Pacific Coast of the continent, bordered by cold water currents, and are cut off to the east by high mountains. The southern boundary in Chile and the northern in Oregon are wetter and more temperate. Inland areas to the east of the main ranges in both North and South America are semi-desertic and characterized by greater temperature extremes. The alternation of wet winters and dry summers, and the relatively moderate summer temperatures in the west coast areas are the result of their location between wet marine climates on the poleward side and dry tropical desert climates on the side of the equator.

Although this region was the center of origin of settled farming and of domesticated plants and animals, because of the prevailing aridity it now has one of the lowest ratios of cultivated land to total area of any major geographic region. For the Mediterranean countries this is less than 10% (Table 14.1). Conversely, it has the highest proportion of wasteland, mainly desert, including most of the Sahara and Arabian Deserts, a large part of the great Australian Desert, and the southern Kalahari. In these areas aridity places absolute limitations on settled farming except in oases, or where irrigation is possible from the few major river systems such as the Nile, Euphrates, or Indus.

Much of the remaining area is virtually treeless grazing land of low productivity (150 to 200 fodder units per hectare per year according to Le Houerou and Hoste 1977). Extensive systems of animal management have developed: nomadism and transhumance in the countries around the Mediterranean where, according to Faulkner (1972), some 9 million nomadic people own nearly 100 million sheep and goats, and large, commercial ranching operations in North and South America and parts of Australia.

Rainfed cropping within the warm temperate semi-arid region is largely confined to the Mediterranean winter rainfall climatic zone, on which

TABLE 14.1. POPULATION AND LAND USE CHANGES IN MEDITERRANEAN COUNTRIES

	Total Population		Agric. Population		Land Use Data				
								Growth Rate 1963-74	
	1970 Millions	1960-70 Growth % p.a.	1970 Millions	1960-70 Growth % p.a.	Total Area	1974 Arable (Million ha)	1974 Irrig.	Arable % p.a.	Irrig. % p.a.
Total	307.0		144.01		1302.0	122.9	24.18	0.1	1.4
Greece	8.79	0.30	3.79	−2.26	13.2	2.98	0.86	0.2	4.9
Spain	33.62	1.04	8.74	−3.72	50.5	15.92	2.82	0.1	2.0
Yugoslavia	20.37	1.02	10.14	−1.44	25.6	7.36	0.13	−0.3	2.2
Italy	53.66	0.66	10.08	−4.19	30.1	9.32	3.60	−2.1	2.1
Portugal	8.63	−0.23	2.88	−2.99	8.9	3.05	0.63	−1.7	0.1
Egypt	33.33	2.54	18.13	1.82	100.1	2.72	2.86	1.0	0.3
Iraq	9.36	3.17	4.36	1.83	43.5	5.10	1.15	0.8	0.9
Israel	2.91	3.25	0.28	−0.68	2.1	0.34	0.18	0.6	1.5
Cyprus	0.63	1.00	0.24	0.21	0.9	0.37	0.94	0.0	−1.2
Iran	28.36	2.78	13.05	1.17	164.8	15.70	5.35	0.5	1.0
Lebanon	2.47	2.89	0.48	−3.76	1.0	0.24	0.09	2.1	3.2
Saudi Arabia	7.74	2.62	5.11	1.80	215.0	0.70	0.19	7.8	5.0
Afghanistan	16.98	2.14	13.87	1.70	64.8	7.92	2.44	0.7	0.8
Algeria	14.33	2.87	8.70	1.89	238.2	6.40	0.32	0.2	2.6
Jordan	2.28	3.01	0.77	0.32	9.8	1.17	0.06	1.3	0.0
Libya	1.94	3.69	0.62	−1.42	176.0	2.40	0.14	0.1	3.5
Morocco	15.13	2.65	8.60	1.69	44.6	7.18	0.44	0.7	2.6
Syria	6.25	3.20	3.19	2.59	18.5	5.68	0.52	−0.7	1.0
Tunisia	5.14	1.98	2.56	0.74	16.4	3.20	0.13	−1.0	1.7
Turkey	35.23	2.50	23.87	1.01	78.0	25.13	1.98	0.7	3.5

Source: FAO (1976C).

this chapter is focused. Although this is often considered to be a rather homogeneous region climatically, De Brichambaut and Wallen (1963) defined no less than 12 Mediterranean sub-climates in the Near East alone, using temperature as the main determining factor, and superimposing rainfall and water balance. Aschmann (1973) sets the approximate annual precipitation levels for the dry boundary of the Mediterranean climate at 275 mm for the coast (350 mm inland), and the humidity boundary at 900 mm, with at least 65% of precipitation being concentrated between October and April in the northern hemisphere. This accords quite well with Emberger's (1955) bioclimatic classification, which adopts 300 mm as the lower limit of precipitation before the climate becomes too arid, and 1000 mm as the lower limit of the humid zone.

The region overall includes parts of some 27 countries with a total population approaching 400 million people. About half of this total live in the "developing" countries to the east and south of the Mediterranean Basin proper (Table 14.1). Per capita income in 1974 ranged from $110 in Afghanistan to $4400 in Libya. Probably no other region on earth exhibits such sharp contrasts of affluence and poverty.

THE MEDITERRANEAN ENVIRONMENT IN RELATION TO AGRICULTURE

Climate

The climate of the warm temperate region, and specifically of the Mediterranean zone, has certain broad characteristics which distinguish it sharply from the semi-arid tropics, and which also largely determine its flora and the character of its agriculture. These are the following:

(1) The Mediterranean climate proper is never found at a great distance from the sea, as indeed its name implies.

(2) There is a marked cool season (October-April/May in the northern hemisphere). Since this is also the rainy season, rainfall is more efficient for crop production in the milder areas of the Mediterranean than in semi-arid summer rainfall areas. However, there are also areas of the Mediterranean where winter temperatures are sufficiently low to be limiting to plant growth. The average low temperature limit for the coldest month in the zone is usually set at 0°C to a −3°C (Koppen and Geiger 1936; Aschmann 1973). These areas exist in the mountainous areas of the maritime countries bordering the seacoast, but are particularly important in the high plateaus of Spain, Turkey, Iran, Iraq,

Afghanistan and, to a more moderate degree, in Algeria. One consequence of low winter temperatures and late springs is that the high-yielding "Mexican" semi-dwarf wheats cannot be grown on the plateaus because they lack cold tolerance. This applies also to certain of the grain legumes and forages—for example to *Medicago truncatulata*, which is so important to the cereal/pastoral system in southern Australia.

(3) Rainfall decreases from north to south (and to some extent from west to east) in the Mediterranean Basin proper, and in all warm temperate areas there is a somewhat abrupt transition from the rainy season to a relatively hot and almost rainless summer. This has led to a preeminence of annual plants and sclerophyllous evergreen trees and shrubs in the flora. Annual plants often constitute half the total species in the Mediterranean regions but rarely as much as a tenth in other climatic regions.

(4) In addition to being concentrated into a period of 6 or less humid months, the rainfall pattern is highly erratic and unpredictable (40 to 50% variability). This results in large year-to-year fluctuations in yields of rainfed crops and restricts the range of annual food crops which can be grown without irrigation mainly to wheat, barley, and pulses. The large component of perennial drought-resistant crops, such as olives, vines, figs, dates, and stone fruits, in the crop mix is a further reflection of high climatic risk in rainfed agriculture, at least in the Mediterranean Basin. Table 14.2 illustrates the importance of these crops in the land-use pattern.

(5) Annual fluctuations of rainfall and its distribution can cause great year-to-year variations in pasture output, and this is reflected in the prevalence of patterns of nomadism and transhumance in traditional animal husbandry, particularly in the drier semi-desert fringes and in the more mountainous and colder areas of the Mediterranean. In many countries this has led to an historic separation of ownership of arable land from that of livestock and to frequent feuds over grazing rights. While these have persisted since the days of Cain and Abel, the situation has been aggravated in recent years by the inexorable growth of both human and animal populations, which have expanded at almost the same rate over the last decade. Increasing competition for land between graziers and cultivators in the marginal areas, aggravated by mechanization of cereal cultivation, has accelerated the forward march of the desert, to which an estimated 100,000 ha of grazing land is being lost every year (FAO 1973).

Topography

The bulk of the Mediterranean climatic zone is both sharply dissected

TABLE 14.2. LAND-USE SYSTEMS ACCORDING TO DISTRIBUTION OF THE CULTIVATED AREA, WESTERN GREECE NOMI OF ILIA AND ACHAIA, 1961

Crop Group	Mountains	Foothills			Plains			Total for Ilia and Achaia
	System A	System A	System B	System E	System B	Systems C & D	System E	
Areas of Principal Crop Groups in Hectares								
Vineyards	1,928	4,763	1,944	4,584	4,593	2,709	15,897	36,418
Plantations	330	2,175	2,052	1,061	1,441	1,927	8,053	17,039
Cereal Grains	21,660	40,221	8,142	1,394	10,927	5,856	9,893	98,093
Other Annual Crops	2,877	5,856	3,169	352	6,908	8,101	3,604	30,867
Alfalfa and Clover	611	810	358	145	566	185	720	3,395
Other Fodder Crops	3,355	8,886	1,644	532	3,454	2,319	2,605	24,795
Fallow	5,420	5,981	642	881	3,930	2,383	2,858	22,095
Total Arable Area	38,181	68,692	17,951	8,949	31,819	23,480	43,630	232,702
Average Arable Land Per Farm (Ha)	3.8	5.4	3.9	3.3	3.8	3.9	2.1	2.6
Percentage of Distribution of Crops								
Vineyards	5.1	6.9	10.8	51.2	14.4	11.5	36.4	15.6
Plantations	0.8	3.2	11.4	11.9	4.5	8.2	18.5	7.3
Cereal Grains	56.8	58.6	45.4	15.6	34.4	24.9	22.6	42.1
Other Annual Crops	7.5	8.5	17.7	3.9	21.6	34.5	8.3	13.3
Alfalfa and Clover	1.6	1.2	2.0	1.6	1.8	0.8	1.7	1.5
Other Fodder Crops	14.0	12.9	9.1	6.0	10.9	9.9	6.0	10.7
Fallow	14.2	8.7	3.6	9.8	12.4	10.2	6.5	9.5
Area of System as	100%	100%	100%	100%	100%	100%	100%	100%
Percentage of Total Arable Area in Region	16.4%	29.5	7.7	3.8	13.7	10.1	18.8	100%
% GDP From Crops	51.3	72.0	74.0	92.8	82.7	88.7	87.9	81.0
% GDP From Animal	48.7	23.0	26.0	7.2	17.3	11.3	12.1	19.0
GDP Per Caput ($)	91	145	132	215	184	205	154	152

Sources: Anon (1961); FAO (1965).

topographically and subject to strong tectonic influences, e.g., all around the Mediterranean Basin, in California, and in Chile. Axelrod (1973) has pointed out that this has had an important influence on shaping the genetic diversity of the flora. However, because of its unstable and mountainous terrain, its relatively young and heterogeneous soils, and the fact that the rainfall often comes in torrential storms, runoff is high and there are some serious erosion hazards in much of the region.

The more mountainous areas of the region were once extensively forested, and Quezel (1977) has stated that the biogeographical, climatic, and topographic heterogeneity of the Mediterranean region has resulted in an extremely diverse range of some 40 major tree species and 50 sub-varieties, compared to at-most 12 and 20, respectively, in the colder regions of central and northern Europe. However, he points out that the forest cover has been progressively pillaged for timber, grazing, and farm land to the extent that three-quarters of the original vegetation had disappeared by the mid 1800s. Mechanized agriculture is generally practiced without conservation measures, and the traditional stone terraces in hilly areas are increasingly being abandoned for lack of labor to maintain them.

In his Criteas, written about 400 B.C., Plato remarks, "There are mountains in Attica which can now keep nothing more than bees, but which were clothed not so very long ago in fine trees—rainfall was not then lost, as it is at present, through being allowed to flow over a denuded surface to the sea!" There is abundant botanical and archaeological evidence that it is the destructive influence of man on a fragile environment, over the long history of settlement in the Mediterranean and Near East, rather than any major shift in the regional climate, which has led to the degradation of its natural resources.

Soils

The soils of the Mediterranean regions are markedly heterogeneous, which is not surprising in view of their transitional nature, varied geological substrata, uneven topography, and tectonic influences. Thus, the soils of California are formed largely from igneous basalt and granite, sedimentary sandstone, and metamorphic rocks; whereas, there are major areas of limestone formations in the Mediterranean Basin from which the extensive "terra rossa" soils are derived. However, the combined influences of topography and elevation often produce catenas which override the nature of the parent material. These can develop a remarkably consistent sequence of soil, pH, and vegetation, which is comparable in situations as far apart geographically as Italy, Greece, and California (Zinke 1973).

Many of the agricultural soils are reasonably easy working, and, except for their extreme dryness at the end of the summer, do not usually present major cultivation problems. They have a predominantly neutral to high pH and a satisfactory base status. There are considerable areas of deep calcareous sands in North Africa, and parts of the Arabian Peninsula, which are low in organic matter, have a low water holding capacity, and are extremely deficient in nitrogen. These are subject to severe wind erosion if cultivated mechanically without conservation measures, or when subjected to excessive grazing pressure. They, therefore, demand especially careful management for sustained cultivation, and, where irrigated, the use of distribution systems which minimize the high losses from permeability.

Fine-textured calcareous soils derived from limestone parent material, including the important "terra rossa" soils, constitute an important part of the agricultural area in the "Fertile Crescent"—Syria, Jordan, Lebanon, northern Iraq, parts of southern Turkey and western Greece, Egypt, eastern Libya (Cyrenaica), Tunisia, and Algeria. These can also suffer from poor physical conditions related to water retention, and where gypsiferous—as in western Egypt, Syria, and Iraq—are difficult to cultivate. According to Specht (1973), the presence of calcium carbonate is a key factor in determining the fertility of the phosphorus-deficient Mediterranean soils of Australia.

Salinity and/or sodium accumulations are serious problems in the Near East. A recent FAO (1975) report states that more than 50% of the lower Rafidain Plain in Iraq, about half of the Euphrates Valley in Syria (some 250,000 ha), almost a third of the total cultivated area in Egypt (750,000 ha), some 15% in Iran, and much of the Indus Basin irrigation system in Pakistan, suffer various degrees of salinity (or saline-sodic conditions), waterlogging, and impeded drainage. In most areas these conditions are the result of poor irrigation practices and lack of proper drainage which together with silting of dams and irrigation channels consequent to overgrazing and lack of protection in the headwaters, have led to the abandonment of once-fertile lands (Pereira 1973).

Pressure of Population on Land Resources and Irrigation

Particularly around the Mediterranean, but to an increasing extent also in Chile and California, there is a critical shortage of new land for cultivation in the warm temperate semi-arid regions. Although many of the pressures leading to the progressive deterioration of the environment can be attributed to the combined effects of the scarcity of arable land and the aridity of the climate, one important and positive consequence of the difficulties facing rainfed agriculture has been the expansion of irrigation.

Table 14.1 shows that for the Mediterranean Basin as a whole, irrigated land averages over 20% of total arable area, the highest proportion of any large geographical region. Although irrigation has a long history in this region, the equipped area has doubled in the last 20 years to a total of over 24 million hectares. However, the growth rate is slowing down, mainly because the best sites have now been utilized, but it is still greater than the growth rate of the total arable area. This reflects the high priority accorded by governments of the region to irrigation as a means of compensating for land scarcity. Irrigation systems vary considerably in their efficiency; these differences are reflected in the wide variability of yields and cropping intensities on irrigated land. Nevertheless, irrigation contributes a significant share of the value of agricultural production nearly everywhere. And because of the relative security it provides from climatic vagaries, and the opportunities for higher yields, cropping intensities, and more diversified cropping patterns, this share is usually much greater than the ratio of irrigated area to total arable area.

AGRICULTURE IN THE ECONOMY OF THE REGION

In recent years there have been major changes in the demography and economy of the countries around the Mediterranean Basin. Industrialization, the migration of people from the land to the cities (or in some cases to other more affluent countries—especially from Spain, North Africa, and Turkey to the EEC), tourism, and the expansion of petroleum production to a wider number of countries have contributed to a decline in the importance of agriculture and widened income disparities both among countries of the region and among agriculture and other sectors within countries.

Changes in Crops

In addition to changes in the factors contributing to overall economic growth of the region, there have also been significant shifts in recent years in the composition of agricultural GDP. In many of the countries listed in Tables 14.1 and 14.3, and particularly those of southern Europe, the importance of cereals and pulses, which are the traditional rainfed crops, occupying some 50% of total arable area (and also that of cotton), has declined relative to that of fruits (especially citrus) and vegetables (Table 14.3). This is largely a result of the expansion of irrigation and the greater diversification of cropping which this permits, but it is also due to the comparative advantage of the countries with a mild Mediterranean climate for producing out-of-season high value crops such as early pota-

TABLE 14.3. AREA, YIELD AND PRODUCTION GROWTH RATE TRENDS FOR SOME MAJOR CROPS (% PER ANNUM) 1961–1975

Country	Wheat Area	Wheat Yield	Wheat PGR[1]	Rice Area	Rice Yield	Rice PGR	Pulses Area	Pulses Yield	Pulses PGR	Seed Cotton Area	Seed Cotton Yield	Seed Cotton PGR	Vegetables & Potatoes Area	Vegetables & Potatoes PGR	Citrus PGR	Olive Oil PGR
Greece	−2.5	3.7	1.14	−1.5	2.1	0.67	−6.6	3.8	−2.82	−2.8	5.7	2.91	0.3	7.25	3.24	2.26
Spain	−2.6	2.8	0.21	−0.1	−0.3	−0.41	−3.0	0.5	−2.5	−10.4	4.0	−6.45	1.0	2.76	3.86	2.44
Yugoslavia	−1.2	4.3	3.1	2.5	1.3	3.82	−2.6	−0.3	−2.89	−0.5	1.9	1.44	−1.9	0.21	13.5	−8.38
Italy	−1.6	2.4	0.76	4.0	0.1	4.01	−9.1	4.6	−4.51	−13.6	−1.9	−15.55	−2.3	0.25	5.43	2.22
Portugal	−3.2	4.5	1.28	0.0	−1.3	−1.32	−3.3	0.7	−2.66	—	—	—	1.7	2.88	0.72	−3.66
Egypt	0.7	1.4	2.06	3.2	0.0	3.18	−2.6	2.7	0.08	−1.4	1.6	0.24	3.2	3.64	8.31	—
Iraq	1.6	1.9	3.49	11.3	5.1	2.76	1.4	−0.4	0.83	−0.4	6.8	6.48	4.7	4.55	6.25	—
Israel	5.8	5.3	11.04	—	—	—	1.1	6.1	7.15	8.4	0.9	9.38	2.6	4.85	8.5	—
Cyprus	−0.8	−2.6	−3.44	—	—	—	−0.9	−4.1	−4.99	—	—	—	0.2	2.82	9.83	−5.6
Iran	3.7	0.9	4.57	2.2	2.2	4.34	−2.0	1.7	−0.3	−0.8	5.8	5.04	2.3	5.04	7.47	−0.91
Lebanon	−2.8	1.7	−1.1	—	—	—	4.1	1.4	2.74	—	—	—	2.6	2.24	3.02	2.27
Saudi Arabia	−3.1	2.0	−1.13	−1.6	1.2	−0.41	4.4	0.2	4.57	—	—	—	9.2	9.94	6.5	—
Afghanistan	0.1	1.6	1.69	−0.2	2.2	1.99	1.6	0.0	1.57	−1.2	5.4	4.25	0.4	0.27	1.19	0.00
Algeria	0.4	−1.0	−0.61	−2.6	−8.0	−10.55	2.8	−0.2	2.66	−8.3	5.3	−2.97	4.0	2.03	2.52	−1.48
Jordan	−4.7	2.0	−2.67	—	—	—	−1.5	2.8	1.33	—	—	—	−5.4	−7.28	−3.08	−10.62
Libya	−2.3	5.1	2.82	—	—	—	−0.2	9.1	8.91	—	—	—	13.6	17.85	3.11	6.23
Morocco	1.8	2.1	3.9	−2.7	0.0	−2.68	2.8	4.4	7.17	4.6	1.8	6.41	1.8	7.34	4.35	4.63
Syria	1.0	0.2	1.2	−4.2	−0.1	−4.34	2.8	−1.2	1.61	2.3	2.4	0.08	2.1	9.98	14.4	2.21
Tunisia	0.8	5.2	6.0	—	—	—	4.4	5.4	9.76	—	—	—	5.0	8.46	1.38	7.7
Turkey	1.1	2.3	3.42	0.4	0.6	1.0	0.4	0.4	0.74	1.1	5.1	6.22	1.7	2.93	8.14	1.29

Source: FAO (1976B).

[1]Production growth rate.

toes, tomatoes, and strawberries. California represents the apogee of this movement. The demand for these crops has risen rapidly in northern Europe as well as in the U.S. and Australia as a result of increasing affluence and the progressive improvements in transport and marketing systems since the end of World War II. The production of wine and table grapes has also expanded significantly in several Mediterranean countries as a result of increasing demand in other countries.

Livestock

A feature of the agricultural economy in the Mediterranean climatic zone is the importance of livestock, particularly ruminants. In 1966 35% of agricultural GDP in value terms in the Near East and 28% in North Africa was accounted for by livestock. Among developing regions, *per capita* consumption of mutton in the Near East and North Africa is higher than anywhere else in the world, and that of beef is surpassed only by the countries of Latin America and East Africa. In Chile, the U.S., and Australia animal production, especially mutton and wool, is also of very great significance in the farming economy of the warm temperate zone.

Despite considerable success in expanding irrigated agriculture and recent efforts to introduce into the Mediterranean the innovative Australian approach to rainfed farming through crop-livestock integration, demand for food is outstripping production in most countries of the region. Cereal and meat production in the populous countries of North Africa and the Near East with the largest dependence on rainfed agriculture has risen at around 2.5% per annum over the last 15 years, barely keeping pace with population growth, and falling increasingly behind a dynamic economic demand averaging 3.7% for cereals and over 8% for meat.

Food Deficits

In the 15 non-industrialized countries around the Mediterranean Basin, food imports have risen substantially in every case during the decade 1963 to 1974, and in most of them cereal imports rose more rapidly than total food imports (Table 14.4). This pattern is not confined to the oil-rich countries which are generally able to finance such imports from non-agricultural exports (although some of these show a particularly fast growth of imports), nor to those which have traditionally been food importers. Countries such as Algeria, Iraq, Turkey, Syria, and Tunisia, which have from time-to-time been large exporters of grain, show a decreasing level of foodgrain self-sufficiency. Projections shown in Fig. 14.2 indicate a food deficit between 19.5 and 33.8 million tons by 1990.

TABLE 14.4. GROWTH RATES OF AGRICULTURAL, FOOD, AND CEREAL IMPORTS, 1963–1974, NEAR EAST AND NORTH AFRICAN COUNTRIES (RANK ORDER OF IMPORTS OF AGRICULTURAL PRODUCTS)

	All Agric. Products Value	Food Products Value	Cereal Products Value	Cereal Products Quantity
		(Rates in % Per Annum)		
Yemen A.R.	25.8	25.0	37.0	36.0
Libya	20.3	20.4	17.7	12.4
Iran	16.1	17.0	21.5	16.1
Syria	14.7	16.0	23.7	20.0
Iraq	13.0	13.2	18.4	9.7
Saudi Arabia	12.7	11.6	12.7	9.8
Cyprus	12.5	12.8	21.8	17.0
Tunisia	10.8	11.5	12.7	7.2
Afghanistan	10.0	7.9	−2.3	−4.4
Algeria	9.3	8.1	16.0	12.9
Morocco	8.7	8.2	17.6	10.1
Jordan	8.2	14.8	14.6	7.3
Turkey	6.2	14.8	14.6	7.3
Lebanon	6.0	5.2	4.6	6.8
Egypt	3.9	2.5	4.2	−1.2

Source: FAO (1974).
Note: A minus sign indicates net exports.

Furthermore, in practically every country demand for meat far exceeds domestic capacity to produce and the gap between demand and availability has been reflected in sharply increased meat prices wherever the latter has not been artificially regulated. A recent World Bank report (1977) postulates a staggering import bill for meat of $2.2 billion for the Near East countries by 1985. An additional complication resulting from the rising demand for meat has been a rapid increase in demand for feed grains, which compete with cereals for food.

In terms of foreign exchange these deficits may to some extent be offset by shifts in land use to a pattern of higher value crops for exports or processing, made possible by irrigation development. However, in most countries, the balance of payments in agriculture is likely to deteriorate progressively unless a radical improvement can be effected in the productivity of cereals and livestock from the rainfed areas, which are also falling behind irrigated and urban areas in income and social amenities.

BASIC FARMING SYSTEMS IN THE REGION

Because of the historical evolution of agriculture in this zone, where there is an extremely long history of human influence on land use and

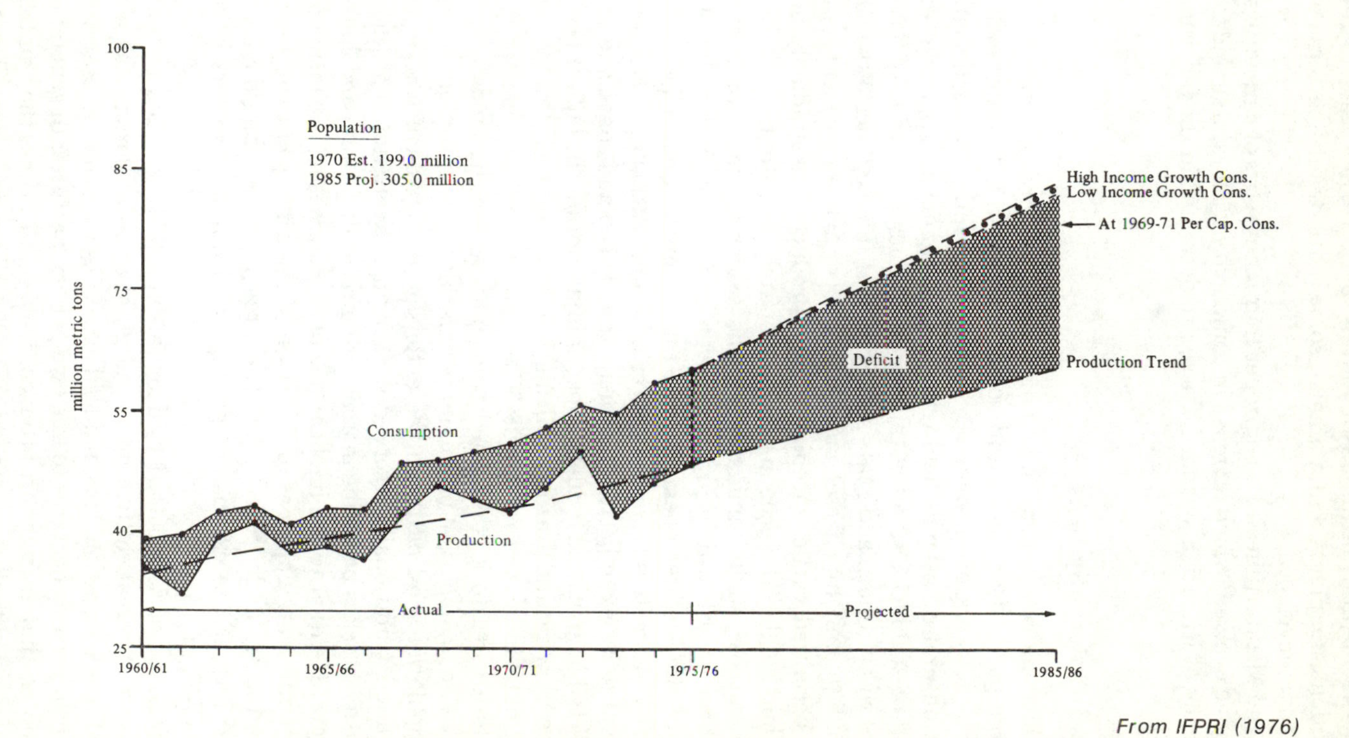

From IFPRI (1976)

FIG. 14.2. NORTH AFRICA/MIDEAST: FOOD DEFICIT DEVELOPING MARKET ECONOMIES. CEREALS: PRODUCTION AND CONSUMPTION 1960–75 AND PROJECTED 1985

misuse in the Mediterranean Basin (dating back 10,000 years to the origins of domestic plants and animals), some quite diverse patterns of land use have developed.

Around the Mediterranean Basin proper four main types of agricultural systems have evolved, which increase in complexity with rainfall and/or the adequacy of irrigation systems. These are described in the following sections.

Traditional Pastoral Systems

Animal husbandry is the main, and often the only, form of agricultural use of large parts of the drier and colder areas of the Mediterranean Region. Flockowners often own no arable land, and animals depend largely on natural rangelands held tribally or in common, supplemented by grazing of stubbles and weed fallows over which graziers may have traditional rights. Sheep and goats are often kept in mixed flocks, which provide milk (mainly for cheese), meat, and wool or hair for garments and carpets. Specialized cattle, sheep, and goat production is gradually increasing, and cattle are becoming proportionately more numerous; but the triple-purpose small ruminant remains extremely important, and milking sheep is still widely practiced even in the more industrialized countries of the Mediterranean Basin—for example, in many locations within a few kilometers from the center of Rome. Although cattle and buffalo tend to be given better treatment than sheep and goats, and are often stall-fed on irrigated fodders, most animals obtain the bulk of their sustenance from natural grazings. Calculations based on FAO Production Yearbook data indicate that about 57% of total feed dry matter requirements for livestock is obtained from natural grazing lands.

Due partly to an explosion of animal population parallel to that of the human population, and partly to the extension of cereal cultivation to marginal areas traditionally reserved for grazing, the existence of these free-range livestock is becoming increasingly precarious. Oram and Jones (1955) drew attention to the implications of the increasing expansion of cultivation into marginal areas of Turkey and the consequent encroachment on grazing lands. Since 1950 the area of wheat has doubled, almost entirely at the expense of grazing, and animal numbers have increased (Wright 1977).

A similar situation is reported by Hopkinson (1974), who states that the extension of cereal cultivation to marginal areas of Jordan, largely because of mechanization, has reduced carrying capacity of pastures from one ewe per two hectares to one ewe per ten hectares. In general it has been noted that grazings are often carrying 3 to 12 times the number of head they can actually support (ALAD 1974). This is not due solely to the extension of cultivation, but also to associated factors such as loss of grazing due to mechanized ploughing of stubbles, chemical weed control, and the elimination of weedy fallows, the enclosure of lands held under common grazing rights for settled cultivation, and the prevention of grazing in forest reserves in order to protect and regenerate forests.

These trends restrict the freedom of movement that has characterized the various migratory systems of nomadism and transhumance designed to provide insurance against drought and cold, and this accounts in part for the failure of animal production to keep pace with demand. The historical importance accorded to animals in the region and enshrined by customary rights (Tomaselli 1977) is progressively being eroded, and as Oram (1956) has stated, "Livestock tends to be the poor relation in the farming pattern of the region." The need to develop more productive systems of settled ruminant animal husbandry is, therefore, urgent.

Nevertheless, it is difficult to be optimistic about the short-run prospects for improving output from the rangelands under the constraints referred to earlier and with existing systems of grazing management. So far this has defeated most "upstream" attempts, such as grazing control, range reseeding, the provision of state-supported fodder reserves, disease control measures, and increasing the number of watering points, largely because these have not been coordinated with "downstream" measures to take the pressure of animals off natural grazings at critical periods in the plant growth cycle.

Measures such as resettlement of nomadic herdsmen on irrigated land, or banning free-range goats have been attempted in some Mediterranean countries; but because they were often based on emotional judgments against nomadism as a way of life or against goats as "black locusts," they have met with opposition and have had only limited success. As Faulkner (1972) has pointed out, nomadism is a rational system in the Mediterranean environment, and the goat is a useful animal within such a system because of the shrubby nature of the vegetation. Banning free-range animals has sometimes led to serious increases in brush fires. Since the only use of a high proportion of the land is for extensive grazing, there is an urgent need for research to establish the carrying capacity of the range, accompanied by the provision of advice and incentives to the flockowners to achieve its more conservative management.

Measures to Restore Natural Balance.—A first measure to help restore the natural balance would be to ban all cultivation in marginal areas where rainfall is below 250 mm, and possibly below 400 mm in more erodable soils. This is particularly important (apart from being a prime cause of erosion) because re-establishment of natural vegetation on land which has been ploughed is a lengthy process (Draz 1974). Certain countries, particularly in the Near East, are moving in this direction, either through compulsion or fiscal measures.

Second, it is important to foster accepted systems of voluntary grazing control within traditional tribal boundaries. In many countries these have been legally abolished in favor of state ownership of land, but because this can rarely be properly enforced, the result has been uncontrolled free-for-all grazing. The need to work with the local people is increasingly being recognized and efforts are being made to restore the traditional Bedouin rights of range ownership designed to protect and conserve the resource.

Third, means must be found to provide more stable outlets for immature animals from the range, surplus to the needs of maintaining the breeding flocks, so as to reduce their mass slaughter at low weights as the natural vegetation dries up. Fattening cooperatives in irrigated areas, the introduction of fodder crops into the cereal system in place of fallow for grazing, improved marketing channels, and prices which give incentives to better finishing of animals are some of the "downstream" measures necessary to provide this stability. The elimination of over-stocking of pastures as an insurance policy against bad years, which these measures would facilitate, would also permit the upgrading of stock through selection and crossing among indigenous breeds. The potential has been clearly demonstrated (Demiruren 1972; Fox 1974; Faulkner 1972) but is frustrated by inadequate nutrition.

What is needed in many countries is, therefore, a systems approach to the problem, involving a package of interrelated measures. The absence of such integration in the past, with the various components of the system being attacked piecemeal and in isolation, has probably been the main reason for the lack of success in raising the growth of ruminant livestock production despite the escalating demand for meat referred to earlier. Table 14.5 indicates the kind of improvement which could be expected from the introduction of such a system.

Rainfed Cereal Production

The possibilities of improving productivity of ruminant livestock are strongly correlated both with changes in the winter cereal production system and with higher yields of cereals, since if these are not achieved even more pasture may be ploughed up for grain production. To a considerable extent feedgrains for pigs and poultry must also depend on success in increasing cereal yields. Although some expansion of irrigated cereal production may be feasible, especially if high yield can be combined with quick maturity so as to fit into rotations with crops such as cotton, the bulk of cereal output must continue to come from rainfed land, principally from winter-sown wheat and barley. It is to this that research and development policies must be directed.

TABLE 14.5. COMPARATIVE PRODUCTION PERFORMANCE OF INDIGENOUS SHEEP UNDER EXTENSIVE AND SEMI-INTENSIVE SYSTEMS OF HUSBANDRY

Production	Extensive System[1]		Semi-intensive System[2]	Possible Increase in Production %
Lamb Production per Ewe	0.8		1.12	40
Average Carcass (meat production), kg	21.0	E[3]	28.0	80
		R[4]	34.0	62
Milk Production Ewe/Lactation, kg	45.0		59.5	32
Wool Production/Ewe, kg	1.3		2.2	69
Flock Off-take, %	33		42	21

Source: Demiruren (1972).
[1]Animals entirely dependent on natural pastures.
[2]Range grazing supplemented with feed to meet normal nutritional requirements.
[3]E=Ewe carcasses.
[4]R=Ram carcasses.

Two main avenues of approach are open: *first*, to try and increase total output and income from the rainfed cereal system by introducing a second crop to replace the fallow, which occupies between 25 and 40% of the land each year, and *second*, to improve the yields of the main winter cereal species whether fallow is maintained or not.

Changes in Use of Fallow.—The role of fallow has long been controversial. Many farmers believe it is essential to the conservation of moisture (mainly through weed control) and to the restoration of fertility, by mobilizing nitrogen for the next cereal crop. The validity of this concept, except in the more marginal areas, can be challenged. Since the bulk of the fallow land is either cultivated inefficiently or not at all, it probably performs neither function, although the weeds and stubbles are an important source of forage for livestock. After an extensive review of the agricultural rotations in the region, Oram (1956) concluded that the only measures for breaking the vicious cycle of overgrazing and low productivity of crops and grazing land is to develop arable rotations which provide more fodder without detriment to cereal production. Statistics show how limited success has been in this direction.

Despite years of subsequent research this issue is still controversial (Oram 1974). Advocates of the Australian dry farming system, in which self-reseeding annual forage legumes replace the traditional fallow break in the cereal rotation, point to the enormous benefits of the introduction of reseeding annual forage legumes in place of fallow in that country. Leeuwrik (1975) has suggested that its adoption would be feasible on at least 3 million hectares annually in the 350 to 500 mm rainfall zone in Morocco, Algeria, Tunisia, Syria, Jordan, Lebanon, and Iraq, where average temperature in the coldest month does not fall below 2°C. In the first 6 countries he estimates that this would add about 174,000 t of nitrogen annually to the soil, resulting in an additional 1.74 million tons of cereals, an increase of 24% over their aggregate 1966-1970 average production. It would provide an increment of 11.6 million tons dry matter to stockfeed, equivalent to about the same increase in numbers of sheep. His conclusions are summarized in Table 14.6.

So far, this promise still shows little sign of being fulfilled, although experimental work is being expanded in Algeria, Tunisia, Syria, and Jordan. There are both technical and social problems to be overcome, even though initial results on plots and some farms are promising (Doolette 1975). Suitable strains of Medicago have to be tested; seed is in short supply; rhizobial nodulation varies considerably from location to location and supplies of inoculum are limited; optimum methods of seeding and cultural practices have to be worked out; and phosphate has to be provided to the legumes. Not only are many of the cultural practices and concepts for the crops quite foreign to the farmers, but even more important, the critical livestock component and experience of grazing management is missing from their traditional systems. The average farm size in comparable areas of South Australia is 1000 ha, with about 200 ha under crop annually (mainly mechanized wheat) and a flock of some 2000 sheep. The average cereal holding (except on state farms) in North Africa and the Near East is probably around 15 ha, with only working animals (Dalrymple 1976).

Because of these difficulties many research workers in the region are in favor of the use of annual forage legumes such as vetches. Although these have to be resown every year and are probably more vulnerable to drought than the self-seeding species used in Australia, they fit better with the traditional farming practices and factor endowments of the small farmer and do not have to be grazed. They are cut usually for hay or fed green. As they are also more tolerant of cold and rough seedbeds, they have a role in upland areas where the Australian system may be inapplicable.

A recent report (ECWA 1977A) states that in Syria a wheat-vetch rotation with a focus on private farms could cover an estimated potential of nearly one million hectares of good rainfed lands with favorable effects on wheat and livestock production as well as reducing seasonal underemployment. There is a similar potential in northern Iraq.

Although in one form or another there seems to be great scope for the

TABLE 14.6. ESTIMATES OF PRESENT FALLOW LANDS TO BE BROUGHT UNDER LEGUME PASTURES AND BENEFITS IN TERMS OF NITROGEN ADDED AND CROP YIELDS

Country	Dryland Under Wheat, Barley & Lentils (avg. 1966–70) 1000 ha	Average Production of Wheat Plus Barley 1966–70[3] tons	Estimates of Fallow to be Sown to Pasture 1000 ha	Nitrogen Added to Soil by Legume Pasture[1] tons	Additional Cereals Production[2] tons	Increase in Cereal Production over 1966–70 Average %	Additional Dry Matter Produced From Pasture[4] 1000 tons	Additional Sheep Carried[5] 1000 head
Jordan	277[6]	152,000	83	4,980	49,800	32.8	332	332
Syria	1,787	1,200,000	536	32,160	321,600	26.8	2,144	2,144
Lebanon	62		19	1			76	76
Morocco	3,746	3,615,000	1,124	67,440	674,400	18.7	4,496	4,496
Algeria	2,635	1,647,000	791	47,460	474,600	28.8	3,164	3,164
Tunisia	1,140	472,000	342	20,520	205,200	43.4	1,368	1,368
Total	9,647	7,151,000	2,895	173,700	1,737,000	24.3	11,580	11,580

Source: Leeuwrik (1975).
[1]Assumed nitrogen is added at an average of 60 kg nitrogen per hectare per year.
[2]An average 10:1 conversion ratio of grain to nitrogen is assumed.
[3]Includes production from high rainfall and marginal areas. If the proportion of crops grown in marginal areas is high, this may lead to overestimation of percentage of increases.
[4]Assumes an average annual additional production of 4000 kg D.M./ha.
[5]Assumes that with proper management 1 ton D.M. will maintain 1 ewe equivalent for a year.
[6]Calculated from FAO Production Yearbook. Corresponds well with statistics from other sources.

replacement of fallow in milder areas where precipitation exceeds 350 mm, a substantial proportion of the winter cereals is produced from drier and/or colder areas. Here it is generally accepted that fallow is essential for moisture conservation; but conventional tillage practices and poor weed control often defeat this objective, while high risk inhibits the use of fertilizer nitrogen to compensate for the absence of legumes in the rotation.

Again there are both technical and social problems. French and Italian colonists in North Africa adopted a system of clean fallowing, using heavy mechanized equipment and deep moldboard ploughing. This permitted timely sowing and controlled weeds, thus enabling reasonably good yields to be attained. However, these practices could not be followed by indigenous farmers lacking mechanical equipment, and needing the stubbles and fallows for grazing. Furthermore, the value of deep ploughing has proved controversial and mechanized cultivation has been pursued often with little regard for soil conservation.

New Technologies.—Recently in Turkey consistent improvements in yields have been demonstrated (especially under the difficult conditions in the Anatolian Plateau), as a result of the use of a relatively low cost "package" of practices based on winter fallow from harvest until the following spring, then sweep chisel ploughing to 18 to 20 cm and harrowing early (in March or April) to form a weed stubble mulch to retard erosion and reduce evaporation. Sweeping is continued at intervals through the summer to control weeds and by the first rains in autumn the seedbed is well prepared. Nitrogen is incorporated at 40 to 60 kg per hectare with the last pre-seeding sweep operation and 40 kg per hectare phosphate is drilled with the seed. A further top dressing of 20 kg of nitrogen can be applied the following spring, and herbicide is applied in spring to control weeds in the wheat. Comparisons between this system and normal cultivation practices (moldboard ploughing, tandem discing, broadcast seeding, grazing of weeds until early summer) on demonstration farms showed a ratio of increased benefits over increased cost of 5:1 (CIMMYT 1976). Yields averaged over 2000 kg per hectare quite consistently. This may be reflected in the impressive upturn in Turkish wheat yields since 1973. Similar results have been obtained in Jordan (Hopkinson 1974; ECWA 1977B). There timing and intensity of early rainfall are critical in influencing decisions about areas to be planted, crop varieties, and fertilizer applications.

In addition to attempts to introduce leguminous crops to replace fallow where rainfall exceeds 350 mm, therefore, attention is being concentrated (1) on genetic improvement of cereals to improve drought and cold tolerance and early maturity, (2) on cultural practices to reduce climatic insecurity, to improve moisture conservation, and to facilitate early sowing (such as deep furrow planting), and (3) on combinations of tillage with herbicides to improve weed control.

Since 1966 an increasing number of warm temperate countries have been introducing and testing genetic material derived initially from the work of the International Maize and Wheat Center (CIMMYT) in Mexico, and adapting it to their own use. Despite the great success achieved by the use of this material in Mexico, northern India, and Pakistan under climatic conditions which are warm temperate semi-arid if not strictly Mediterranean, their adoption by farmers and their impact on production in the Near East and North Africa have been somewhat patchy. Thus, by 1975, 15.7 million hectares were sown to the "high-yielding wheats" in 4 countries of South Asia, compared to 3.5 million hectares in 11 Near East/North African countries. The percentages of HYV wheat area were 64.4% and 13.4%, respectively.

There are several reasons for this relatively disappointing rate of adoption of a promising technological development. First, a large proportion of the wheat area in the Mediterranean and Near East countries is durum wheat grown for macaroni or semolina rather than bread wheat. Until fairly recently there were no semi-dwarf, high-yielding "Mexican" equivalents of durum wheat and although some breeding had been undertaken, notably in North Africa, many of the local varieties are long strawed and relatively late maturing. Their high quality and price premium over bread wheats compensate for their lower yield, but their response to fertilizer is much less and they are more susceptible to rust diseases. There also appears to be a negative correlation between yield and grain quality.

Second, much of the success of the bread wheats in Mexico, India, Pakistan, and increasingly in Bangladesh can be attributed to irrigation, and in South Asia particularly to the rapid spread of private tubewells and low-lift pumps. Nothing comparable has occurred in the Mediterranean and Near East, where the majority of the wheat is not irrigated. The high-yielding varieties have not demonstrated a conclusive superiority over local cultivars in rainfed areas.

Third, certain diseases—notably *Septoria tritici*—are much more serious in the Mediterranean area than in Mexico or South Asia. Fourth, much of the wheat in Spain, Algeria, Turkey, Iran, and Afghanistan is grown under much colder conditions than in Mexico or South Asia. The Mexican wheats are not cold tolerant, and both durum and bread wheat varieties are needed that will stand short cold periods, as well as those that are truly winter hardy (requiring vernalization to flower) for the high plateaus.

Most of what has been said of wheat is also true of barley. Some improved varieties have been produced; for example, in Cyprus and

Morocco some indigenous material has found its way to the U.S. and then returned after improvement (for example, the California selection of the Egyptian Mariout six-row barley). But much of the barley grown in the region is susceptible to mildew in more humid conditions, and so weak in the straw that very little fertilizer can be used before it lodges. It is in fact best suited to marginal areas too dry or infertile for wheat. Varieties imported from western Europe have proved to be too late maturing.

Clearly, climatic risk is an important reason for the relatively slow rate of adoption of modern varieties, herbicides, and other improved practices, the low rates of fertilizer use (Table 14.7), and the persistence of fallow. Undoubtedly cereal varieties eventually will be produced with greater tolerance of drought and cold, better response to fertilizer under more favorable conditions, and lower susceptibility to disease. Indeed, CIMMYT together with ICARDA (International Center for Agricultural Research in Dry Areas, with locations in Syria and Iran), the well-equipped Italian institute at Bari, and national stations in other countries of the Mediterranean are making significant progress in their development. The decision of the Consultative Group on International Agricultural Research to establish ICARDA at two locations reflects the differences between the needs of rainfed agriculture on the plateaux and the milder, low-land areas of the Mediterranean.

Modern varieties and other innovations discussed above require greater draft power, different types of cultivating equipment, seed drills, sprayers, herbicides, fodders to replace grazed weedy fallow, and other departures from traditional low-cost cereal culture. As indicated in Table 14.7 mechanized equipment is increasing. These developments need to be accompanied by intensive farmer education and demonstration campaigns, credit for purchase of equipment and recurrent inputs, price stabilization measures (including the provision of adequate storage), and crop insurance to cushion farmers against risk. Experience in Turkey, Syria, and Tunisia shows that the success of such measures depends greatly on total government commitment, backed by careful planning and organization including joint research and extension operations. It is also clear that they will demand greatly increased investment in rainfed agriculture than has been the case in the past, when the bulk of public funds has gone to irrigation.

Mixed Rainfed Farming

Mixed rainfed farming involves a wide variety of tree species, vines, and annual crops, including leguminous forage for sheep and cattle, and is typified by 5 different systems described in Table 14.2 for western

TABLE 14.7. USE OF PRODUCTION FACTORS, 1975, AND GROWTH RATES, 1961–1975, FOR MEDITERRANEAN AND NEAR EAST COUNTRIES

	Fertilizer Consumption (NPK)		Tractor Utilization		Harvester Threshers		Animal Units (Equivalents)[2]		
	kg/ha[1] Nutrients	Growth Rate %P.A.	Number of Units	Growth Rate %P.A.	Number of Units	Growth Rate %P.A.	Units per sq km[3]	Million head	Growth Rate % P.A.
Total		6.8	2,027,611	9.4	122,225	7.0	2.0	69.78	0.74
Greece	119	6.6	87,000	9.4	4,700	5.2	2.5	2.22	0.21
Spain	69	5.5	379,070	10.8	39,674	12.3	1.2	5.29	0.15
Yugoslavia	90	5.2	225,524	14.2	11,045	1.5	3.7	5.46	−0.09
Italy	121	4.0	819,334	7.5	27,774	9.6	4.4	8.36	−0.77
Portugal	68	3.7	44,452	10.8	3,815	16.1	3.3	1.30	0.65
Egypt	175	5.0	21,500	3.9	—	—	15.1	3.25	2.6
Iraq	6	24.1	20,222	12.4	5,084	6.1	4.4	2.92	3.19
Israel	171	6.1	21,650	7.4	520	−4.9	2.1	0.23	1.44
Cyprus	61	4.3	9,800	7.9	300	−1.8	2.0	0.08	1.87
Iran	21	24.6	29,000	8.2	2,150	7.0	3.6	8.89	1.05
Lebanon	37	−2.3	3,000	6.2	90	5.6	3.5	0.15	−1.53
Saudi Arabia	12	6.3	950	11.9	280	8.5	3.6	1.07	2.31
Afghanistan	4	33.8	700	11.3	—	—	0.2	5.13	0.00
Algeria	23	10.4	51,000	5.2	7,000	3.1	0.5	1.64	2.72
Jordan	4	5.3	3,400	8.1	190	11.4	1.3	0.20	−0.84
Libya	14	19.8	4,250	3.4	—	—	0.7	0.64	0.22
Morocco	21	12.4	20,000	7.9	2,700	−0.4	2.6	4.22	1.98
Syria	12	11.3	15,303	7.7	1,812	2.3	0.8	0.85	1.77
Tunisia	11	7.8	29,000	7.8	3,250	1.5	1.7	1.00	2.24
Turkey	30	19.4	242,456	14.4	11,841	5.5	3.2	16.87	0.34

Sources: FAO (1976A, 1976C).
[1]Per ha arable land and permanent crops.
[2]Animal unit equivalents: camel=1.1; buffalo=1.0; cattle=0.8; sheep and goats=0.1.
[3]Includes arable, permanent crops, and permanent pasture.

Greece: (1) cereal livestock association (75% or more of arable area in cereal/fodder/fallow rotation); (2) mixed farming (at least 3 major crop groups, none except cereals occupying over 30% of arable area); (3) industrial or vegetable crops (occupying 40% or more of arable area); (4) intensive arable farming (industrial, vegetable, vines, and fruit); and (5) vines and plantations (occupying 40% or more of arable land). These types of varied land use patterns are most commonly found in the more humid and less extreme climatic areas of the Mediterranean climatic zone, especially in the mountains and valleys of northern and eastern Mediterranean and the larger islands such as Cyprus, Crete, Cephalonia, Corfu, Corsica, Sicily, and Sardinia.

Because these areas have been isolated both geographically and by poor internal communications, semi-subsistence types of economy have arisen with a considerable proportion of each family's produce being retained for its own use, and with small, inefficient milling, olive oil, wine, and cheese processing industries geared to serving small local communities.

Further consequences of the geography, topography, and subsistence-oriented farming in such areas are a wide diversity of enterprises (the author has recorded 16 different species of fruit on one 2-ha holding), and a high degree of fragmentation of holdings and dispersion of plots within holdings. Situations where average farm size is under 5 ha, fragmented among 8 to 10 plots each under .5 ha, some of which may be several kilometers from the homestead, are widely reported in FAO (1965) and World Bank studies (1966A,B) from Greece, Turkey, Spain, and Morocco.

Perennial Crops.—In these areas of varied topography, soil and slope are as important as rainfall in determining farming patterns. Land use tends to be strongly influenced by altitude, with (1) forest interspersed by mountain pastures at higher elevations, (2) vines, olives, and fruit trees becoming increasingly important below 1000 m down to sea level, and (3) cereals, grain and forage legumes (especially alfalfa) being grown on terraces with perennial crops or on lower slopes and valley bottoms. Forest tree species, especially pines, evergreen oaks, and chestnuts are also common on steeper slopes at intermediate altitudes. Sclerophyllous scrub (variously described as maquis, garrigue, matorral, or chaparral) is widely distributed at lower altitudes where temperatures are higher, especially where soils are thin (limestone) or unstable (conglomerates). The extent, biological composition, and history of degradation of the Mediterranean forest and maquis is well described by UNESCO (1977) and by Naveh and Dan (1973).

An important point of difference in the use of natural vegetation between the Mediterranean Basin and many other regions of the world is the extent to which it is harvested for products other than timber and firewood. These include (1) resin from pines, (2) oil from eucalyptus, (3) chestnuts, pistachios and other nuts and berries, (4) cork from *Quercus suber*, and (5) acorns for fattening pigs from this and other species of oak. Had this not been the case, the rate of destruction of the forest and maquis might have been even faster.

Among the cultivated perennial species in this mixed economy, vines and olives are of crucial importance. Both are grown as monocultures in parts of the Mediterranean, sometimes on a large scale—olives in Portugal, Spain, Italy, Greece, Turkey, Syria, Tunisia, and Morocco, and vines in Portugal, Spain, Italy, Yugoslavia, Greece, Algeria, and, to a lesser extent, Turkey, Tunisia, and Morocco. However, whereas specialized viticulture predominates in California, Australia, South Africa, Chile, and Argentina, both vines and olives are widely grown on small farms in scattered plots throughout the countries surrounding the Mediterranean Basin. Other fruit trees (figs, apricots, almonds, peaches) are both grown alone and in mixed plots; while in some areas (for example, in Jordan, Libya, and Tunisia), cereals or other annual crops may be sown between widely spaced rows of trees (usually olives or almonds). In the drier areas olives with intercrops may be irrigated, more rarely so in the northern countries where rainfall is higher and more reliable. Vineyards, except in the more remote areas, are rarely intercropped because this compounds the difficulties of pruning, inter-row cultivation for weed control, spraying, and harvesting. In the wetter areas, forage (clover, alfalfa, vetches, lupins, and, very rarely, grasses) may be sown under trees for dairy cattle, sheep, or goats; other legumes such as broadbeans (*Vicia faba*), chickpeas, or lentils may be grown for human consumption.

Fragmentation of Farms.—Although gross income per hectare is generally higher from mixed farming than from cereal culture, because of the greater range and value of commodities which can be produced, small farm size and fragmentation of holdings often result in higher costs and very low farm income. While no simple prescription fits the diverse systems found throughout the Mediterranean, action to improve agrarian structure may well be the first priority to improve the economic situation of these small farmers. Thomson (1963) cites one case of a farmer whose annual travel distance between plots averaged almost 5000 km, and another, where a 20-minute spraying operation required several hours travel. The U.N. Special Fund Regional survey for the Peloponnesus singles out fragmentation as the principal limiting factor to agricultural expansion in western Greece.

Attempts to rectify this problem have been made in several Mediterranean countries, as well as in Chile where a somewhat similar small-farm problem has developed due to population pressures on land in the coastal mountains. These efforts have ranged from government-assisted voluntary consolidation efforts in Spain and Greece and pension schemes to entice farmers from the land in southern France, to more draconian land reform measures in Chile and Iran, including legal minimum and maximum holding sizes and redistribution of the large holdings, and the establishment of state farms or agrarian cooperatives in North Africa and Israel.

A major impediment is inadequate cadastral survey. Disputes over land ownership and tenant rights are frequent, and are aggravated by lack of title registration and complicated inheritance laws. An FAO (1966A) study showed that 40% of the land ownership in the Antalya region of western Turkey was in dispute in the mid 1960s. Land reform can open up opportunities for technical progress, but it will not transform traditional agriculture unless the other conditions for such progress are also created!

Strategies for Improved Land Use.—The technical blueprint for improvement should be a careful study of existing land use, farm structure, and the composition of farm income as a basis for the development of an integrated plan for future land use—including provision of necessary social and economic infrastructures to ensure its viability. Rational resource management may require substantial shifts in historical patterns of land use imposed by isolation, which has led to cultivation of cereals and vines on steep slopes and at high altitudes where yields are low and costs of cultivation and harvesting are high, and to destruction of the natural cover by uncontrolled free range grazing. Rationalization implies specialization according to ecological comparative advantage, and may or may not require land consolidation and a more uniform cropping calendar before it can be successful. To the extent that irrigation or mechanization is the objective, or a reliable flow of produce to an economic sized modern processing or exporting plant is the objective, some form of aggregation of farming may be essential.

One example of an attempt to redesign the traditional production pattern in the mountain and foothill areas of southwestern Greece is shown in Fig. 14.3. The basic objective was to manage the resource in such a way that total income and employment from the land would be increased, and the forces of physical and social disintegration as a result of its destruction through mismanagement and erosion would be reversed. This would be achieved by forest and managed watersheds expanding at the expense of unimproved rangeland at higher altitudes, and marginal cereal land being replaced by improved pasture and terraced fruit trees or viticulture on intermediate slopes. Irrigation would be developed through drainage and empoundment in valleys and upland plains for fruit and vegetables, annual cash crops, and forage production or for quick growing trees such as poplars in wetter areas.

A combination of incentives and legislation was planned to bring about the desired changes; for example, free range grazing would be prohibited in the high forests and in areas planned for new forest plantations. But compensation for loss of grazing land would be paid, and grants would be given for purchase of improved animals for tethering, for pasture improvement, and for fodder crop cultivation. Assistance would be provided for cooperative management of flocks and herds, and for flock recording and improvement, as well as for marketing of dairy produce and meat, with appropriate price support to discourage slaughter at excessively light weights and provide incentives to fattening.

These measures would be accompanied by (1) the improvement of transport and communications; (2) the development of local market towns as poles for social events, improved schooling and health services, and agricultural servicing and processing industries; and (3) where possible, the stimulation of tourism. The various works associated with afforestation, terracing, torrent control, irrigation, drainage, transportation, and tourism would in themselves generate significant increases in income and employment. This has been discussed at some length to illustrate the complex nature of the problem, but basically similar combinations of social and technical innovations are being attempted in comparable areas of southern France, Turkey, Jordan, Syria, Morocco, Portugal, Spain, Tunisia, Italy, and Yugoslavia.

Irrigated Agriculture

The warm temperate regions are the cradle of irrigated agriculture and represent a microcosm of its problems. Irrigated perimeters vary from a few hectares fed by shallow wells or springs to vast projects covering hundreds of thousands of hectares. Every source of surface and groundwater resource is utilized, and all forms of delivery sytems and water distribution devices can be found: from Persian wheels, Archimedes screws, or other ancient waterlifting devices to sophisticated drip-feed and rotary sprinkler systems. The general principles for improving irrigated farming are discussed in Chapters 8 and 15.

The largest known ground water aquifer lies at the eastern edge of the warm temperate region, covering an estimated 18 million hectares in the Indus Basin. This supplies over 100,000 public and private tubewells and 200,000 Persian wheels, providing about 30% of the water for the 14 million hectares of crops in Pakistan. The remaining water is derived from one of the world's largest canal systems with a total length of 48,000 km (FAO 1977). In the Mediterranean area a much larger proportion of irrigation is from surface water, mainly from government

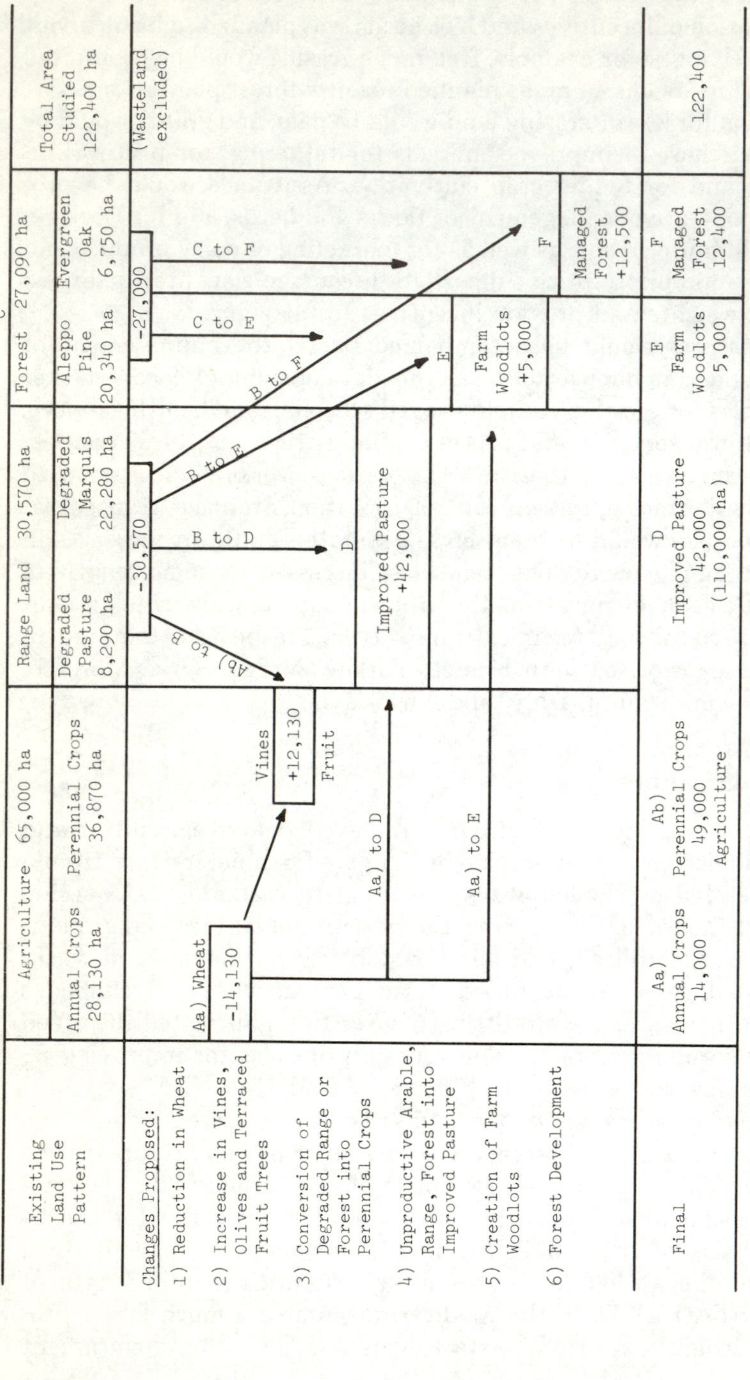

	A		B		C		
	Agriculture 65,000 ha		Range Land 30,570 ha		Forest 27,090 ha		
Existing Land Use Pattern	Annual Crops 28,130 ha	Perennial Crops 36,870 ha	Degraded Pasture 8,290 ha	Degraded Marquis 22,280 ha	Aleppo Pine 20,340 ha	Evergreen Oak 6,750 ha	Total Area Studied 122,400 ha
Final	Aa) Annual Crops 14,000	Ab) Perennial Crops 49,000 Agriculture	D Improved Pasture 42,000 (110,000 ha)		E Farm Woodlots 5,000	F Managed Forest 12,400	122,400

Changes Proposed:

1) Reduction in Wheat
2) Increase in Vines, Olives and Terraced Fruit Trees
3) Conversion of Degraded Range or Forest into Perennial Crops
4) Unproductive Arable, Range, Forest into Improved Pasture
5) Creation of Farm Woodlots
6) Forest Development

Aa) Wheat -14,130
Vines +12,130 Fruit
Aa) to D
Aa) to E
-30,570
Ab) to B
B to D
B to E
B to F
-27,090
C to E
C to F
D Improved Pasture +42,000
E Farm Woodlots +5,000
F Managed Forest +12,500
(Wasteland excluded)

Example taken from actual land-use survey data in typical coastal foothill area, southwestern Greece. Conversion of 14,130 ha of poor quality wheat-land; 30,570 ha of degraded pasture and scrub; and 27,090 ha of degraded forest to provide an additional 12,130 ha of vines, olives, and fruit; 42,000 ha of improved pasture (≡ 20,000 animal units) 5,000 ha of farm woodlots for shelter, firewood, and timber, and 12,500 ha of managed productive forest.

FIG. 14.3. MIXED AGRICULTURAL SYSTEMS-FOOTHILL AREAS: IMPROVEMENT THROUGH SHIFTS IN LAND USE

projects, and often derived from impounding rivers. In some countries private irrigation is important; for example, it amounted to 780,000 ha out of a total of roughly 2 million hectares of irrigated land in Spain in 1965. In a few cases, such as Libya, most of the area is irrigated from wells. However, in its Indicative World Plan, the FAO (1966B) does not foresee ground water playing a major role in the Near East or Mediterranean as compared to South Asia.

Changes in cropping patterns as a result of the pull of export markets in the industrialized countries of eastern and western Europe have led both to increased diversification of land use and in some areas to greater specialization, particularly with respect to citrus and horticultural crops. Better water control, new varieties with shorter maturity, and the use of plastics to accelerate growth and restrict weeds have contributed to these trends.

Nevertheless, there are greater risks as well as potentially higher profits in growing crops which are perishable and subject to sharp seasonal gluts and price fluctuations outside the control of producers and even of their countries. Some efforts have been made to alleviate these problems by the development of local freezing and canning plants. Governments have intervened to ensure stable food supplies for local markets by manipulating prices, subsidies, area and input quotas, and other measures aimed at providing cheap food to urban consumers or creating employment in agro-industries as much as at protecting farm incomes. Were it not for these influences, the share of high income crops relative to cereals in irrigated farming systems in the Mediterranean would probably be much higher, as it is in California.

Low Productivity of Farms.—Despite encouraging trends towards intensification and modernization of production, the productivity of irrigated agriculture is still very low in many warm temperate countries. According to FAO (1975), cropping intensities average only 57% for the countries of the Near East and North Africa with the exception of Egypt with a high figure of 168 and also high average yields of most crops.

The growth of irrigated area in the region is also slowing down. Some trends are shown in Table 14.1. FAO (1977) projects the rate of expansion from 1975 to 1990 to be about 1.5% per annum compared to an average of 3.5% between 1960 and 1975. This mainly reflects the rising costs of construction for dams, irrigation systems, and related infrastructure (now as high as $6000 per hectare for major reservoir-fed systems) and the fact that many of the better and more accessible sites are used up in an area with such a long history of irrigated agriculture. However, it may also indicate some disenchantment by governments and funding agencies with low returns to costly investments in water resource development, which in several countries has represented 70 to 80% of total public expenditure on agriculture.

Need for Improved Water Management.—Poor performance in irrigation projects in most Mediterranean and Near East countries compared, for example, to Egypt, Israel or California, is due to numerous factors shown schematically in Table 14.8. Essentially these include (1) weaknesses in planning, (2) weaknesses in design, and (3) weaknesses in management and implementation.

Faulty planning is often due to the use of irrigation mainly as a means of alleviating poverty and creating employment. This has led several countries to locate major projects in unfavorable ecological areas whose comparative advantage for the products they can grow is lower than for those same products in more favorable regions of the same or competing countries. Sometimes water has been spread too thin in an attempt to "benefit" as many people as possible or farm sizes have been established which are smaller than optimum for a decent living standard. Other consequences have been heavy subsidies on water (resulting in its wasteful use), and the establishment of inefficient high-cost processing industries linked to the projects and distortions of prices or subsidies which have affected the whole economy of a country in an attempt to make marginal producers viable. This is not to say that irrigation cannot or should not serve social objectives, as well as being a vehicle for increasing individual incomes or achieving national production or export targets, but there are pitfalls in attempting to do both which several governments in warm temperate regions have failed to avoid.

While these criticisms apply particularly to more recent projects, design weaknesses affect both modern and traditional systems, but more especially the latter. They include lack of adequate storage for year-round irrigation, canal systems which do not have the capacity to support high cropping intensities, inflexible water delivery to farms, inefficient land-shaping, and failure to extend irrigation and drainage systems to farms. Over-extending canal systems results not only in low irrigation intensities, but in high transportation losses and increased waterlogging and salinity.

Weaknesses in irrigation management and on-farm use of water are partly due to inadequate training of irrigation staff and farmers, but they are greatly compounded by weaknesses in planning and design. A common cause of failure is to devote insufficient resources to the institutions and services essential to the development of a modern, highly productive agricultural system. This applies both to the development and dissemination of new technology through research and extension, and to the provision of the inputs essential to sustain its application.

Future economic planning in many warm temperate areas will certainly have to give continuing high priority to investment in water development

TABLE 14.8. THE EXTENT TO WHICH VARIOUS CONSTRAINTS APPLY TO THE INTENSIFICATION OF IRRIGATED LAND-USE IN THE REGION

Constraints	Sudan	U.A.R.	Fertile Crescent	Central Iraq & Euphrates in Syria	Lower Iraq Khuzistan	Arabian Peninsula	Iran and Afghanistan		Total Area in Region Affected[3] %
							Traditional	Modern	
Seasonality of Flow	VS	VS	M	M	S	S	M	M	22
Drainage and Salinity	S	VS[1]	S	G	G	S	S	M	40
Climate (frost)	VS	VS	VS	S	M	VS	G	M	29
Capacity of Canals	S	VS	S	M	M	VS	G	VS	51
Plant Nutrients	VS	VS	G	G	G	G	G	G	55
Labor	M	VS	S(Syria)	S(Syria)	VS	VS	S	M	24
Draft Power	VS	VS	M	M	G	VS	G	M	55
Operating Capital on Farms	VS	S	S	G	G	G	G	G	61
Skills and Knowledge	VS[2]	VS	M	G	G	S	M	G	47

Source: FAO (1966B).
[1]Improvements in drainage in the U.A.R. contribute significantly to yield increases, but are judged as not being limited to intensification.
[2]In the Sudan the centrally managed Gezira and Managil schemes represent a very large percentage of total irrigated area.
[3]The figures in this column are weighted averages for the region (where for each country the area under irrigation is used as weight).

Key to the letters used in the table: G—great constraint S—slight constraint
M—moderate constraint VS—very slight or no constraint

as the main motor for change and growth in the rural sector, but especially to increasing its productivity rather than merely to expanding area. Groundwater resource development, the combination of ground and surface resources, the modernization of traditional systems, the increasing use of improved technology and high-yielding varieties, and the flexible use of drip-feed or sprinkler systems to increase efficiency in water use or to irrigate land which cannot be covered effectively by normal furrow or basin systems are all likely to play an increasingly important role compared to the construction of large, new reservoir-based systems.

A restraint both on the planning and implementation of water development in the southern Mediterranean and Near Eastern countries is the scarcity of trained manpower. FAO points to weaknesses in research and extension services and low levels of farmers' skills as a main reason for failure to use knowledge acquired over time, which if properly applied could have upgraded irrigated agriculture to a much greater extent than has actually been the case (FAO 1973). FAO has estimated a need for 35,600 skilled workers, 8400 technicians, 4500 extension workers, and 1100 professional staff to manage irrigation developments planned in the region during 1975 to 1990 (FAO 1977). Considerable strides have been made in recent years; but training, and especially training related to the improvement of on-farm water use and management, remains a high priority for a region where water scarcity is frequently the main constraint on agricultural progress.

THE NEW DYNAMISM IN THE MEDITERRANEAN REGION

The systems of land use described above were each logical within a given set of physical and socioeconomic parameters, otherwise they would have long since disappeared. However, in most Mediterranean countries they are no longer adequate in isolation from one another to meet the combined stress of unparalleled population and rising income growth. This is being reflected in increasing food deficits, greater fragmentation of holdings, rapid migration to the cities, and aggravated conflicts in land use. Yet, the conjunction of economic and social forces, which has created this situation, has also opened up unique opportunities to reshape farm structures and has provided new flexibility to repair the damage caused by centuries of agricultural stagnation and misuse of resources. The increasing availability of indigenous finance from sources such as the Arab and Kuwait funds, as well as from the World Bank, the European Investment Fund, and the new International Fund for Agricultural Development should provide the wherewithal to grasp those opportunities. And the rapid spread of literacy and higher education is raising the absorptive capacity of many countries to use the funds more efficiently.

This is not to say that the pressures described in this chapter can be relieved easily, especially in the poorest countries and more difficult environments; the constraints imposed by climate, topography and the limited scope for the expansion of cultivated area are too severe. Success will depend on improving yields and yield stability of crops and livestock, on intensifying land use, and, perhaps most of all, on developing complementarities between land use systems in different agroclimatic zones which will maximize their comparative advantages and replace the conflicts of interest which have been so destructive in the past by symbiosis.

The evolution of agriculture in the warm temperate areas remote from the Mediterranean Basin, as well as in some countries in that region in recent years, suggests that, despite the constraints, much can be done both to raise the productivity of the traditional farming systems and to introduce new methods. The possibilities have been demonstrated and the conditions for dynamic change exist.

REFERENCES

ALAD. 1974. Report of the regional sheep and forage production workshop. Arid Lands Agricultural Development Center, The Ford Foundation, Beirut, Lebanon. Feb. 11-14.

ANON. 1961. Agricultural Census. National Statistical Service, Athens, Greece.

ASCHMANN, H. 1973. Distribution and peculiarity of Mediterranean ecosystems. *In* Ecological Studies Vol 7. F. di Castri and H.A. Mooney (Editors). Chapman and Hall, Ltd., London.

AXELROD, D.I. 1973. History of the Mediterranean ecosystem in California. *In* Ecological Studies Vol. 7. F. di Castri and H.A. Mooney (Editors). Chapman and Hall, Ltd., London.

BAGNOULS, F., and GAUSSEN, H. 1957. Biological climates and their classification. Ann. Geogr. *355*,193-220. (French)

CIMMYT. 1976. Turkey's wheat research and training project. CIMMYT Today *6*. International Maize and Wheat Improvement Center, Mexico.

DALRYMPLE, D.G. 1976. Development and spread of high-yielding varieties of wheat and rice in the less developed nations. Foreign Agric. Rep. *95*. Econ. Res. Ser. USDA/USAID, Washington, D.C.

DE BRICHAMBAUT, G., and WALLEN, C. 1963. A study of agroclimatology in semi-arid and arid zones of the Near East. Tech. Note. *56*. WMO, Geneva.

DEMIRUREN, A.S. 1972. Sheep production potential in the Eastern Mediterranean. World Anim. Rev. *2*.

DOOLETTE, J.B. 1975. The strategy of establishing a crop rotation program using annual forage legumes. Third Regional Wheat Workshop, Tunis, Apr.

28-May 2, 243-253. Sponsored by FAO, Rockefeller Foundation, CIMMYT, Ford Foundation and the Cereal Board of Tunisia. CIMMYT, Mexico.

DRAZ, O. 1974. Management and rehabilitation of the natural grazing lands. Proceedings, ALAD Regional Sheep and Forage Production Workshop, Beirut, Feb. 11-14.

ECWA. 1977A. Short-term possibilities for increasing food production in selected countries of the ECWA region. Sub. Prog. *3.* U.N. Economic Commission for West Asia, Amman, Jordan.

ECWA. 1977B. A pilot study on food security planning—a case study of wheat in Jordan. U.N. Economic Commission for West Asia, Amman, Jordan.

EMBERGER, L. 1955. A biogeographic classification of climates. Recl. Trav. Lab. Bot. Geol. Zool. Univ. Montpellier 7, 3-43. (French)

FAO. 1965. Economic survey of the western Peloponnesus, Greece. *FAO/SF: 8/GRE.* United Nations Special Fund/FAO, Rome.

FAO. 1966A. Pre-investment surveys of the Antalya region, Turkey. *FAO/SF: 14/TUR.* United Nations Special Fund/FAO, Rome.

FAO. 1966B. Indicative world plan for agricultural development 1965-85. Near East subregional study No. 1,1. FAO, Rome.

FAO. 1973. Research review mission to the Near East and North Africa. Report to the Technical Advisory Committee of the Consultative Group on International Agricultural Research. FAO, Rome.

FAO. 1974. Trade Tape. FAO, Rome.

FAO. 1975. Land and water use program in the Near East and North Africa. Project document. FAO, Cairo.

FAO. 1976A. Annual Fertilizer Review. FAO, Rome.

FAO. 1976B. Production Tape. FAO, Rome.

FAO. 1976C. Production Yearbook, Vol. 30. FAO, Rome.

FAO. 1977. Water for agriculture. Prepared by FAO for U.N. World Water Conference. Mar del Plata, Argentina. FAO, Rome.

FAULKNER, D.E. 1972. Near East regional animal husbandry study. FAO, Cairo.

FOX, C.W. 1974. The ALAD program for sheep and forage improvement. Proceedings, ALAD Regional Sheep and Forage Production Workshop, Beirut, Feb. 11-14.

HOPKINSON, D. 1974. The development of dryland farming in Jordan. Proceedings, First FAO/SIDA Seminar on Improvement and Production of Field Food Crops for Plant Scientists from Africa and the Near East. Sept. 1-20, 1973. Cairo, *TF/INT 88 (SWE).* FAO, Rome.

IFPRI. 1976. Research Report *1.* International Food Policy Research Institute, Washington, D.C.

KÖPPEN, W., and GEIGER, H. 1936. Handbook of Climatology. Gerruder Borntrager, Berlin. (German)

LEEUWRIK, D.M. 1975. The relevance of the cereal-pasture rotation in the Middle East and the North African region. Proceedings, Third Regional Wheat

Workshop, Tunis, Apr. 28-May 2, 266-291. Sponsored by FAO, Rockefeller Foundation, CIMMYT, Ford Foundation and the Cereal Board of Tunisia. CIMMYT, Mexico.

LE HOUEROU, H., and HOSTE, C. 1977. Relationships between rangeland production and average annual rainfall. Int. Livestock Center for Africa (ILCA), Addis Ababa, Ethiopia.

NAVEH, Z., and DAN, J. 1973. The human degradation of Mediterranean landscapes in Israel. *In* Ecological Studies Vol. 7. F. di Castri and H.A. Mooney (Editors). Chapman and Hall, Ltd., London.

ORAM, P.A. 1956. Pastures and fodder crops in rotations in Mediterranean Agriculture. Agric. Dev. Pap. *56*. FAO, Rome.

ORAM, P.A. 1974. The potential for increasing sheep production in the Near East and North Africa through crop-livestock integration within a stratified system. Proceedings, Regional Sheep and Forage Production Workshop, Beirut, Feb. 11-14.

ORAM, P.A., and JONES, D.K. 1955. The agricultural revolution in Turkey. World Crops *7*, 137-142, 278-283.

PAKISTAN (FAO). 1977. Report of the seminar on agricultural perspective planning, Jan. 10-19. *ESP/PAK/75/003*. FAO/Govt. Pakistan, Islamabad. FAO, Rome.

PAPADAKIS, J. 1975. Climates of the world and their potentialities. J. Papadakis, Av. Cordoba 4564 Buenos Aires, Argentina.

PEREIRA, H.C. 1973. Land use and water resources. Cambridge University Press, England.

QUEZEL, P. 1977. Forests of the Mediterranean Basin. MAB Tech. Notes *2*. UNESCO, Paris.

SPECHT, R.L. 1973. Structure and functional response of ecosystems in the Mediterranean climate of Australia. *In* Ecological Studies Vol. 7. F. di Castri and H.A. Mooney (Editors). Chapman and Hall, Ltd., London.

THOMSON, K. 1963. Farm fragmentation in Greece. Res. Monograph *5*. Center for Economic Research, Athens, Greece.

TOMASELLI, R. 1977. Degradation of the Mediterranean maquis. MAB Tech. Notes *2*. UNESCO, Paris.

TROLL, C., and PAFFEN, KH. 1965. Seasonal climates of the earth. *In* World Maps of Climatology, 2nd Edition. H.E. Landsberg *et al.* (Editors). Springer-Verlag, New York.

UNESCO. 1977. Mediterranean maquis: ecology, conservation and management. MAB Tech. Notes *2*. UNESCO, Paris.

UNESCO-FAO. 1963. Bioclimatic map of the Mediterranean zone. Arid Zone Research XXI. UNESCO, Paris-FAO, Rome.

WORLD BANK. 1966A. The development of agriculture in Spain. International Bank for Reconstruction and Development and FAO, Rome. Washington, D.C.

WORLD BANK. 1966B. The Economic Development of Morocco. Johns Hopkins Press, Baltimore.

WORLD BANK. 1977. Meat consumption, production and trade in the Near East and North Africa. Commodity Pap. *26*. The World Bank, Washington, D.C.

WRIGHT, B.C. 1977. A brief review of the Turkish wheat research and training project. Wheat Research and Training Project. (CIMMYT-Government of Turkey). Ankara, Turkey.

ZINKE, P.J. 1973. Analogies between the soil and vegetation types of Italy, Greece and California. *In* Ecological Studies Vol. 7. F. di Castri and H.A. Mooney (Editors). Chapman and Hall, Ltd., London.

15

Irrigated Farming in Arid and Semi-Arid Temperate Zones

D. Wynne Thorne

The steady and at times dramatic increases in irrigated land are often due as much to the desire for stabilizing yield and ensuring against the disasters of drought as it is to the added potential for attaining increased yields. In the developed countries, however, the large investments required to establish irrigation systems necessitate that substantially higher yields be obtained than were achieved under rain-fed conditions.

The Irrigation Journal makes an annual survey of areas being irrigated by various methods and types of equipment in the U.S. The 1972 and 1976 data for the 17 western states are summarized in Table 15.1. This survey indicates that about 26% more land is being irrigated in these states than the 1974 Census of Agriculture indicates. About 42% of the irrigated land is in the 5 states having primarily warm temperate climates. Nearly 28% of the irrigated land is watered by some type of sprinkler equipment and about 15% of the gravity irrigated land has underground pipe and valve distribution systems. Much of the remaining has levelled land and water controlled by headgates and syphon tubes. The most dramatic increases in irrigated land since 1949 have been in the southern two-thirds of the Great Plains region, including Nebraska, Kansas, Oklahoma, and Texas. A 450% increase occurred in these states in the 27-year period.

The trend toward irrigation is not limited to arid regions. In the U.S. during the 30-year period from 1939 to 1969, irrigated farm land in the arid and semi-arid western states increased by 7,120,000 ha, or just over 100%. During the same period irrigated land in the other more humid states increased by 2,241,000 ha, or about 335%.

The most extensive irrigated areas in temperate zones are in North America, 23 million ha; the Near East, 18 million ha; and southern Europe, 12 million ha. There are additional areas in Turkey, Pakistan, Korea, Japan, and the temperate parts of China.

TABLE 15.1. IRRIGATED LAND IN THE SEVENTEEN WESTERN STATES, 1972 AND 1976

States	1972 1000 ha	1976 1000 ha
Arizona	464	466
California	3,515	3,685
Colorado	1,329	1,255
Idaho	1,559	1,653
Kansas	762	1,228
Montana	1,296	1,256
Nebraska	1,836	2,552
Nevada	527	586
New Mexico	418	433
North Dakota	39	42
Oklahoma	267	381
Oregon	746	793
South Dakota	86	120
Texas	3,443	3,524
Utah	567	772
Washington	608	654
Wyoming	721	741
Total 17 States	18,183	20,141
Total in Other States	2,047	2,908
Total United States	20,230	23,049

Source: Anon. (1972, 1976).

Problems of Water Supply

Irrigation has expanded to utilize most of the available water in the warm temperate parts of California, Arizona, New Mexico, and Texas and in the adjacent areas of Mexico. The waters of the Colorado and Rio Grande and other rivers of the area are almost fully used and litigation over the allocation of remaining waters and over increasing salinity from current and anticipated uses are matters of national concern. Also, underground waters have been developed extensively and in many areas, such as the Salt River Valley of Arizona and the High Plains area of Texas, the ground water level is falling to such an extent as to threaten long-continued use at present rates. New tube wells are being developed in much of the Northwest and in the central and northern parts of the High Plains, resulting in increased irrigated areas over Idaho, Oregon, Washington, and Nebraska.

In the cool temperate arid and semi-arid zones water supplies are being utilized largely in the Great Basin area of Utah and Nevada, and only modest increases in irrigated land can be expected in the upper Colorado River Basin areas of Wyoming, Colorado, Utah, and New Mexico. The diversion of Colorado River water to the central valleys of Arizona is still in development stages. The water situation is less restrictive in the Snake River/Columbia River region and in headwaters region of the Missouri River in Montana, Wyoming, Nebraska, and the Dakotas. Except for some remaining underground waters, however, the costs of storing and transporting unused surface waters is very high and the extent of future development for irrigation is being intensively debated.

The irrigation development picture in western U.S. is that of approaching the limits of available water supplies. Unless large regional or interregional water storage and transport schemes are developed, such as transporting large volumes of water southward from Alaska and Canada, the rate of expansion of irrigated land will be slowed and there will be increasing pressure on water users to attain higher levels of irrigation efficiency. Also, competitive demands for water for urban uses, industry, and recreation are increasing the costs of water and decreasing the supplies available for irrigation. Future development of energy resources in western U.S. such as electric steam generating plants and coal gasification could also divert large quantities of water away from irrigation.

Further complication is legislation and threatened litigation over water quality. Irrigation is the major consumer of water and return flows from irrigated lands are increasing the salinity of most major streams. Irrigators are facing limits on the quantities and nature of pollutants in waste or return-flow waters that enter ground waters or that enter interstate streams. These public concerns and legislation are broadening the responsibilities of irrigators and forcing many types of changes in water use and drainage.

CHANGES IN METHODS OF IRRIGATION

Events leading to changes in irrigation practice include those described in the previous section plus the increasing difficulty in employing experienced irrigators and the rising cost of irrigators and of all types of labor. These problems are forcing changes in irrigation systems so water can be applied more efficiently and uniformly, and so the quantity of labor can be reduced. The substantial shift to sprinkler-type systems in western U.S. is shown in the data of Table 15.2.

The manpower situation is changing in two major ways. First, there is a need for technical skills for designing, installing, and maintaining irrigation systems. In many cases large commercial farms are handling this by contract with consulting firms. These firms offer comprehensive services in designing and overseeing the installation of systems, the development and operation of irrigation schedules, determining the need for and supervising the application of fertilizers, and diagnosing and

TABLE 15.2. IRRIGATION METHODS AND EQUIPMENT IN THE SEVENTEEN WESTERN STATES HAVING PRIMARILY ARID AND SEMI-ARID WARM TEMPERATE OR COOL TEMPERATE CLIMATES, 1976

Irrigation Method	Warm Temperate States[1] 1000 ha	Cool Temperate States[2] 1000 ha	Total 17 Western States 1000 ha
Sprinkler Systems			
Tow line and side roll	465	994	1,459
Center pivot	340	983	1,323
Hand move	678	1,094	1,772
Solid set	134	70	204
Trickle or drip	61	2	63
Gun and other	38	71	109
Total Sprinkler	1,716	3,214	4,930
Gravity Systems			
Open ditch and syphons	2,961	2,607	5,568
Underground with valves	2,072	904	2,976
Gated pipe from source	1,602	273	1,875
Gravity, unclassified	—	4,792	4,792
Total Gravity	6,635	8,576	15,211
Total	8,351	11,790	20,141

Source: Anon. (1976).
[1]Includes the states of Arizona, California, New Mexico, Oklahoma, and Texas.
[2]Includes the remainder of the 17 western states.

prescribing treatments for the control of insects and diseases. Second, there is a decreasing need for technical skills in carrying out irrigation practices since rather simple schedules of turning the water on and off are being developed. Difficulties in finding dependable labor as needed is cited by many farmers as reason for investing in automated irrigation systems.

Modifications in facilities and practices to achieve the above goals are highly varied and are involved with nearly all types of irrigation methods. Some of the types and modifications in irrigation methods include the following.

Surface Irrigation

Irrigation in furrows, borders, or basins is still the most extensively used method of irrigation (Table 15.2). Through government assistance and private initiative, improvements are being made continually through land levelling or grading, furrowing, and bedding the soil. These are accompanied usually by improvements in ditches, controls for water entering fields, and in other facilities. One alternative being adopted by some farmers is levelling fields to zero slope and filling basins, borders, or furrows in these flat fields as rapidly as possible to attain uniform depth in soil wetting.

Another development is automation in handling water. Headgates are being constructed to be automatically raised or lowered with time clock controls so the water will move from one location to another at prescribed intervals. Such headgates are operated by electricity or air pressure, and can be activated by clocks or soil moisture sensors (Haise *et al.* 1969). Self-starting syphons are an added technology that makes automatic turning of water practical where the water is in turn syphoned out into furrows. Gated pipe and regulating spiles are also useful in such situations.

The use of field or tail-water reservoirs to catch and retain water flowing from ends of furrows and borders is helpful in improving water application efficiency. The water is either pumped back to the head ditch or is reused in irrigation at a lower level.

Sprinkler Irrigation

Over the past two decades there has been a significant shift to sprinkler types of irrigation. Between 1972 and 1976 the area irrigated by sprinkler systems in the 17 western states increased from 3,439,000 ha to 4,930,000 ha, or by 43% (Table 15.2). Equipment modifications and improvements are frequent. If properly designed and managed, most types of sprinkler systems can be used to attain water application efficiencies of 70 to 80%. The principal modifications in sprinkler equipment are, therefore, directed toward lower costs, convenience, and reduced labor and maintenance. Fertilizer materials and other agricultural chemicals can be applied in irrigation water through sprinkler equipment. Four major types of sprinkler systems are briefly described in Chapter 8.

The costs of installing, maintaining, and operating irrigation facilities are substantial. The operating costs for 4 systems under Nebraska conditions in 1976 are itemized in Table 15.3. Fixed costs remain the same each year, regardless of the amount of time the system is used. Such costs include interest, depreciation, and insurance. Operating costs vary year by year depending on the number of hours of operation. The systems shown in Table 15.3 include conventional gated pipe without reuse of tail water, such systems with reuse, an automated gated pipe system with reuse, and a center pivot system. The expected life of a center pivot system is usually considered to be 10 to 15 years. Stationary systems may last 25 years. Difficulties in obtaining reliable labor and high water application efficiencies often make the automated gated pipe and center pivot systems attractive. As fuel costs increase, the high pressure center pivot and related systems may become less attractive.

The development of irrigation in semi-arid and humid zones is increas-

TABLE 15.3. ESTIMATED OPERATING COSTS AND TOTAL COSTS PER HECTARE FOR IRRIGATION WITH VARIOUS SYSTEMS IN NEBRASKA IN 1976

	Gated Pipe Without Reuse	Gated Pipe With Reuse	Automatic Gated Pipe With Reuse	Center Pivot
Water Applied/year, cm	43	36	30	30
Water Application Efficiency	60	75	85	85
Operating Pressure, bars	0.55	0.55	0.55	4.83
Fuel, 38¢/gallon[1]	$ 19.22	$ 17.30[2]	$ 14.83[2]	$ 30.32[2]
Oil, $2.30/gallon	3.13	1.75	1.51	3.36
Maintenance and Repairs	6.40	6.77	7.78	
Labor, $4.00/hour	10.20	10.20	4.94	4.94
Total Operating Costs/ha/year	$ 37.95	$ 36.02	$ 29.06	$ 40.02
Total Fixed Costs/ha/year	62.44	66.67	108.67	118.06
Total Irrigation Costs/ha/year	$100.39	$102.69	$137.73	$167.08

Source: Eisenhauer and Fishbach (1977).
[1]Based on 27.4 m (100 ft) well lift and pumping plant operating at Nebraska Standard.
[2]Includes cost of pumping reuse water at 3¢/kw-hr.

ingly an economic issue. This is illustrated by the data of Sheffield (1977) in Nebraska in which the costs of producing irrigated maize was $677.18 per hectare (Table 15.4), and the cost per 100 kg varied by 60% depending on the yield obtained. Obviously, the relative advantage from irrigation rests with securing high and stable yields.

TABLE 15.4. RELATIONSHIPS BETWEEN YIELD AND BREAK-EVEN PRICE FOR MAIZE IN NEBRASKA IN 1976

Yield		Break-even Price	
kg/ha	bu/a	Per 100 kg	Per bu
7,847	125	$4.38	$2.19
8,475	135	4.06	2.03
9,416	150	3.66	1.83
12,555	200	2.74	1.37

Source: Sheffield (1977).
NOTE: Total cost of producing irrigated maize = $677.18/ha ($274.05/acre). Includes land cost of $1,112/ha ($450/acre), with return computed at 6%.

CROPPING SYSTEMS

As with many other types of agriculture in western U.S., the general trend in cropping patterns has been toward specialization and varying degrees of monoculture. Areas in the Northwest have concentrated on apples or other fruits or nuts. Relatively frost-free areas in California and Arizona have developed an intensive citrus culture. Regions with long growing seasons in California, Arizona, and northern Mexico have specialized in melon and vegetable crops. Cotton has also become a major cash crop in these and adjacent areas. The mountain and range states have featured hay, feed grains, and silage to supplement grazing on the open range lands. In the Great Plains region maize for grain, sorghum, alfalfa hay, and some small grains are popular choices. In the Texas Panhandle cotton, sorghum, and wheat are widely grown under irrigation.

While the trend has been away from repeated crop sequences in fixed rotations, a somewhat flexible cropping system with a limited number of crops is usually adhered to, particularly on farms emphasizing livestock. Alfalfa, because of bacterial wilt and other diseases, is commonly grown no longer than 3 to 5 years on irrigated land. It is often followed by maize for silage or grain or potatoes. A small grain such as barley is commonly used as a nurse crop for establishing alfalfa. Sugar beets is also a common crop in such sequences. Potatoes and sugar beets are usually grown in rotations with other crops that do not favor their disease-causing organisms.

Economics is a major factor in the selection of crops. Prices paid for potatoes, onions, dry beans, cotton, and other crops may change significantly from year to year. In consequence, the planted areas of these and other crops fluctuate appreciably from year to year depending on the price or the anticipated market. After having purchased expensive specialized equipment for handling a crop there may be substantial costs, however, in shifting to another crop with different equipment requirements. Such considerations restrain changes by many farmers.

Multiple cropping is primarily limited to the warm temperate climatic zone. One of the most common combinations is the growth of wheat during the winter months and maize or sorghum in summer. Because of its long growing season, cotton is usually not grown in multiple cropping sequences except for a winter cover crop such as vetch and oats that can be grazed and plowed under early in the spring.

The diversity of crops grown on irrigated farms in the warm temperate climatic zone is tremendous. California is reported to grow 200 different crops. Many of these have restrictive requirements for climate, soil, water, and farm machinery. Many also need unique handling and processing treatments before being marketable. Such factors favor numerous rather small areas devoted to somewhat unique crops.

In some areas soil and water conditions restrict the choice of crops. Salinity affects large areas of irrigated land. In western U.S. relatively salt tolerant crops such as barley, alfalfa, sugar beets, and cotton are favored for such situations. Farms obtaining irrigation water from the direct diversion of mountain streams often have greatly reduced flows during the latter part of the growing season. Under such conditions early maturing crops such as barley and wheat are commonly grown. Forage crops such as alfalfa are also adapted to limited water because the deep rooted systems enable the plants to survive relatively severe drought and to resume growth when soil moisture is replenished.

Fruit production is a major enterprise in many parts of the region. According to the 1974 Census of Agriculture, California has about 720,000 ha of orchards. This is essentially all on irrigated land and is about equally divided between deciduous fruits and citrus, avocados, and vineyards. Arizona and the Rio Grande Valley of Texas also specialize in citrus. California, Washington, and Oregon and favorable climatic areas in Idaho, Colorado, and Utah feature deciduous fruits. Often frost and other weather variables are major hazards to fruit production.

CROP PRODUCTION SYSTEMS

Along with improvements in irrigation technology has been a general improvement in skills in utilizing other inputs to maximize yields and give higher production per unit of water.

Fertilizer

Fertilizer applications have been identified as the treatment factor contributing most to increased crop yields. With adequate well-managed irrigation, fertilizer needs can be fairly accurately estimated based on soil tests and a knowledge of crop requirements.

For many crops such as sugar beets, maize, sorghum, fruit trees, and cotton, annual fertilizer applications often exceed 200 kg of nitrogen per hectare. Most often phosphate and potash materials and all, or in heavy treatments about two-thirds, of the nitrogen, are plowed under or worked into the soil before planting. Supplemental treatments with nitrogen are side-dressed or added in irrigation water, particularly if sprinkler or other carefully controlled irrigation systems are used. The least expensive nitrogen material is anhydrous ammonia. It can be injected directly into the soil or applied in irrigation water. Application in irrigation water requires monitoring to determine that the pH of the water does not exceed about 8.5. At about this point losses of ammonia by volatilization may become excessive.

Other nutrients such as zinc, manganese, and boron are also frequently deficient in western soils. The small quantities needed can often be mixed with other fertilizer materials. They often give better results, however, with foliage applications. They can be added in sprinkler irrigation water near the end of the irrigation treatment, or they can be applied with pesticides by special spray treatments.

Crop Varieties

Crop varieties are carefully selected for irrigated agriculture. The first semi-dwarf wheats (Gaines and New-Gaines) were developed in Washington for high yielding potential without lodging under irrigation. New varieties of maize, sorghum, barley, cotton, and many other crops also have been bred for high yielding potential under irrigation. Many of such varieties have limited areas of adaptation so selection should be made on the basis of local tests.

Plant Populations

With adequate water and high soil fertility, higher plant populations can be attained through either closer row spacings or through higher densities of plants in the row. Thus, the number of plants per hectare in maize has been doubled to 85,000 per hectare or more under favorable conditions with substantial increases in total yields.

Weed Control

Weed control is especially important under irrigated conditions. In addition to weed seeds carried over in the soil from previous seasons, seeds are often brought in or scattered around a farm in irrigation water unless precautions are taken to screen or filter them out. Herbicides are extensively used in irrigated farming. The herbicide materials and methods and rates of treatment must be carefully designed to be compatible with both current and succeeding crops and with climate and other conditions.

Cultivation and crop rotations are also important in weed control. In general, however, repeated cultivations for weed control are avoided because of costs and soil compaction from tractor wheels.

Insect and Disease Control

The control of such pests is a necessary part of crop production programs. Since insects carry diseases as well as feeding directly on crops, they receive special attention. In such crops as fruits and cotton and a number of the vegetables, insect control is one of the most significant and difficult management practices in terms of producing marketable products. The pests involved are so numerous and of such critical importance that public officials such as County Farm Advisers and special private consultants are often employed to keep farmers advised of pending problems and of actions to be taken.

MAJOR FACTORS LIMITING CROP YIELDS

Limiting factors for crop yields vary appreciably among the various farming areas. In consequence, no priority list can be given that would be true for all of western U.S. Several items can, however, be identified as of broad importance.

Salt Problems

Salinity and associated sodic soil problems are estimated to affect crop yields on as much as 25% of the irrigated land in the 17 western states. Also, 15% of the irrigated land in the Alberta Province of Canada is similarly affected. These problems are receiving increased public attention. The principles involved in reclaiming and managing these problem soils are discussed in Chapter 8.

Currently the problems involved in managing salt in surface waters and return flows from irrigated lands are receiving intensive study in the U.S., but no general solutions have been found. The principles of reclamation of salty soils are well understood; the difficult problem for the future may be the disposal of salts in drainage waters.

Water Supply

Water supply and efficient management is a general problem and affects all irrigated farms to some degree. Probably more than half of all irrigated farms suffer reduced yields from lack of skilled management of irrigation waters. Water may be applied unevenly over the fields. The scheduling of irrigation is often not properly timed to keep crops in vigorous states of growth. Frequently too much water is applied, particularly with surface methods, causing leaching of nutrients and wasting of water. Water application efficiencies on farms are often as low as 30 to 50%. Seepage from unlined canals and ditches wastes water and contributes to soil waterlogging and salinity.

Irrigation Scheduling

Proper scheduling of irrigation is particularly important to achieve high yields when water supplies are inadequate to keep the soil well supplied with available moisture throughout the crop growth season. This is illustrated by a field irrigation experiment conducted in the High Plains of Texas (Shipley and Regier 1975). In this two-year test 1, 2, 3, and 4 irrigations per season were applied in various combinations to land planted with sorghum. The timing of irrigation was based on the physiological stage of crop development. Some of the data obtained are shown in Table 15.5. The tests were conducted in 1969 and repeated in 1972. All of the plots were given one preplant irrigation. The most effective irrigations were those applied at mid to late boot stage and at the heading and booting stage. Two irrigations properly timed resulted in nearly as high yields as the field with four irrigations. Many crops have comparable stages of growth at which moisture stress is particularly harmful to the harvested yields.

TABLE 15.5. SUMMARY OF YIELDS AND IRRIGATION TREATMENT DATA OF GRAIN SORGHUM RESPONSE TO IRRIGATION APPLICATIONS AT DIFFERENT STAGES OF GROWTH

Irrigation Application at Stages of Growth				Grain Sorghum Yield, kg/ha		
6–8 Leaf	Mid to Late Boot	Heading and Flowering	Milk to Soft Dough	1969	1972	2-year Average
Preplant only				1615	3123	2369
Preplant plus one 10 cm irrigation						
X	—	—	—	896	3186	2041
—	X	—	—	4505	4763	4634
—	—	X	—	3550	5502	4526
—	—	—	X	1279	3663	2471
Preplant plus two 10 cm irrigations						
X	X	—	—	4102	4380	4241
X	—	X	—	4687	6401	5544
X	—	—	X	1412	4709	3060
—	X	X	—	5871	6257	6064
Preplant plus three 10 cm irrigations						
X	X	X	—	7170	6715	6942
X	X	—	X	4166	6247	5206
—	X	X	X	6677	6681	6679
Preplant plus four 10 cm irrigations						
X	X	X	X	7623	7603	7613

Source: Shipley and Regier (1975).

Excessive pumping from underground water often exceeds the rate of annual recharge and the resulting falling water tables are making pumping costs marginal for irrigation in some areas. These problems of inadequate water supply are increasing in severity in parts of the Great Plains, in the Salt River Valley of Arizona, and in other restricted areas.

Soil Fertility

Soil fertility maintenance is a varied problem. Fertilizer materials and soil and crop testing services are available. The problems are most often with the farmer. Either he lacks capital for investing in plant nutrients or he is not motivated toward achieving high production levels. Often there is lack of managerial capability for putting the many production factors together in an effective management system.

Soil Physical Properties

Soil physical properties often reduce plant emergence, water movement, and root development. Crusting associated with irrigation or spring storms often interferes with seedling emergence. Soil compaction from tractor wheels has been repeatedly identified as contributing to reduced root and crop growth. Reduced tillage and varied spacing of crop rows, with irrigation being restricted to rows not traversed with tractor wheels, are examples of practices being adopted to reduce this problem. The conservation and management of crop residues is another factor that affects the physical condition of soils.

INTEGRATION OF CROP PRODUCTION PRACTICES

Several developments have assisted and promoted the integration of crop production factors at the farm level toward attaining efficient farm operations and high levels of production. The principles involved have been discussed throughout this book. The steps leading to the present state of farming in the U.S. and most developed countries are much the same.

Agricultural research is among the first and most important of modern innovations that have led to a productive agriculture. Initiated in the U.S. and several European countries as a public responsibility about 100 years ago, research findings have provided crop varieties with high quality and yield capabilities plus resistance to many insects and diseases. Research has provided a knowledge of soils and plant nutrition and the technology to produce and apply correct fertilizers as needed. It has generated principles of soil and water management. New chemicals and practices to control weeds, insects, and disease organisms have been evolved. Research has provided the knowledge and much of the technology on which modern farming is based.

Closely associated with research advances, and in part a product of research, has been the development and improvement of all types of farm machinery, including the internal combustion engine, the tractor, the moldboard plow, and the many other types of tillage equipment. Specialized harvest equipment ranging from the early grain reaper to the strawberry harvester has helped remove the drudgery of farming and has reduced the costs of agricultural products. The adoption of new mechanical equipment has been encouraged by shortages and increasingly higher costs of farm labor. Mechanization makes possible more timely planting, fertilizing, pest control treatments, and harvesting.

The approval and development of government supported agricultural extension programs with agents or advisers in every county has been a companion to research. These extension workers have taken new research findings to farmers and have demonstrated how they could be used to increase yields and make farming easier. A centralized staff of subject matter specialists keeps up-to-date on latest research and provides technical information to farmer advisers. They also work with industry representatives serving agriculture.

A related effort in the U.S. was the organization of the Soil Conservation Service. The first and primary mission of this agency was to control soil erosion. It became evident, however, that the causes of erosion and the means for its control were intertwined with nearly all farm practices. Even the adoption of special erosion control practices was found to be practical primarily when they became part of the total farm management and crop production system. From this concept evolved the principles and steps in farm planning. Farm planning emphasizes the uniqueness of each farm. It involves the development by the farmer, assisted by various specialists, of a plan to improve his farm facilities, to control erosion and other soil degradation problems, and to bring together the numerous factors necessary to an efficiently operated and highly productive farm.

Another incentive and source of great assistance has been the government farm support program. Through such assistance the government has provided a substantial part of the cost for many types of farm improvements, including land levelling, installing drains, and improvement in irrigation facilities. Financial assistance has also been given to help promote the adoption of new practices. Among such promotion schemes have been the application of phosphate fertilizer where needed, the liming of acid soils, the reclamation of saline and sodic soils, and the control of noxious weeds.

The nearby availability of farm supplies and equipment offered for sale by informed sales representatives who are able to help in assuring their proper use has been an added and essential part of the revolution in farming.

Yet in the final and ultimate sense, the farmer is the key to development. He makes the decisions. Without his cooperation, research, extension, farm planning, and the marketing of farm equipment and supplies would have no impact on the advancement of farm technology.

REFERENCES

ANON. 1972. Irrigation survey. Irrigation Survey Supplement. Irrig. J. *22*(6).

ANON. 1976. Irrigation survey. Irrigation Survey Supplement. Irrig. J. *26* (6).

AYERS, N.W., MENZIE, C., and JACOBS, J. 1977. Sulfuric acid in Arizona agriculture. Ariz. Agric. Exp. Stn. Tech. Bull. *231.*

CANTOR, L.M. 1970. A World Geography of Irrigation. Praeger Publishers, New York.

EISENHAUER, D.E., and FISHBACH, P.E. 1977. Cost of conventional and improved irrigation systems. Paper Presented at Irrigation Short Course, Lincoln, Nebraska, Jan. 24-25. Univ. Nebraska, Lincoln.

HAGOOD, M.A. 1972. Automated irrigation is here. Agric. Eng. *53*, 16-18.

HAISE, H.R., KUSE, E.G., and ERIL, L. 1969. Automating surface irrigation. Agric. Eng. *50*, 212-217.

PAIR, C.H., HINZ, W.W., REID, C., and FROST, K.R. 1969. Sprinkler Irrigation, 3rd Edition. Sprinkler Irrigation Association, Washington, D.C.

SECKLER, D.W., SECKLER, F.H., and DE REMER, E.D. 1971. Sprinkler irrigation. Agric. Eng. *52*, 374-376.

SHEFFIELD, L.F. 1977. The economics of irrigation. Irrig. J. *27*(1) 18-22.

SHIPLEY, J., and REGIER, C. 1975. Water response in the production of irrigated grain sorghum, High Plains of Texas. Tex. Agric. Exp. Stn. Misc. Publ. *1202.*

16

Crop Production Systems on Acid Soils In Humid Tropical America

John Nicholaides, III

Humid tropical America is one of the areas which offers mankind perhaps his best hope in the raging war between food supply and population growth. The largest continuous block of unused and potentially arable land in the world lies in humid tropical America on the vast savannas and forests of acid, highly weathered, low base status, infertile soils known as Oxisols and Ultisols. Potentially arable are 391 million hectares of Oxisols and Ultisols in humid tropical America alone. This figure represents 241 million more hectares than the entire harvested cropland in the U.S. in 1976.

Traditionally man has settled on the more fertile, higher base status soils, due in large part to their relatively higher agricultural production. Attesting to this fact is the consequent demographic density around the higher base status soils except in arid regions where water is the limiting factor to crop production. In humid tropical America, man presently is reaching the obvious limits for agricultural development in the areas of his traditional settlement on the coasts and on high base status mountainous soils. There is no question that mounting population pressures will force settlement of the acid, infertile soils of humid tropical America. In fact, this is occurring today and several Latin American countries have made priority commitments to develop agriculturally these areas. Less than 20 years ago, Brazil's capital city was moved from the coast to an Oxisol savanna location. A similar move for the capital city of another Latin American country to an Ultisol forest location is in the discussion stage.

Recent soil and water management research for increased crop production on these low base status soils of humid tropical America is laying to rest the myth that these soils are non-productive. Several U.S. universities, foreign, national, and international institutions are finding that with proper management these soils can be among the most productive in the world.

This chapter will center on the acid, infertile savannas and forests of humid tropical America since these areas dominate the region, are those currently under- or unutilized, and are those which offer the greatest potential for increased crop production through improved soil and water management. Certainly other soil expanses, such as the seasonally flooded plains and high base status mountain soils, are important, though they occupy smaller areas of humid tropical America; however, these will not be covered in this treatise. Also overlooked, by design, will be the plantation-type agriculture of sugar cane, bananas, coffee, etc. Chapter coverage will describe, as related to the Oxisol and Ultisol savannas and forests of humid tropical America, (1) the environment of the humid tropics, (2) land-clearing methods, (3) current agricultural production systems, (4) constraints to high yields, and (5) potential improvements in these systems which make this area one of the best hopes of mankind in the food versus population race.

ENVIRONMENT OF THE HUMID TROPICS

Geographic Area

The humid tropics are those areas of the tropics which have 7 or more humid months and receive more than 1500 mm precipitation annually (Fig. 16.1). This area covers approximately 50% of the tropics or roughly 2450 million hectares of land. In tropical America this humid region occupies about 780 million hectares, which is slightly greater than 50% of the land area. Within this area exists a range of climatic and soil conditions which affect the crop production systems employed.

Climate

Rainfall.—The moisture regimes of the humid tropics encompass the udic and part of the ustic soil moisture regimes. As defined by the Soil Survey Staff (1975), the udic soil moisture regime is in those areas where the rooting zone of the soil is dry for no more than 90 cumulative days during the year. The ustic environment occurs where the soil's rooting zone is dry for more than 90 but less than 180 cumulative days or 90 consecutive days during the year. The humid tropics generally do not suffer from moisture limitations, although erratic rainfall can occur, even in the udic environments, sufficient to restrict plant growth. Potential evapotranspiration usually is exceeded by rainfall, though the reverse can occur in some areas of the humid tropics (Fig. 16.2).

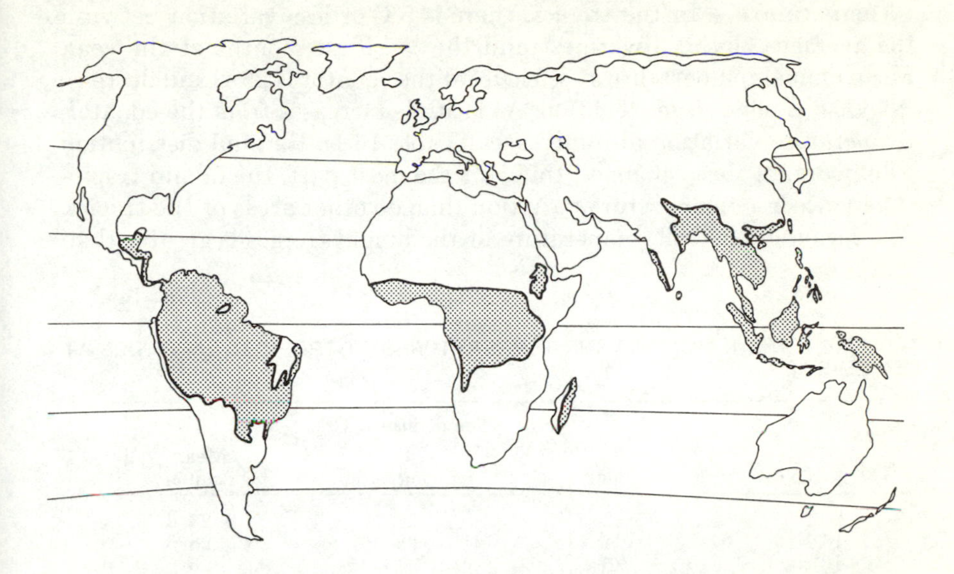

FIG. 16.1. MAP OF THE HUMID TROPICS AS DEFINED BY THOSE AREAS OF THE TROPICS WHICH RECEIVE 1500 mm RAINFALL AND HAVE 7 HUMID MONTHS

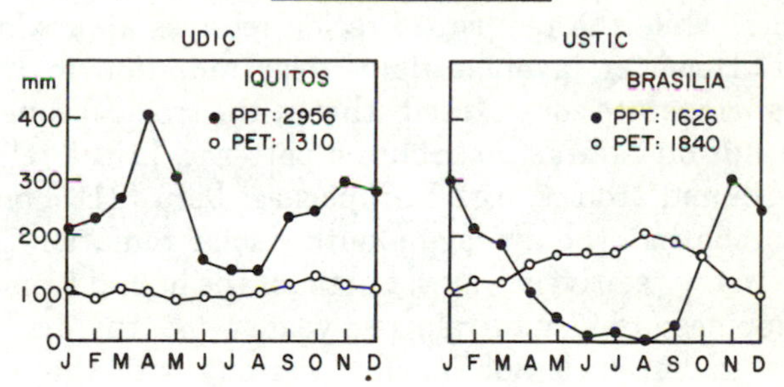

FIG. 16.2. AVERAGE MONTHLY PRECIPITATION (PPT) AND POTENTIAL EVAPOTRANSPIRATION (PET) OF UDIC AND USTIC SOIL MOISTURE REGIMES AT LOCATIONS IN PERU AND BRAZIL, RESPECTIVELY. ANNUAL TOTALS ARE NUMBERED

Temperature.—In the tropics, there is 5°C or less variation between the averages of the 3 warmest and the 3 coldest months of the year. Mean annual temperature at sea level at the equator is 26°C and decreases by 0.6°C per 100 m elevation. As latitude increases from the equator, temperature variation also increases (Table 16.1). Rainfall distribution will moderate these changes; thus, for the most part, the humid tropics experience less temperature variation than do other areas of the tropics. The mean annual soil temperature of the humid tropics is greater than 22°C.

TABLE 16.1. MEAN MONTHLY AIR TEMPERATURES AT TROPICAL LATITUDES AT SEA LEVEL (°C)

	Temperature (°C)		
Latitude	January	July	Mean Annual
20°N	22	28	26
15°N	24	28	26
10°N	26	27	26
5°N	26	24	26
0°	26	25	26
5°S	26	25	26
10°S	26	24	25
15°S	26	23	24
20°S	25	20	23

Source: Sanchez (1976).

Solar Radiation.—The average solar radiation is 400 langleys per day for the tropics, while the temperate region receives approximately half this amount. However, average daily solar radiation in the growing season of the temperate zone exceeds that of the tropical zone. The daily average depends on rainfall distribution patterns. In the relatively wet areas of the humid tropics, e.g., Yurimaguas, Peru (2137 mm rain per year, no dry months), the average monthly solar radiation is 306 langleys per day. In the relatively less wet areas of the humid tropics, e.g., Los Banos, Philippines (1847 mm rain per year, 4 months dry), the mean monthly solar radiation is 366 langleys per day with the dry and wet months respectively averaging 417 and 341 langleys per day. High year-round solar radiation when combined with sufficient rainfall and moderately high year-round temperatures stresses the important solution which agriculture in the humid tropics offers to the world food versus population battle.

Daylength.—Annual daylength variation ranges from zero at the equator to 2 hours and 50 minutes at 23.5° latitude. Daylength variation in the temperate regions is much greater. However, there is little correlation between solar radiation and day length in the tropics as contrasted with the temperate region.

Soils

The geology of the humid tropics of Latin America is influenced primarily by the Guyana and Brazilian Shields and the Orinoco, Amazon, and Paraná Basins (Fig. 16.3). The soils resulting from these geologic formations are of two main orders, Oxisols and Ultisols, both of which are characterized as being extremely weathered, acid, and infertile. The criteria for their classification are based on subsoil characteristics and may or may not reflect the surface soil characteristics.

Oxisols.—The Guyana and Brazilian Shields represent the oldest land surfaces in tropical America, having originated in the Archean and Paleozoic periods. Parent materials of these shields are granite, gneiss, and some sandstones. Oxisols are the most widespread soils originating from these shields, though other soil orders occur. These soils usually are deep and well-drained, having uniform properties with depth, firm granular structure, low fertility, and red or yellow color. The Oxisols' excellent physical properties allow cultivation almost immediately following rainfall; the granular soil structure permits rapid water infiltration and consequently these soils are not rapidly eroded except during periods of intense rainfall. However, available soil moisture contents of the Oxisols are lower than what would normally be indicated by their high clay content as compared with less weathered soils of temperate areas.

Ultisols.—The Orinoco and Amazon Basins derived their sediments from the shield areas and the Andean uplifts. Tertiary deposits with some recent alluvium cover these areas. The Paraná Basin was derived primarily from basalt deposits. All three basin areas are dominated by Ultisols though other soils are present. Generally, these soils are deep, do not have uniform properties with depth, are low in weatherable minerals and base saturation, have relatively low fertility, and are of red or yellow color. Ultisols have coarser textured surfaces and slower permeability than Oxisols.

Oxisols are the most widespread soils of the tropical world and tropical America, covering 22.5 and 43.3%, respectively. The Ultisols are the fourth and second most common of the tropical world and tropical America, covering 11.0 and 19.1%, respectively. When combined, the Oxisols and Ultisols dominate the humid tropics and especially humid tropical America (Fig. 16.4).

Though Oxisols and Ultisols dominate the humid tropics, other soils are

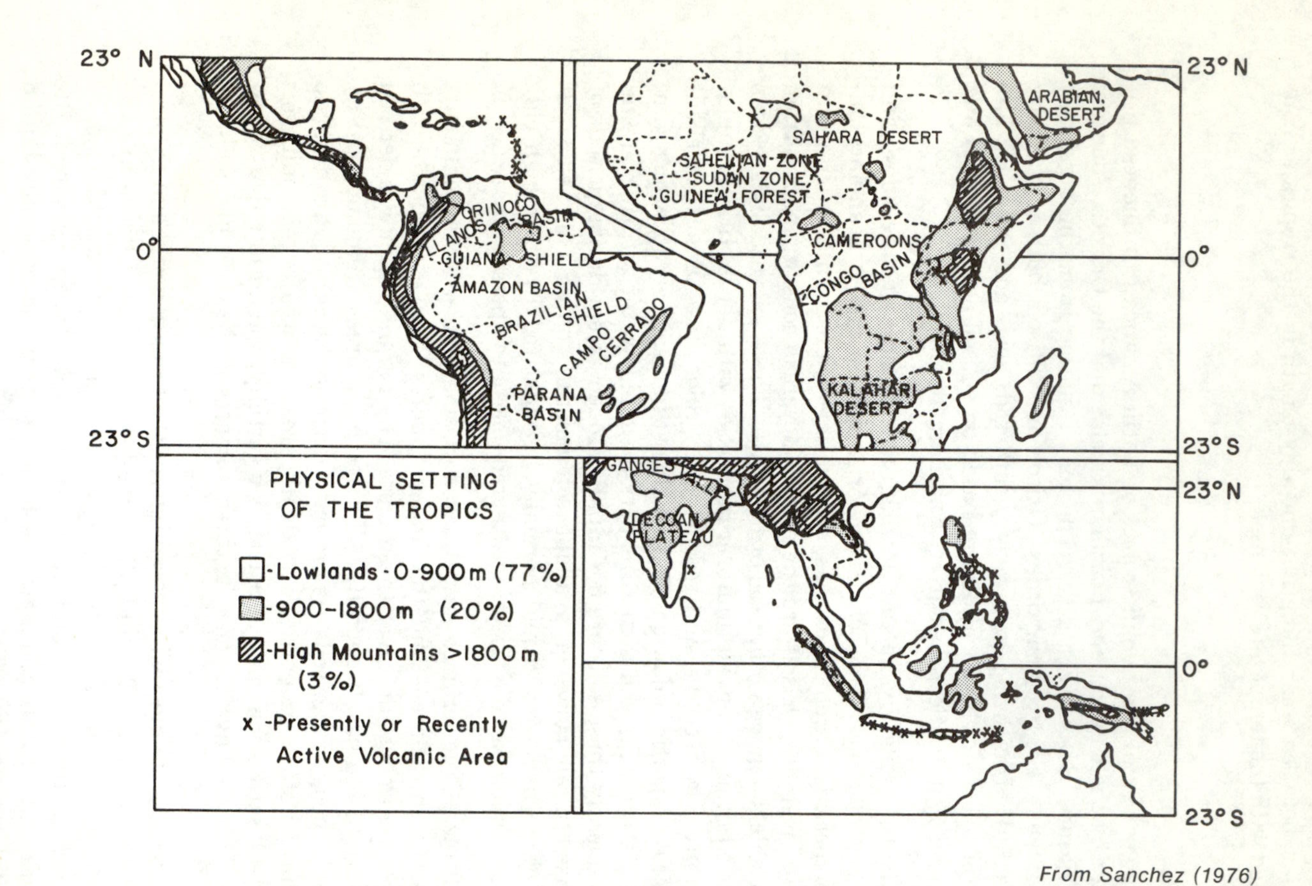

From Sanchez (1976)

FIG. 16.3. PHYSICAL SETTING OF THE TROPICS WITH REFERENCE TO SHIELDS AND BASINS INFLUENCING SOIL DEVELOPMENT

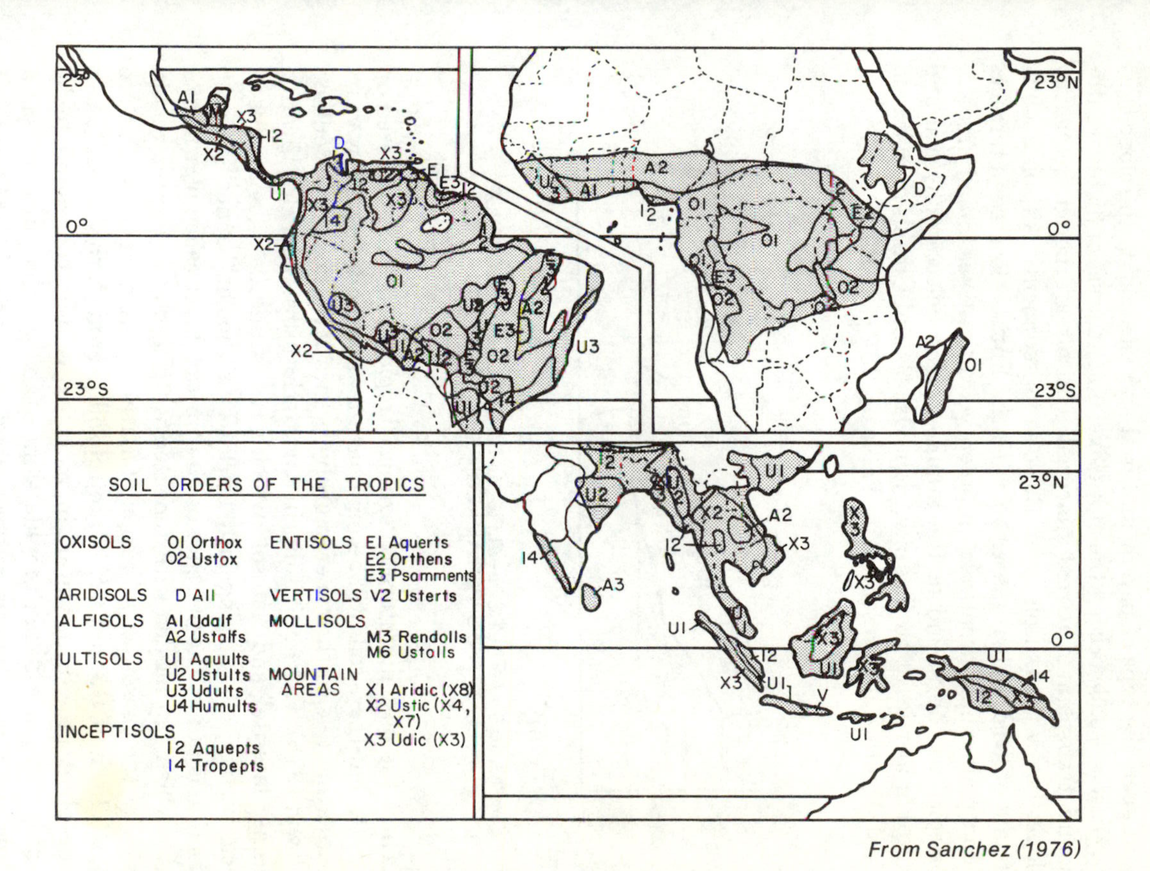

From Sanchez (1976)

FIG. 16.4. SOIL ORDERS OF THE TROPICS

present and usually in predictable locations in the landscape. In the humid tropics, intermixtures of Oxisols, Ultisols, and Inceptisols and of Ultisols, Alfisols, and Entisols are found. Their combinations and extent depend upon the type of parent material and location on the slope (Fig. 16.5). Oxisols usually occupy the more level upland areas of highly weathered and transplanted parent material. Ultisols are noted frequently on slopes below the Oxisols or on these same sediments. Alfisols usually are found on basic parent materials with the Entisols located in terraces or flood plains and Inceptisols on steep volcanic slopes. Small areas of other soils also are found in the humid tropics. Selected physical and chemical properties of profiles of various Oxisols and Ultisols from humid tropical America are presented in Table 16.2.

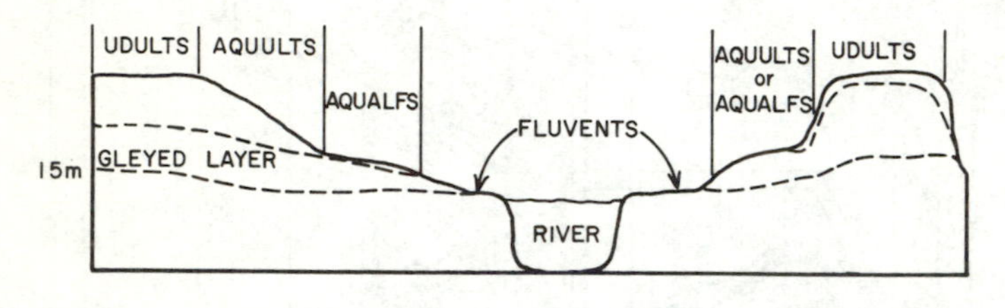

From Sanchez (1976)

FIG. 16.5. AN ULTISOL-ALFISOL-ENTISOL TOPOSEQUENCE OF THE UPPER AMAZON JUNGLE OF PERU

Much of the past literature regarding "tropical soils" has centered on the myth that these soils will become barren wastelands only a few years after clearing due to their metamorphosis into lateritic rock (Goodland and Irwin 1975). This myth is due in part to the use of older classification terms as "laterite" and "latosol" which gave erroneous connotations of the cultivated soils turning into brick. This "major barrier" to crop production is due to the presence of an iron-rich layer called plinthite or laterite which can harden with repeated wetting and drying. However, only 7% of the Oxisol-Ultisol region of Latin America contains plinthite within 1.25 m of the surface. If the topsoil is eroded, the plinthite can impede plant growth; however, in most areas it is considered to be advantageous since it can provide slope stability when it occurs as a hardened outcrop at scarps between erosion surfaces. In addition, it is considered to be an inexpensive and suitable material for road beds in areas where rock is at a premium. In summation, these soils can be among the most productive in the world when proper soil-water management and conservation techniques are effective.

TABLE 16.2. PROFILE ANALYSIS OF SELECTED OXISOLS AND ULTISOLS FROM EXPERIMENT STATIONS IN LATIN AMERICA

Horizon (cm)	Clay %	Sand %	pH (1:1 H_2O)	Org. C %	Exchangeable Cations (meq/100 g) Al	Ca	Mg	K	CEC	Al satn. %
Oxisol, Brasilia, Cerrado of Brazil. Typic Haplustox, fine, kaolinitic isohyperthermic. Dark Red Latosol Profile 1, Centro de Pesquisa Agropecuaria dos Cerrados.										
0-10	45	36	4.9	1.8	1.9	0.4		0.10	2.4	79
10-35	48	33	4.8	1.2	2.0	0.2		0.05	2.2	89
35-70	47	35	4.9	0.9	1.6	0.2		0.03	1.8	88
70-150	47	35	5.0	0.7	1.5	0.2		0.01	1.7	88
150-260	42	39	4.6	0.3	0.7	0.2		0.02	0.9	76
Oxisol, Carimagua, Llanos Orientales of Colombia. Tropeptic Haplustox, fine clayey mixed, isohyperthermic. Profile Ca-4, Centro Nacional de Investigaciones Carimagua.										
0-20	37	6	4.9	3.1	2.8	0.2	0.2	0.10	3.4	82
20-51	39	5	5.0	1.5	2.0	0.1	0.1	0.10	2.3	85
51-82	40	5	4.8	0.8	1.9	0.1	0.1	0.10	2.2	84
82-117	40	5	5.4	0.6	1.1	0.1	0.1	0.10	1.6	69
117-132	48	5	5.8	0.4	—	0.2	0.2	0.30	0.8	—
132-152	52	4	5.9	0.3	—	0.2	0.2	0.30	0.7	—
Ultisol, Jusepin, Mesas Orientales de Venezuela, Oxic Paleustult, fine loamy, isohyperthermic. Maturin series Profile 1. Universidad del Oriente.										
0-28	24	69	4.7	0.9	0.5	0.5	0.1	0.01	1.1	43
28-65	29	61	4.6	0.6	0.6	0.4	0.1	0.01	1.1	54
65-96	37	53	4.7	0.3	0.6	0.4	0.1	0.01	1.1	50
96-220	40	51	4.9	0.4	0.7	0.4	0.2	0.01	1.3	52

TABLE 16.2. (*Continued*)

Ultisol, Yurimaguas, Amazon Jungle of Peru. Typic Paleudult, fine loamy, siliceous, isohyperthermic. Profile Yu-13, Estación Esperimental de Yurimaguas.

0-5	6	80	3.8	1.3	2.0	0.84	0.37	0.20	3.4	59
5-13	10	70	3.7	0.8	2.6	0.05	0.03	0.04	2.7	96
13-43	15	61	3.9	0.4	3.1	0.05	0.03	0.03	3.2	96
43-77	17	57	4.0	0.3	3.1	0.03	0.02	0.02	3.2	97
77-140	25	51	4.1	0.2	4.5	0.03	0.01	0.03	4.6	98
140-200	24	54	4.4	0.2	3.8	0.06	0.03	—	3.9	96

Ultisol, Pucallpa, Amazon Jungle of Peru. Aquic Paleudult, clayey, mixed, isohyperthermic. Profile Pu-2. Instituto Veterinario de Investigaciones del Trópico y Altura.

0-3	27	25	5.2	6.3	0.2	4.2	2.1	0.52	7.1	3
3-21	45	17	4.3	1.9	4.0	2.2	1.2	0.40	7.9	51
21-62	59	15	4.2	1.0	8.7	0.8	0.9	0.32	10.8	81
62+	59	21	4.1	0.5	11.6	0.4	0.7	0.24	13.1	89

Source: Sanchez (1977).

Vegetation

Climate and soil factors influence the native vegetation of the tropics. Two different native vegetation types, savannas and forests, are found in the area of the humid tropics with which this chapter is concerned (Fig. 16.6). Generally speaking, the savannas are dominated by the Oxisols and occur in the ustic moisture regime, while Ultisols are more prevalent in the udic moisture regimes of the forested areas. Quite obviously, a continuum between these two extremes exists. Oxisols occur in forested areas and savannas can be located on Ultisols. As earlier discussed, other soil orders also occur in the savanna and forest areas of Latin America.

Savannas.—The savannas occur in 43 and 28%, respectively, of the tropical world and American tropics. In Latin America, about 180 million hectares of savanna are found in the Cerrado of Brazil alone, with additional savannas being located in the Llanos of Colombia and Venezuela, some areas of the Pacific coasts of Central America and Mexico, and most of Cuba. Many of these grassland- and shrub-dominated areas are located in the ustic moisture regime and barely receive the 1500 mm annual precipitation with 7 humid months required to be classed as humid tropics. Other savannas occur in aridic soil moisture regimes and will not be discussed herein.

Three savanna types, tall grass-low tree, Acacia-tall grass, and Acacia-desert grass, are recognized. The Acacia-tall grass savanna is the dominant type in humid tropical America and consists of tussock grasses which grow to 1 to 2 m during the rainy season with interspersed deciduous or evergreen trees. In Brazil, the savannas range from "Campo Limpo," primarily grass-dominated areas, to "Cerradao," containing large trees with some grass vegetation (Fig. 16.7). More information regarding savanna vegetation is supplied by Eyre (1962).

Savanna genesis probably varies with location and can occur even in udic soil moisture regimes after the rain forests have been cut. The "naturally occurring" savannas can be caused by such diverse factors as low nutrient availability, high levels of exchangeable aluminum, laterite formation, extreme alternating waterlogging and drying, or fire. Certainly the savanna vegetative type is present on diverse soil conditions.

Forests.—Both rain forests and some semideciduous and deciduous forests are located in the humid tropics. The rain forests are found in the udic moisture regimes and cover 30 to 52% of the tropical world and the American tropics, respectively. The Amazon Basin, the Pacific coast of Colombia, and some of the Atlantic coast of Central America are dominated by 550 million hectares of rain forests. Diversity of tree species characterizes tropical rain forests, with approximately 2500 (few of

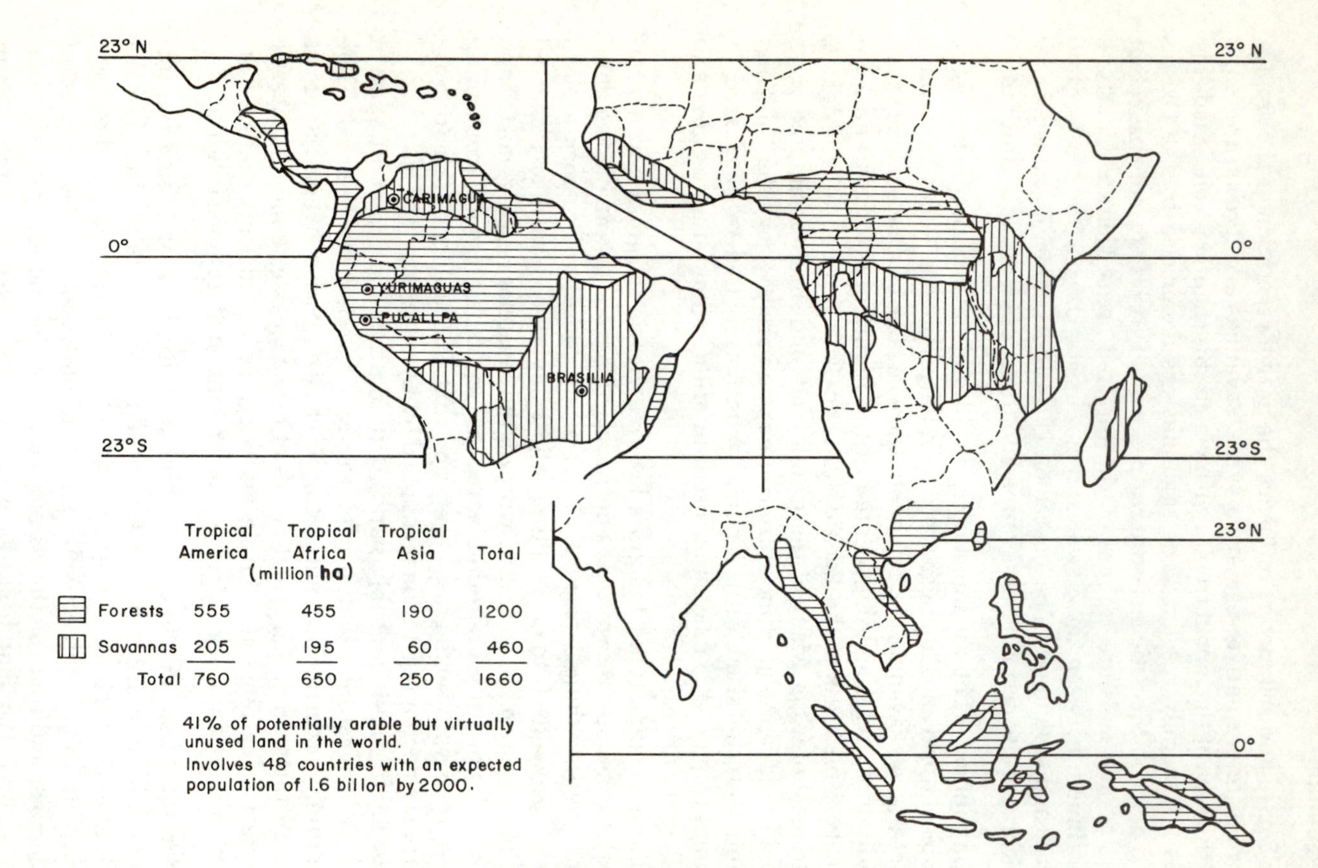

FIG. 16.6. SAVANNAS AND FORESTS IN THE TROPICS

which are deciduous) being found both in the Amazon jungle and in Malaysia, while only about 12 species are noted in temperate forests.

FIG. 16.7. NATIVE SAVANNA VEGETATION RANGING FROM CAMPO LIMPO (OPEN AREA) TO CERRADAO (MORE FORESTED AREAS) ON BRAZILIAN OXISOLS

The almost closed nutrient cycle of the rain forest and its soil is maintained by a constant litter fall and decomposition. Solar radiation reaching the soil surface under the customary 3-layer canopy is about 15% of that which reaches a cleared area. As expected, when compared with cleared areas, air temperatures are lower and humidities are higher in the rain forests. A more complete study of tropical rain forests is presented by Odum (1972).

Semideciduous and deciduous seasonal forests of ustic environments are found in some areas of the humid tropics, including tropical America, where the dry season is sufficient to prevent rain forest establishment, but where few grasses are found. Usually two-layer canopies are present. The nutrient cycle of these forests is different from that of the rain forest in that during the dry season solar radiation reaching the substantial leaf fall slows decomposition of the litter.

LAND-CLEARING METHODS

The Latin American savannas and forests which are to be cropped or put into pasture usually are subjected to the same traditional land-clearing technique, the "slash and burn" practice. In this method, the larger trees and shrubs are cut by hand during low rainfall periods, allowed to dry for about ten days, and burned either in place or after localizing the smaller cut trees and shrubs (Fig. 16.8).

Bulldozer clearing of the forests and the more heavily vegetative savannas is practiced in some of the larger land holdings. In this system the vegetation and part of the topsoil is pushed by the large equipment to the edge of the cleared area (Fig. 16.9) and burned or preserved as windbreaks. Mechanized clearing of Oxisol savannas is not usually as detrimental as is that of Ultisol forests. This is because the Oxisol savannas possess less vegetation to supply fertilizer ash and are less subject to compaction than are the Ultisol forests.

The crop production systems that are established on these cleared savannas and forested soils are often as varied as the vegetation which preceded them.

OXISOL SAVANNAS

Current Production Systems

Although some of the savanna areas have been cleared and fertilized for more intensive crop production, the most widespread and traditional use of these areas is pasture for extensive livestock production. More detailed information regarding both systems for Oxisol savannas is found in Bornemisza and Alvarado 1975; North Carolina State University (Anon.) 1973, 1974, 1975, 1976; Sanchez 1976, 1977; CPAC 1976 (Anon. 1976B); and CIAT 1976 (Anon. 1976A).

Pastures.—The livestock in tropical America are primarily beef and some dairy cattle; sheep and goats account for about one-third of the forage-consuming animals in this area. Used to feed these animals are native grass pastures of *Paspalum*, *Axonopus*, *Themeda*, *Trachypogon*, *Heteropogon*, or *Hyparrhenia*. Some native legumes can be noted in the pastures, but they usually are not prominent. Pastures for livestock production comprise 11% of the tropics with a large portion of humid tropical America's Oxisol savannas dedicated to this practice. However, much of these savannas, especially in the Cerrado of Brazil, are sparsely populated and, consequently, are little utilized by man for either livestock or crops.

Periodic burning of the savannas is a universal management practice

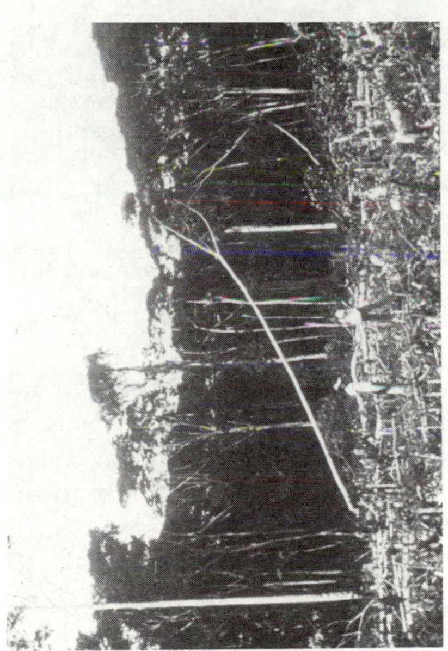

FIG. 16.8. THE “SLASH AND BURN” OR MANUAL CLEARING OF AN AMAZON RAINFOREST NEAR YURIMAGUAS, PERU

for pastures. This periodic burning destroys dry, unpalatable grass and promotes regrowth of the young, more palatable shoots. Animal and pasture pests are reduced by the rapid burn. Most of the nitrogen, sulfur, and carbon are lost to the atmosphere; the hotter the fire, the greater the loss. Phosphorus, potassium, calcium, and magnesium are returned in the ash to the soil. Many of the native grasses and some introduced species, such as guinea grass (*Panicum maximum*), are benefited by this burning, while others, such as molasses grass (*Melinis minutiflora*), are not. In high rainfall areas, annual burning is required; in lower rainfall regions, burning each 2 to 3 years is sufficient to destroy the mature unpalatable grasses and to promote fresh regrowth.

FIG. 16.9. BULLDOZER CLEARING OF AN AMAZON RAINFOREST NEAR YURIMAGUAS, PERU

Fertilization of these native grasses has increased annual dry matter production in some locations but usually not economically so. The native grasses seem to be relatively well-adapted to the low native fertility of the soils and generally are inefficient in their response to added nitrogen and phosphorus. Fertilization of the native pastures of the Oxisol savannas is not a common practice.

What have these traditional practices produced as far as beef production on these native pastures is concerned? Annual beef liveweight gains range from 10 to 50 kg per hectare, with weight gains coming during the rainy season and severe weight losses and mortality occurring during the dry season. Stocking rates of 0.05 to 0.30 animals per hectare are as common as are calving rates of 40 to 50% under the traditional pasture management of the Oxisol savannas in tropical America. This low productivity means that beef cattle require 5 to 8 years to reach marketable weight. Thus, although livestock is an important component of tropical agriculture, meat and milk production in the tropics is extremely low, being one-third and one-fifth, respectively, of what is produced with the same number of animals in the temperate areas.

Crops.—A wide array of crops can be found on the savannas of the humid tropics. Growing in monocultures and, at times, in various combinations are maize, soybeans, beans, upland rice, sorghum, groundnuts, wheat, cowpeas, cotton, cassava, sweet potato, sugar cane, plantains, and others. The production systems of the above crops, with the exception of cassava and plantains, when grown on savanna soils in the tropics, are quite similar to those systems employed for these crops in the temperate zone. Productivity of these crops under intensive management can be very good, while productivity under low management conditions is understandably low.

Constraints to High Crop and Pasture Yields

Many factors have been related to the low productivity of both livestock and crops grown on the savanna Oxisols. Feed supply for the livestock is accepted as the most limiting factor, though environmental stress and disease also enter into the picture. Both quantity and quality of forages used for livestock production, as well as crops, are directly dependent upon the soil. The most limiting soil factors to pasture and crop production in the Oxisol savannas are moisture and fertility related.

Moisture Limitations.—Crops are produced on the Oxisol savannas during the rainy season, which usually occurs during October through April. However, short term droughts, called "veranicos," of up to two weeks in duration (Table 16.3) can occur during the rainy season each year and considerably affect crop production, especially when occurring at critical growth stages. The low water holding capacity of the Oxisols (Table 16.4) further intensifies this problem because soil moisture storage is insufficient for crops to withstand the drought. A drought of only 7 days during the rainy season can reduce corn yields by more than 50% when it occurs at silking stage.

TABLE 16.3. FREQUENCY OF OCCURRENCE OF DROUGHTS DURING THE RAINY SEASON FOR BRASILIA, BRAZIL BASED ON 42 YEARS OF RECORDS

Duration of Drought Days	Frequency (Per Year)
8	3
10	2
13	1
18	0.29 (1 per 3.5 years)
22	0.13 (1 per 7 years)

Source: Anon. (1974).

Soil Fertility Limitations.—Compounding the soil moisture limitation problem even more are problems with aluminum toxicity (Table 16.2), phosphorous deficiency and fixation (Table 16.4), and deficiencies of nitrogen. In some instances, calcium, magnesium, potassium, and zinc deficiencies are noted (Table 16.5).

TABLE 16.4. CLAY EFFECT ON AVAILABLE SOIL WATER AND PHOSPHORUS FIXED (TO GIVE A 0.2 ppm P IN SUPERNATANT SOLUTION) OF 44 OXISOL TOPSOILS FROM THE CERRADO OF BRAZIL

% Clay	% Available H_2O (0.1-15 bars)	P Fixed (ppm)
<18	4.9	110
18-35	8.5	219
35-60	9.8	377
<60	9.1	442

Source: Lopes (1977).

The aluminum toxicity problem, which becomes more pronounced in the subsoil, inhibits crop rooting depth. When short term droughts occur, the shallow rooted crops are unable to draw moisture from the subsoil and thus, are even more adversely affected than they normally would be. Even during brief droughts, serious water stress occurs as a result of the rooting depth being limited by high exchangeable aluminum levels in the subsoil.

Native phosphorus levels in these Oxisols range from 0 to 3 ppm, which indicates extreme deficiency of this element. Maize will not grow under these conditions (Fig. 16.10). Due to the relatively high phosphorus retention capacity of the higher percentage of clay soils, large initial amounts of phosphorus must be added to provide the necessary amount of soil solution phosphorous for plant growth (Table 16.4). In fact, 2 representative Brazilian Oxisols containing 60% clay in the surface soil required applications of 730 and 1070 kg of phosphorus per hectare to maintain 0.05 ppm of phosphorus in the soil solution (Anon. 1973).

TABLE 16.5. MEAN TOPSOIL NUTRIENT AVAILABILITY INDICES IN CERRADO TOPSOILS (0-20 CM) OF 510 SAMPLES DIVIDED ACCORDING TO PHYSIOGNOMIC SAVANNA TYPES (NUMBER OF OBSERVATIONS IN PARENTHESIS)

Soil Property	Campo Limpo (64)	Campo Cerrado (148)	Cerrado (255)	Cerradão (45)	Forest (16)	Critical Level[2]
Organic matter (%)	2.21	2.33	2.33	2.32	3.14	—
Exch. K (me/100 g)	0.08	0.10	0.11	0.13	0.17	0.15
Exch. Ca (me/100 g)	0.20	0.33	0.45	0.69	1.50	—
Exch. Mg (me/100 g)	0.06	0.13	0.21	0.38	0.55	0.50
Avail. Zn[1] (ppm)	0.58	0.61	0.66	0.67	1.11	1.00
Avail. Cu[1] (ppm)	0.60	0.79	0.94	1.32	0.95	1.00
Avail. Mn[1] (ppm)	5.4	10.3	15.9	22.9	24.1	5.00
Avail. Fe[1] (ppm)	36	34	33	27	32	—

Source: Lopes (1975).
[1]Extracted with the North Carolina dilute double acid.
[2]According to soil test recommendations of the State of Minas Gerais.

Potential Improvements in Production Systems

Agronomically feasible solutions to overcome factors limiting to crop production on the Oxisol savannas have been found. Improvements in the pasture production systems are being investigated, but at present are not as conclusive as are those for crop production systems. Economic feasibility of these agronomic improvements will depend upon various factors, including availability and cost of the needed fertilizers and lime, transportation costs, and availability of markets. This economic feasibility will vary with the individual farmers of the areas.

Crop Production Systems.—A package of technological improvements is necessary to produce high crop yields on the Oxisol savannas since several factors restrict crop production. Basically, the solutions include the judicious use of lime and fertilizers to overcome soil fertility barriers, the use of crop species and varieties tolerant to high aluminum and low phosporus levels in the soil, gravity irrigation where possible, and the combination of deep liming and use of grass mulch to reduce the effects of the "veranicos."

FIG. 16.10. PHOSPHORUS MUST BE SUPPLIED FOR CROP GROWTH ON SAVANNA OXISOLS

Left, Corn with no P other than that supplied by the soil itself. Right, adequate crop growth is obtained with sufficient P.

Overcoming Aluminum Toxicity.—The high levels of subsoil aluminum which limit rooting depth and adversely affect crop growth can be overcome by deep incorporation of lime or use of aluminum-tolerant crop species and varieties.

Incorporation of lime to 30 cm rather than the traditional 15 cm lowers the high exchangeable aluminum in the subsoil which, in turn, promotes deeper root development. Consequently, crops can not only grow in these soils but they are also better able to withstand the effects of drought. Immediate and residual beneficial effects of this deep lime incorporation in soils with high percentage of aluminum saturation are apparent (Fig.

16.11). The amount of lime recommended is 1.65 t per hectare for each milliequivalent exchangeable aluminum per 100 g soil. An in-depth study of the effects of deep lime incorporation on corn in Oxisols is presented by Gonzalez-Erico (1976). However, current research shows that calcium and magnesium move into the soil profile with time. Accordingly, deep incorporation of lime is of benefit to only the initial crops.

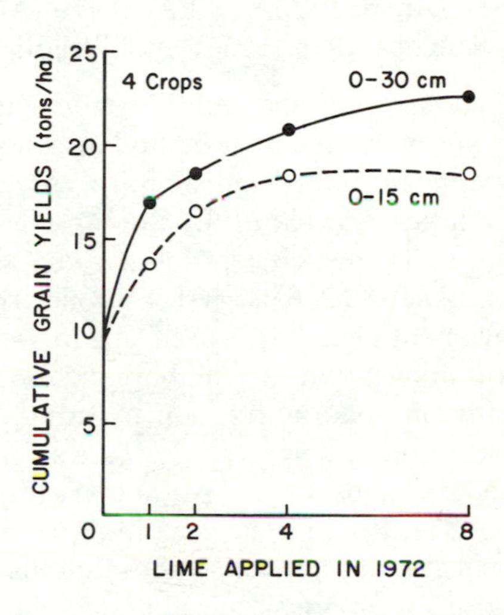

FIG. 16.11. CUMULATIVE GRAIN YIELDS OF FOUR CONSECUTIVE CROPS HARVESTED AFTER INITIAL LIME APPLICATIONS

Adapting the crop to the soil instead of modifying the soil to fit the crop is an alternative strategy which is regarded as one of the keys to agricultural production on these low base status soils. Species and varieties of plants which evolved under acid soil conditions will tolerate higher levels of exchangeable aluminum than those developed on higher base status soils. For instance, on an unlimed Colombian Oxisol having 80% aluminum saturation a native upland rice variety, Colombia 1, yielded 2.3 t per hectare while an improved variety, IR5, developed on high base status Philippine soils, yielded only 1 t per hectare (Anon. 1976A). On a Brazilian Oxisol with 50% aluminum saturation, an aluminum-tolerant sorghum variety, Taylor Evans Y-101, yielded 6 t per hectare after application of 1 t of lime per hectare 3 years previously. To produce the same yield, an aluminum-susceptible variety,

RS-610, required 4 t of lime per hectare and a reduction of aluminum saturation to 20% (Anon. 1975). Many varieties of various crop species developed in the Cerrado of Brazil or the Llanos of Colombia exhibit joint tolerance to aluminum toxicity and low available phosphorus levels. Under natural conditions or at low levels of lime and phosphorus, yields of these "adapted" varieties and species are usually much better than those of varieties developed on higher base status soils. Supplementary detailed information can be found in Bornemisza and Alvarado 1975; CIAT 1976 (Anon. 1976A); CPAC 1976 (Anon. 1976B); Papendick *et al.* 1976; Sanchez 1976, 1977; and Wright 1977.

Overcoming Phosphorous Deficiency and Fixation.—In addition to employing crop species and varieties tolerant to low soil phosphorus levels, several other management improvements are currently used in the savannas of Latin America. Oxisol and Ultisol topsoils with greater than 35% clay and a free Fe_2O_3 per clay ratio exceeding 0.15 exhibit high phosphorus fixation (Anon. 1976). In these soils, the two successful management improvements currently used are increased residual efficiencies of phosphorus applications by combining various rates and placements, and decreasing phosphorus fixation by other soil amendments.

Broadcasting 140 kg of phosphorous per hectare at planting of the first crop plus banding 35 kg of phosphorous per hectare for each of the next 5 successive crops produced 81% of the maximum corn yield given by a single broadcast application of 560 kg of phosphorous per hectare (Yost 1977). This combination of broadcast plus banded phosphorous fertilizer is the most promising method of phosphorus fertilization shown by recent research on infertile Oxisol savannas of Brazil. Soybean yields obtained after broadcasting as much as 264 kg of phosphorus per hectare were profitable at 1976 Brazilian phosphorus and soybean prices; however, 80% of the maximum yield was obtained with only 132 kg of phosphorus per hectare.

Lime or calcium silicate applied at rates sufficient to neutralize the exchangeable aluminum decreased the amount of phosphorus required to provide 0.03 ppm of phosphorus in the soil solution by 41 or 46%, respectively. Thus, this practice is expected to lower the amounts of phosphorus which are required to produce good crop growth on the Oxisol savannas.

Overcoming Other Nutrient Deficiencies.—In addition to the primary growth limiting factors for the Oxisol savannas described above, other nutrient deficiencies may also exist (Anon. 1976B; Anon. 1973, 1974, 1976). Although maize on Cerrado Oxisols will respond to nitrogen fertilization, yields of maize adequately fertilized with all necessary elements except nitrogen usually ranged from 3.2 to 4.2 t per hectare.

With the application of 100 kg of nitrogen per hectare to maize, maximum yields of 7.0 t per hectare can be obtained.

Applications of only 3 kg of zinc per hectare increased maize yields from 0.5 to 6.0 t per hectare when other nutrients were non-limiting. The residual effect of this fertilization was noted as 4 consecutive maize crops responded positively, with the fourth crop yields being 7.0 t per hectare.

Application of 63 kg of potassium per hectare raised maize yields from 2.3 to 4.0 t per hectare when other nutrients were non-limiting. These potassium applications were quite economical, giving a $9.30 rate of return for each $1.00 invested in potassium fertilizer at 1976 Brazilian fertilizer and maize prices.

Overcoming Short-duration Droughts.—Approximately 5% of the Cerrado soils close to Brasilia are easily irrigated by gravity after channeling or impounding streams. This practice allows year-round farming of these soils. However, most of the savanna Oxisols are not so easily irrigated.

Without irrigation, the use of deep lime placement and mulching increased maize yields by more than 25% (from 5.6 to 7.1 t per hectare) when an 18-day "veranico" occurred at a critical growth stage (Bandy 1976). Due to deeper root development in the deep limed areas, maize took up more soil water than did maize in the shallow limed plots, thereby decreasing water stress. The use of mulches reduced surface soil-water evaporation by 4 to 7 mm per day during the drought. Intensive crop production in these Oxisol savannas may require this combined practice of deep liming and mulching. Additional information is furnished by Wolf 1975; CPAC 1976 (Anon. 1976B); Lopes 1977; and Sanchez 1976, 1977.

Pasture Production Systems.—Although improvements in the pasture production systems of the Oxisol savannas have not been as conclusive as those of the crop production systems, certain beneficial advancements have been realized. These include use of less expensive phosphorus sources, the use of adequate fertilization in improved grass or grass/legume pastures which are tolerant to the environmental stresses of the area, and the use of fertilized crops prior to pasture establishment.

Use of Less Expensive Phophorus Sources.—The reaction of rock phosphate under the native acid conditions of unlimed Oxisol savannas generally makes these phosphorus sources more available. Hence, application of these materials, which are one-fourth the cost of superphosphates per kg of total phosphorus can be more profitable than superphosphates. For instance, the relatively soluble North African rock phosphate, "Hiperfosfato," and superphosphate produced equal growth on a *Brachiaria decumbens* pasture on an Oxisol savanna of Brazil's Cerrado (Anon. 1974, 1975).

Use of Adequate Fertilization in Improved Grass or Grass/Legume Pastures.—Application of 0.5 t basic slag per hectare to a *Brachiaria decumbens* pasture on an Oxisol in Carimagua, Colombia increased grazing cattle's annual liveweight gains 5-fold over that produced on native savannas. The liveweight gain for cattle on this *Brachiaria* pasture for the second year was three-fold over that of native savannas. These results were attributed to the improved grass and fertilization which overcame nitrogen, potassium, and magnesium deficiencies of the Oxisol savannas (Anon. 1976A).

Establishment of a legume in the grass pasture with modest fertilization produced even greater liveweight gains. For example, a pasture of *Stylosanthes guyanensis* and molasses grass (*Melinis minutiflora*) on the same Oxisol described above increased annual liveweight gains to 250 kg per hectare. The *Stylosanthes* died out due to insect and disease infestations thereby indicating that more effective means of controlling pests of the legume are needed before this system can be used.

Use of Fertilized Crop Prior to Pasture Establishment.—Adequate fertilization of annual crops followed by pasture establishment makes use of residual fertilizer nutrients. Current research efforts indicate that this could be an economical and effective practice for establishment of crop-pasture systems in the Oxisol savannas of Latin America.

ULTISOL FORESTS

The current and improved soil-crop management systems for the jungle areas of Latin America are vastly different from those of the savannas. Several factors contribute to this situation. Among these are the following: (1) land clearing is generally more difficult due to the large amount of vegetation, (2) higher annual rainfall with no marked dry season, (3) relatively higher native fertility of newly cleared forests due to previous recycling of nutrients by the lush vegetation, and (4) less developed infrastructure of supplies, transportation, and markets in the jungle areas as compared with the savannas.

Current Production Systems

The above factors have combined to establish "shifting cultivation" as the most widely employed management practice in the jungle areas not only of Latin America but also of most of the humid tropics throughout the world (Fig. 16.12). Shifting cultivation is discussed by Ruthenberg

1976; National Academy of Sciences (NAS-NRC) 1972; Grigg 1974; Manshard 1974; and Sanchez 1973, and is known by local names in the diverse regions where its many variations are employed. Examples are "barbecho," "coamile," and "milpa" in Mexico and Central America and "chaco," "roza," "monte," "conoco," and "roca" in different locations of South America. The term "shifting cultivation" encompasses the many variations practiced around the world and is considered to include any system under which the soil remains fallow for a longer time period than it is cropped.

FIG. 16.12. SHIFTING CULTIVATION IN THE UPPER AMAZON JUNGLE BASIN OF PERU

Most systems of shifting cultivation employ the "slash and burn" technique as previously described. Other variations, such as that used by farmers in rain forests on the Pacific coast of Colombia, follow the practice of broadcasting crop seed, cutting the undergrowth, and then, instead of burning, allowing the ashed vegetation to remain as a mulch.

Crops.—In the most common system of shifting cultivation, crops such as rice, beans, maize, cassava, sweet potatoes, and plantains are planted among the remaining debris left after cutting and burning the vegetation. The farmers use a "coa" or stick to make a hole in which the seed or vegetative portion of the crops are placed. Intercropping of two or more species simultaneously on the small "chacras," "chaqueados," or clearings is quite common. This practice reduces, but does not eliminate, manual weeding of the plantings. After only one or two harvests, crop yields begin declining drastically due to soil fertility depletion. The land subsequently is abandoned to rapid forest regrowth which is allowed to fallow for periods of 4 to 20 years. After accumulation of essential plant nutrients in the forest vegetation and the upper layer of the soil, the practice is repeated.

Pastures.—Although crop production on newly cleared areas of the rain forests is the most common practice, pastures are becoming more noticeable. In the last decade alone more than one million hectares of pasture have been established in the Amazon jungle region of Brazil, Peru, and Colombia.

The current pasture system followed after burning these forests is generally planting of guinea grass (*Panicum maximum*) where moisture is available year-round (udic) and yaragua (*Hyparrhenia rufa*) where there is an annual dry season (ustic). Vegetative propagation is common as is seed planting. Neither legumes nor fertilizers are used in most pastures. The guinea grass and yaragua usually are noncompetitive with the coarse, unpalatable *Paspalum conjugatum* and *Axonopus compressus* which begin to dominate the pastures after a few years. Native legumes such as *Centrosema* and *Desmodium intortum* begin to appear in the pastures and provide some nitrogen to the grasses. With proper management of the established grass pastures, the natural inclusion of the native legumes is quite beneficial. Where there are no annual dry seasons, pastures of this type have persisted for more than 20 years without reestablishment when not overgrazed. Where dry seasons are more intense, as near Pucallpa, Peru, yaragua grass pastures established after mechanical clearing usually become non-productive within 4 to 5 years.

Constraints to High Crop and Pasture Yields

Land-clearing Operations.—One of the primary constraints to high yields in either crops or pastures grown on rain forest Ultisols can be the mechanized clearing by heavy equipment as earlier described. Seubert *et al.* (1977) found that crop and pasture grass yields on bulldozer-cleared land were reduced substantially when compared with yields of these plants on land cleared by the traditional slash and burn technique (Fig. 16.13). In crops which were (1) unfertilized, (2) nitrogen, phosphorus, and potassium fertilized, and (3) nitrogen, phosphorus, and potassium fertilized and limed, bulldozer-cleared land produced ⅓, ½, and ⅘, respectively, of what was produced on burned land (Table 16.6). Three consecutive rice crops grown in 17 months on newly bulldozed land pro-

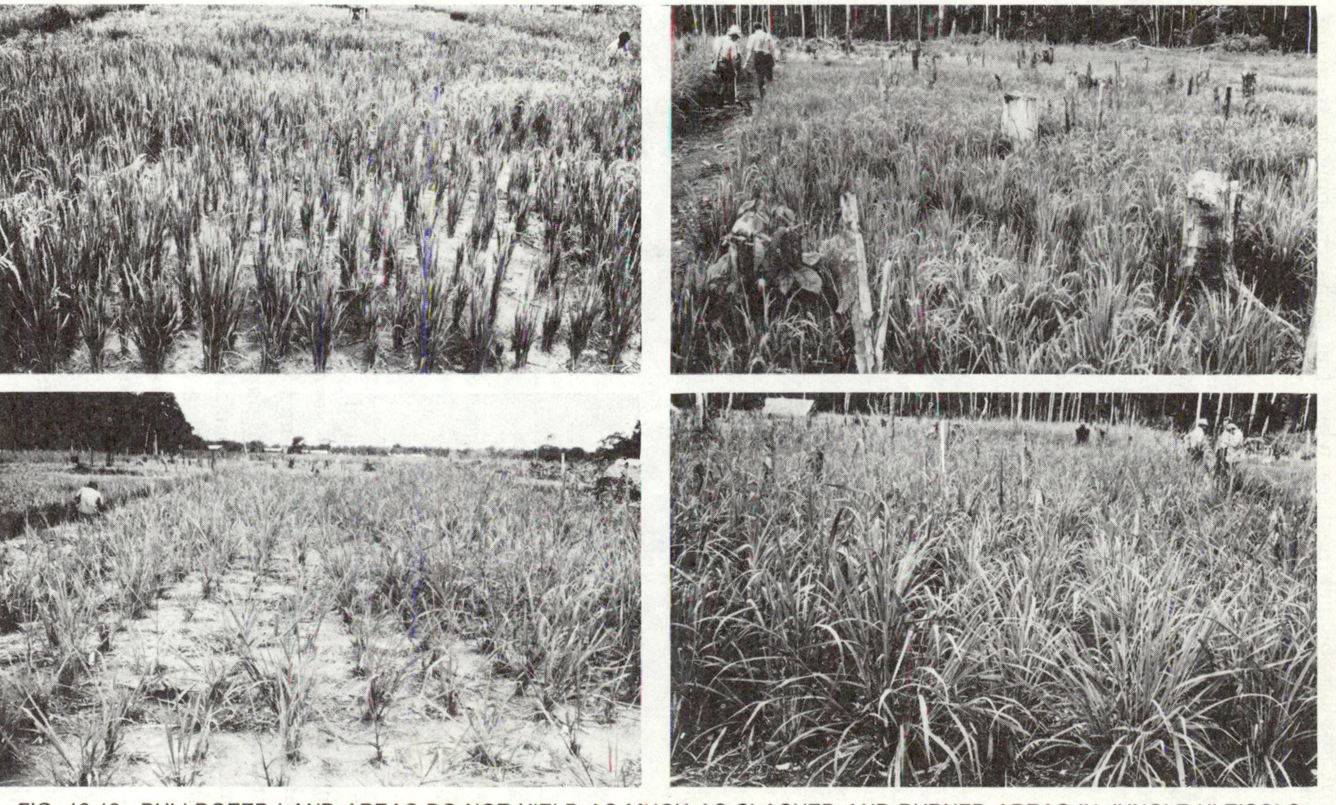

FIG. 16.13. BULLDOZED LAND AREAS DO NOT YIELD AS MUCH AS SLASHED AND BURNED AREAS IN JUNGLE ULTISOLS

Upper left, rice on a bulldozed area. Upper right, rice on a traditionally cleared area. Lower left, *Panicum maximum* pasture on a bulldozed area. Lower right, *Panicum maximum* pasture on a cut and burned area. Areas shown are at Yurimaguas, Peru.

duced a loss of $66 per hectare while the same cropping system on slashed and burned land gave a profit of $648 per hectare.

TABLE 16.6. EFFECTS OF LAND CLEARING METHODS ON CROP YIELDS AT YURIMAGUAS, PERU. YIELDS ARE THE AVERAGE OF THE NUMBER OF HARVESTS INDICATED IN PARENTHESIS

Crop	Fertility Level[1]	Slash and Burn (t/ha)[2]	Bulldozed (t/ha[2])	$\frac{\text{Bulldozed}}{\text{Burned}} \times 100$ (%)
Upland Rice (3)	O	1.3	0.7	53
	NPK	3.0	1.5	49
	NPKL	2.9	2.3	80
Corn (1)	O	0.1	0.0	0
	NPK	0.4	0.04	10
	NPKL	3.1	2.4	76
Soybeans (2)	O	0.7	0.2	24
	NPK	1.0	0.3	34
	NPKL	2.7	1.8	67
Cassava (2)	O	15.4	6.4	42
	NPK	18.9	14.9	78
	NPKL	25.6	24.9	97
Panicum maximum (6)	O	12.3	8.3	68
	NPK	25.2	17.2	68
	NPKL	32.2	24.2	75
Mean Relative Yields	O			37
	NPK			47
	NPKL			88

Source: Anon. (1974).
[1]50 kg N/ha, 172 kg P/ha, 40 kg K/ha, 4 t/ha of lime.
[2]Grain yields of upland rice, corn, and soybeans; fresh root yields of cassava, annual dry matter production of *Panicum maximum.*

These large differences in yields and economics are due to (1) the fertilizer value of the ash, (2) the severe soil compaction and soil removal caused by bulldozer clearing, and (3) the more rapid soil fertility decline on bulldozed land. The ash and partially burned vegetation from the Amazon jungle basin of Yurimaguas, Peru added the following kg of "available" nutrients per hectare: 67 N, 6 P, 38 K, 75 Ca, 16 Mg, 8 Fe, 7 Mn, and 0.3 of Cu and Zn. The calcium and magnesium added to the burned plots were equivalent to 240 kg per hectare of dolomitic lime.

Soil Compaction.—The sandy surface layers of the Ultisols of the Amazon jungle basin are as subject to compaction by heavy equipment as are these same soils in the coastal plain of the southeastern U.S. Water infiltration on bulldozed land has been found to be reduced 20-fold (from 20 cm to 1 cm per 2 hours) when compared with that on land which was cleared by burning. The implications of this finding for crop production and erosion susceptibility are apparent. Such compaction is a long-lasting consequence of the mechanized clearing by heavy equipment and is reflected by the reduced growth of secondary jungle vegetation of bulldozed land when compared with that of burned land five years after clearing. Even burned land can begin to suffer from compaction problems after one year of continuous cultivation. This is due to the impact of the intense rainfall on the bare soil surface, repeated cultivation to control weeds, and weight of human laborers walking through the fields during weeding operations.

Soil Fertility Decline.—Soil fertility decline with its consequent crop yield decrease is the primary reason for shifting cultivation being the most common agricultural practice in the Amazon jungle basin of Latin America. This decline occurs in both burned and bulldozed land, but is much more rapid in areas cleared by bulldozing because the bulldozed areas, which begin at a lower level of fertility than do the burned plots, are more rapidly depleted of their nutrients than are the burned areas.

When the jungle is burned but not fertilized, yields decrease markedly with time. For example, a second crop of rice with 2 plantings per year yielded only about half of the 2.9 t per hectare which the first crop yielded. The fourth crop yielded only one-third of the first year's crop yield, while the sixth crop produced no yield at all (Anon. 1975). Similar decreases have been noted in yields of *Panicum maximum* pasture, yucca, and rotations of rice, corn, and soybeans.

This fertility decline with continued cropping of the forest Ultisols can be attributed to a decrease of soil levels of nitrogen, phosphorus, potassium, calcium, and magnesium. In addition, as exchangeable calcium and magnesium contents are decreasing, exchangeable aluminum content is increasing. Consequently, the percentage of aluminum saturation of the soil increases to levels toxic to most crop species and cultivars.

Due to the consequent yield decline, most farmers of these rain forest areas abandon their cleared land to forest regrowth after producing only one crop each of rice and yucca. The fact of life for the jungle areas of Latin America is that without proper fertilizer and lime amendments, newly cleared land will out-yield by 2- to 3-fold land which has been previously cropped without sufficient fallow.

Potential Improvements in Production Systems

Adequate year-round rainfall, temperature, and solar radiation combine to make the rain forest Ultisol areas some of the most attractive in the world for crop or pasture production. With proper soil-crop management, these areas can be extremely productive (Fig. 16.14).

FIG. 16.14. CONTINUOUS CULTIVATION OF JUNGLE ULTISOL AREAS AFTER TRADITIONAL SLASH AND BURN CLEARING IS POSSIBLE WHEN PROPER SOIL-CROP MANAGEMENT IS FOLLOWED

The rice, peanuts, and corn shown are in fertilized plots of a several year old clearing in the Amazon jungle basin of Peru.

Land Clearing Method.—The traditional slash and burn method of clearing the jungle has been found to be superior to that of heavy mechanization. Hence, the first soil-crop management strategy component for either crop or pasture production on the jungle Ultisols is continuing the traditional land clearing method. Once cleared, the land should remain in continuous production, be it crops or pastures, to prevent erosion and to justify the expenditures for lime and fertilizer.

Crop Production Systems.—Continuous, fertilized intensive cultivation of the Amazon jungle areas, which are primarily Ultisols, has been shown to be both agronomically and economically feasible by research conducted by the Peruvian Ministry of Food and North Carolina State University (Anon. 1976). Three and four crops can be produced each year on these soils. In addition, intercropping which is a traditional system in these cleared areas, can be intensified in such a way as to make it more productive, especially where manual labor is available. Whether monoculture or mixed cropping is employed, the system will make use of the ash and partially burned vegetation from the burn. In a short time this fertility is depleted and fertilizer and lime must be used to overcome the soil fertility decline which forces the natives to shifting cultivation.

Continuous Monoculture.—Many rotational schemes are possible with proper use of lime, fertilizer, and crop residues. One such system, rice-maize-soybeans, when properly mulched to prevent soil compaction and correctly fertilized, produced a net profit of $1539 per hectare per year with a $2.66 return on each dollar invested in fertilizer and lime. This system included fertilization for each crop at 117 kg N, 70 kg P, 110 kg K, and 54 kg Mg per hectare and a total of 4.5 t of lime per hectare. In addition, copper, boron, and zinc were applied at 3 kg per hectare and molybdenum at 0.3 kg per hectare for the first crop. Total costs of production, including transportation of fertilizers to farm and crops to market, were $1140 while total income was $2679. Yields of rice (3.6 t per hectare) and soybeans (2.6 t per hectare) were considered excellent for the region, while those of maize (3.6 t per hectare) were good but limited by disease and insect infestation. The need for maize varieties resistant to these pests is apparent.

The beneficial effects of mulching and green manuring with a legume adapted to the area such as kudzu (*Pueraria phaseoloides*) are great. Yields of soybeans, cowpeas, maize, peanuts, and rice, when unfertilized but mulched or incorporated with kudzu, were 80 to 90%, respectively, of yields of the same crops which were heavily fertilized. In this system, monetary expenditures for fertilizer are replaced by labor expenditures. Additionally, there are land requirements for the production of kudzu.

Soil physical properties improved as a result of kudzu mulch or incorporation. Bulk density and penetrometer resistance decreased with kudzu incorporation. Kudzu mulch decreased day soil surface temperatures by 2°C, reduced soil water loss during dry periods, and inhibited weed competition. Thus, these practices are considered to be important components of improved crop production systems for the rain forests of Latin America.

Intensive Mixed Cropping.—Intercrops with modest, but sufficient fertilization, can provide higher farm income per unit land per unit time than monoculture. In the system outlined in Table 16.7, 9 crops were produced within 21 months on the same unit of land in Yurimaguas, Peru. Marketable value of this intercropped system exceeded that of the combined monocultures by about 25%, though individual crop yields were lower. The advantage of intercropping over monoculture was attributed to a reduction in weed, insect, and disease infestations and improved utilization of sunlight, water, and nutrients.

Thus, the common intercropping systems used by the farmers of the Amazon jungle can be kept continually productive when an adequate fertilization program is used. Other improvements in the system will include crop species and cultivars which are more tolerant to the environmental stresses of the area than those currently used.

TABLE 16.7. INTENSIVE INTERCROPPING SYSTEMS PRODUCING 4 TO 5 CROPS A YEAR AS COMPARED WITH GROWING THE SAME CROPS UNDER MONOCULTURES

Tall crops spaced at 2 m rows; fertilizer application: 1 t lime/ha year, plus 45 N, 44 P, 40 K, 10 S, 0.5 B and 0.5 Mo kg/ha/yr.

Year 1:	Maize	Soybeans	Cassava	Cowpeas	Total Market Values	% Over Monocultures
	Grain or Tuber Yields (t/ha)				(US$/ha)	
Intercropped	1.54	0.83	11.7	0.54	1055	20
Monocultures	3.35	1.15	16.8	1.05	879	—

Year 2:	Rice	Maize	Cassava	Peanuts	Cowpeas	Total Market Value	% Over Monocultures
	Grain or Tuber Yields (t/ha)					(US$/ha)	
Intercropped	2.01	0.51	8.0	2.62	0.24	1996	28
Monocultures	2.38	1.19	22.9	3.05	0.47	1558	—

Source: Anon. (1974, 1975).

Pasture Production Systems.—Many newly cleared areas in the Amazon rain forests are put into grass pastures following only one rice harvest. Establishing an improved grass/legume pasture with adequate fertilization and mineral supplements to the cattle has proven beneficial in these areas. On an Ultisol located in Pucallpa, Peru, an improved pasture of *Hyparrhenia rufa* and *Pueraria phaseoloides* when fertilized with 19 kg of phosphorus per hectare and mineral supplements increased liveweight gains over an unfertilized grass pasture and no mineral supplements from 79 to 350 kg per hectare per year. This improvement produced an 18% return on investment while the grass alone resulted in only a 0.9% return. Other pertinent information regarding forage potential in these areas is presented by Spain (1975).

Development and use of legumes resistant to insect and disease attack and tolerant of overgrazing is a large gap in the use of these improved mixed pastures. Work by the Instituto Veterinario de Investigacion del Tropico y Altura (IVITA) in Pucallpa, Peru has shown that with time these mixed pastures lose their productivity due to the aforementioned pest attacks on the legumes, animal compaction of the soil, and nutrient deficiencies. These researchers suggest to crop once again with adequate fertilization and then reestablish pastures to make use of the residual fertilizers. Essential nutrients can be supplied to the pastures without reverting to cropping, but economic considerations often warrant interchanging crops and pastures after clearing these jungle Ultisols.

For single-family settlers in these areas, the available research suggests the following practice: annually slash and burn an area of land which can be managed by the family (usually 1 to 4 ha); take advantage of the relatively high initial fertility level by planting a cash crop, such as rice; then establish adequately fertilized intensive monocultures or intercrops on part of the land, placing the remainder in improved grass/legume pastures; gradually increase the area under pasture with time to economically produce beef cattle (Sanchez 1977).

POTENTIAL OF THE ZONE

The humid tropics offer an enormous challenge and hope to man in his attempt to prevent the neo-Malthusian predictions from becoming reality. Current populations are located on naturally fertile soils with little opportunity for expanded acreage. The highly weathered, acid, infertile soils dominating the major portion of the sparsely populated humid tropical America were thought to be non-productive in the not too distant past. Current research and consequent improved farm practices on these soils are showing that they perhaps can be the most productive in the world. The National Academy of Sciences (NAS-NRC 1977) estimates that food production in the humid tropics can be from 150 to 200% greater than that produced in the temperate zone on a per hectare per year basis. Contributing to this estimation is the potential year-round cropping in 70% of the humid tropics and the improved crop and pasture management systems covered in this chapter. These improved systems include some traditional practices, such as land clearing by slash and burn, combined with agronomic technology found to be suitable for specific locations in the humid tropics. The technological packages previously discussed are undergoing continual evaluation and revision by researchers and farmers. The evaluations and revisions will make the technological packages for the humid tropics even more productive, agronomically and economically, and now are beginning to enable developing countries to supply more of their populations' demand for food. The future could show some of the developing countries located in the humid tropics to be the new grain and beef baskets of the world.

REFERENCES

ANON. 1973, 1974, 1975, 1976. Agronomic-economic research on tropical soils. Annual Reports. Soil Science Dep., North Carolina State Univ., Raleigh.

ANON. 1976A. Annual Report. Centro Internacional de Agricultura Tropical, Cali, Colombia.

ANON. 1976B. Annual Report. Centro de Pesquisa Agropecuaria dos Cerrados, EMBRAPA, Brasilia, Brazil. (Portuguese)

BANDY, D.E. 1976. Soil-plant water relationships as influenced by various soil and plant management practices on Campo Cerrado soils in the Central Plateau of Brazil. Ph.D. dissertation. Cornell Univ., Ithaca, New York.

BORNEMISZA, E., and ALVARADO, A. 1975. Soil Management in Latin America. Univ. Consortium on Soils of the Tropics, Feb. 10-14, 1974, CIAT, Cali, Colombia. North Carolina State Univ., Raleigh.

EYRE, S.R. 1962. Vegetation and Soils—A World Picture. Aldine Press, Chicago.

GONZALEZ-ERICO, E. 1976. Effect of depth of lime incorporation on the growth of corn in Oxisols of Central Brazil. Ph.D. dissertation. North Carolina State Univ., Raleigh.

GOODLAND, R. J.A., and IRWIN, H.S. 1975. Amazon Jungle: Green Hell to Red Desert? Elsevier Publishers, Amsterdam.

GRIGG, D.B. 1974. The Agricultural Systems of the World: An Evolutionary Approach. Cambridge University Press, London.

LOPES, A. 1975. A survey of the fertility status of soils under "Cerrado" vegetation in Brazil. M.S. Thesis. North Carolina State Univ., Raleigh.

LOPES, A. 1977. Available water, phosphorus fixation and zinc levels in Brazilian Cerrado soils in relation to their physical, chemical and mineralogical properties. Ph.D. dissertation. North Carolina State Univ., Raleigh.

MANSHARD, W. 1974. Tropical Agriculture: A Geographical Introduction and Appraisal. Longman Group Ltd., London.

NAS-NRC. 1972. Soils of the Humid Tropics. National Academy of Sciences-National Research Council, Washington, D.C.

NAS-NRC. 1977. World Food and Nutrition Study. National Academy of Sciences-National Research Council, Washington, D.C.

ODUM, H.T. 1972. A Tropical Rainforest. U.S. Atomic Energy Commission, Washington, D.C.

PAPENDICK, R.I., SANCHEZ, P.A., and TRIPLETT, G.B. 1976. Multiple Cropping. Am. Soc. Agron. Spec. Publ. *27*. Madison, Wisconsin.

RUTHENBERG, H. 1976. Farming Systems in the Tropics, 2nd Edition. Clarendon Press, Oxford.

SANCHEZ, P.A. 1973. A review of soils research in tropical Latin America. North Carolina State Univ., Soil Sci Dep., Tech. Bull. *219*.

SANCHEZ, P.A. 1976. Properties and Management of Soils in the Tropics. John Wiley and Sons, New York.

SANCHEZ, P.A. 1977. Advances in the management of Oxisols and Ultisols in tropical South America. *In* Proceedings of the International Seminar on Soil Environment and Fertility Management in Intensive Agriculture, Oct. 10-13, Tokyo. The Society of the Science of Soil and Manure of Japan, Tokyo.

SEUBERT, C.E., SANCHEZ, P.A., and VALVERDE, C. 1977. Effects of land clearing methods on soil properties and crop performance in an Ultisol of the Amazon jungle of Peru. Trop. Agric. (Trinidad) *54*,307-321.

SOIL SURVEY STAFF. 1975. Soil Taxonomy. U.S. Dep. Agric. Handb. *436*.

SPAIN, J.M. 1975. Forage potential of allic soils of the humid lowland tropics of Latin America. *In* Tropical Forages in Livestock Production Systems. E.C. Doll and G.O. Mott (Editors). Am. Soc. Agron. Spec. Publ. *24*. Madison, Wisconsin.

TROLL, C., and PAFFEN, KH. 1965. Seasonal climates of the earth. *In* World Maps of Climatology, 2nd Edition. H.E. Landsberg *et al.* (Editors). Springer-Verlag, New York.

WOLF, J.M. 1975. Water constraints to corn production in Central Brazil. Ph.D. dissertation. Cornell Univ., Ithaca, New York.

WRIGHT, M. 1977. Plant adaptation to mineral stress in problem soils. Proceedings of Workshop held at National Agriculture Library, Beltsville, Md. Nov. 22-23, 1976. Sponsored by USAID/TAB. Cornell Univ., Ithaca, New York.

YOST, R.S. 1977. Effect of rate and placement on availability and residual value of P in an Oxisol of Central Brazil. Ph.D. dissertation. North Carolina State Univ., Raleigh.

17

Crop Production Systems in Semi-Arid Tropical Zones

B. A. Krantz and J. Kampen

Rainfed agriculture has failed to provide even the minimum food requirements for the rapidly increasing population of many developing countries in the Semi-Arid Tropics (SAT). Although the reasons for this are many, a primary constraint to agricultural development in the seasonally dry tropics is the lack of suitable technology for soil and water management and crop production systems under the undependable rainfall conditions. The SAT includes those areas within the tropics where precipitation exceeds potential evapotranspiration at least 2 and at most 7 months (Fig. 17.1). The severity of the constraints is amplified by the generally high evaporative demands and, in many areas, by soils of shallow depth with limited water holding capacity.

Agricultural research in the developing countries of the SAT has until recently been aimed primarily at increasing production in irrigated and/or dependable rainfall areas. However, in recent years there has been an increasing concern about agriculture in unirrigated areas with erratic rainfall and an increasing number of people have realized that the potential of the SAT to produce food for a hungry world far exceeds present production levels. This widespread concern about SAT and the new confidence in the productive potential of these areas resulted in the creation of the International Crops Research Institute for the Semi-Arid Tropics (ICRISAT) in 1972.

THE SETTING

In several regions of the SAT, the average annual precipitation appears to be sufficient for one, or in many cases, two good crops per year. However, the rainy seasons are brief, the rainfall patterns erratic, and brief or extended droughts are frequent. Much of the rain occurs in high intensity storms, resulting in runoff and erosion even on relatively flat lands. During the rainy season, the storage capacity of the root profile is usually exceeded and water percolates to deeper layers or ground water. Thus, only a portion of the precipitation is available for crop use. This situation has resulted in unstable food production and continued low crop yields. Because of the ever present risk of droughts, farmers of the SAT have been reluctant to adopt the use of high yielding varieties, fertilizers, and other inputs even when available.

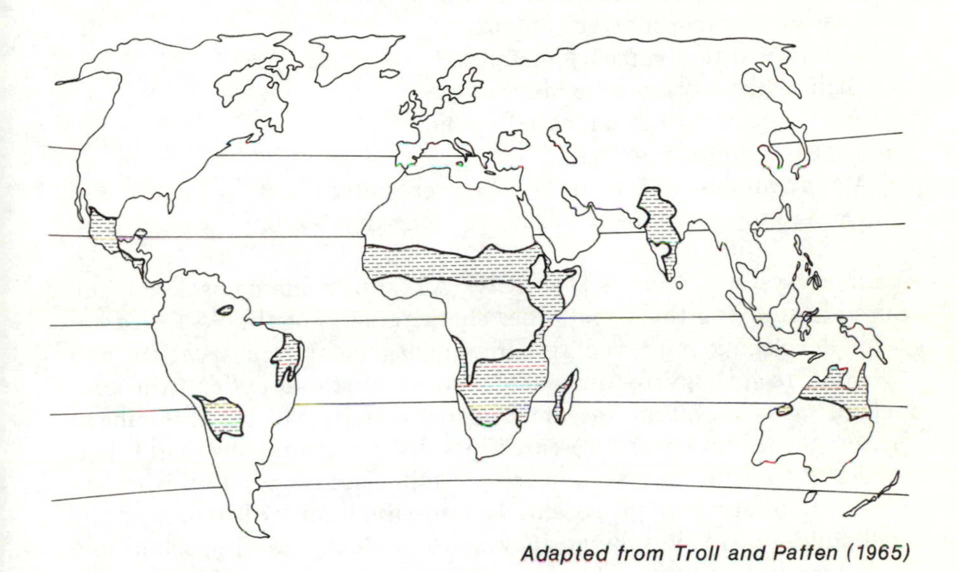

Adapted from Troll and Paffen (1965)

FIG. 17.1. MAP OF SEMI-ARID TROPICAL ZONES

During the past 30 years, populations in many areas of the SAT have doubled; farmers have attempted, therefore, to double agricultural production. Since there has been no substantial increase in per hectare yields during this period, the result has been a tremendous increase in cropped area and livestock numbers. The proportion of geographical area cropped is high, reaching 80% or more in some districts in India (Ryan 1976). Thus steep and erodable lands are frequently being overcropped and overgrazed. Forest lands have been denuded, causing permanent damage to vast areas. The decreasing productivity of the land increases the quest for more land. To break this vicious circle, more stable forms of land use which preserve, maintain, and better utilize the productive capacity of available resources are urgently needed. Two major problems face the small farmers and landless laborers in the SAT: (1) the low and unstable crop yields which have persisted throughout the years and (2) the accelerating rate of soil erosion which occurs in most areas of the SAT.

Some of the most relevant characteristics of the SAT include the following:

(1) Intensive rainfall interspersed with unpredictable droughts
(2) Relatively short rainy seasons
(3) Highly variable rainfall during the rainy season
(4) Low soil organic matter content
(5) Low infiltration capacity of soils
(6) High water erosion hazard
(7) Small farms with fragmented holdings
(8) Limited capital resources
(9) Mainly animal or human labor power sources
(10) Severe unemployment during the long dry season.

Farming systems scientists at ICRISAT and in many national programs are studying these and other characteristics of the SAT as a basis for developing improved resource management, conservation, and utilization technology for increased and more stable agricultural production. Farming systems research is a new activity which is emerging in conceptual form at several research institutions around the world. It is envisioned that farming systems research in close cooperation with improvement of the genetic potential of specific crops will result in the rapid development of substantially more productive and economically viable farming systems. This approach is especially important in the developing world where there is an urgent need for more food.

Water is the major natural constraint to improved systems of farming in the SAT. Alleviation of the effect of this barrier should be the ultimate aim and focus of the farming systems research programs.

PAST APPROACHES TO SOIL AND WATER CONSERVATION MANAGEMENT

Four approaches to ameliorating the problems faced by farmers in the rainfed SAT have been:

(1) Emergency programs to meet droughts and food crises
(2) Developing irrigation facilities
(3) Fallowing the land during the rainy season to accumulate a moisture reserve in the profile for the subsequent crop season
(4) Constructing bunds for soil and water conservation

Crises and Emergency Programs

During droughts and the associated food crises, large sums of money are frequently spent for hastily conceived programs in various types of famine relief. Food aid is often provided and "crash" resource conservation and development schemes are designed and implemented. However, after the calamity is over and life returns to "normal," the acute problem may be forgotten till the next crisis occurs. Severe droughts in the Indian subcontinent in 1965-1967 and 1972-1973, as well as the recent disastrous droughts in the Sahel region of Africa, provide examples of this type of effort. These types of activities seldom result in improving the stability and long term productive potential of the environment.

Conventional Irrigation

Irrigation facilities in the SAT include large gravity irrigation projects, small runoff storage reservoirs (called tanks in India), and wells. Large irrigation projects were initially envisaged as supplemental water facilities for upland crops. In practice these projects usually provide continuous water on a seasonal basis to crops with high water requirements and do not presently have the flexibility required to act in a supplemental fashion. Where there are inadequate drainage facilities irrigation water and rainfall have resulted in waterlogging and salinity problems.

Wells are owned mostly by individuals; however, the water may be drawn from an area exceeding the individual property boundaries. Full irrigation applied to small areas of wetland crops such as rice, rather than supplemental irrigation of rainfed crops over an extended area, is frequently the result.

Existing small runoff storage reservoirs (tanks) are characterized by shallow depths, large evaporation and seepage losses, inefficient use of water, and considerable loss of land. While substantial areas of productive land are occupied, siltation due to lack of erosion control in the contributing watershed has significantly reduced the storage capacity of many of these tanks. A sample survey of the Indian states of Karnataka, Tamil Nadu, and Andhra Pradesh (Anon. 1966) showed that the ratio of tankbed to service area was only 1 to 1.3 due to evaporation, siltation, and shallow depth, use of water for rice cultivation, and inefficient management.

Thus, it appears that present water resource development in the SAT often results in the creation of "small islands of relative wealth" in "a sea of poverty." The fact that the total water resources are insufficient to cover any substantial portion of the cultivated land through conventional irrigation has been overlooked. At the same time, few serious efforts have been made to explore the question of how the limited water resources can be used to stabilize and support large proportions of rainfed agriculture. Further research is urgently needed to answer this question.

Cultivated Fallow

In many of the temperate semi-arid regions, the total moisture storage capacity of the soil exceeds the normal effective annual rainfall. Under these conditions, cultivated fallowing during one or more years will often substantially increase yields due to a larger capacity of moisture available to the crop (Pengra 1952). However, in the SAT, total seasonal rainfall is often several-fold that of the capacity of the root zone to store moisture. Also, high intensity rainfall frequently exceeds the soil infiltration rate.

On many deep Vertisols, especially in India, cultivated fallow is practiced during the rainy season with cropping only during the post rainy season. The reasons for not cropping during the rainy season include such factors as poor drainage, difficulty of tillage and weed control, undependability of monsoon rains, and inadequate soil and crop technology (Kampen *et al.* 1974). However, the consequences of traditional fallowing on these deep Vertisols are serious with regard to soil erosion. Jacks *et al.* (1955) noted that a few minutes of high intensity rainfall on some bare soils is sufficient to cause surface sealing and drastically reduce infiltration. Arnon (1972) stated that frequent cultivation to produce a soil mulch often results in an impaired soil structure, thus increasing runoff and soil erosion. Thus, the logic of cultivated fallow in the SAT is questionable from the standpoint of soil erosion and frequently results in only a small portion of the rainfall actually being used for crop production.

Traditional Bunding

Contour bunds or graded bunds are used in many areas with the aim of decreasing soil erosion and conserving water above the bunds. In the semi-arid tropical environment, well designed and maintained systems of bunds have been found to decrease soil erosion on a watershed basis. However, substantial erosion and sedimentation may occur in the areas between bunds. The long term effect of this interbund erosion results in a large elevation difference upslope and downslope of the contour or field bund (Fig. 17.2). Although large quantities of water are held back by the bunds, the increased infiltration often affects only a small portion of the land in the watershed.

FIG. 17.2. DIFFERENCE IN ELEVATION (1.1 m) UPSLOPE AND DOWNSLOPE OF A FIELD BUND DUE TO SOIL EROSION THAT HAS TAKEN PLACE OVER THE YEARS IN A RAINY SEASON FALLOW, POST-RAINY SEASON CROPPING SYSTEM

Often farmers will not allow the contour bund to pass through their fields. Thus the bunds do not actually follow the contour (Fig. 17.3A) or the graded contour (Fig 17.3B) but follow the field boundary which creates additional drainage and erosion problems. Graded beds and furrows (Fig. 17.3C), which will be discussed later, can avoid this problem.

A study of the effect of bunding on the yields of monsoon crops on heavy soils was conducted in India during the 1960-1967 seasons (ICAR 1970). The average rainfall in the area of study is about 550 mm. It was found that contour bunding decreased the average yield of rainy season sorghum and cotton by 25 and 30%, respectively. The decrease of yields was explained by the delay in cultural operations and crop damage due to stagnant water above the bunds (Fig. 17.4). The effect of bunding on wheat yields following rainy season fallowing on deep Vertisols in central India was studied in 1972 (Krantz and Kampen 1977). These data indicated no increase in wheat yields due to the water trapped near the bunds. There was a decrease in wheat yields in the pit just above the bunds due to removal of surface soil during the construction of the bund. This effect, plus the area removed from cropping by the contour bund, resulted in a negative effect on total production from bunding under these conditions. Similar reduced yields on bunded lands were found at ICRISAT during 1974-1977 with maize, pigeonpea, and chick-

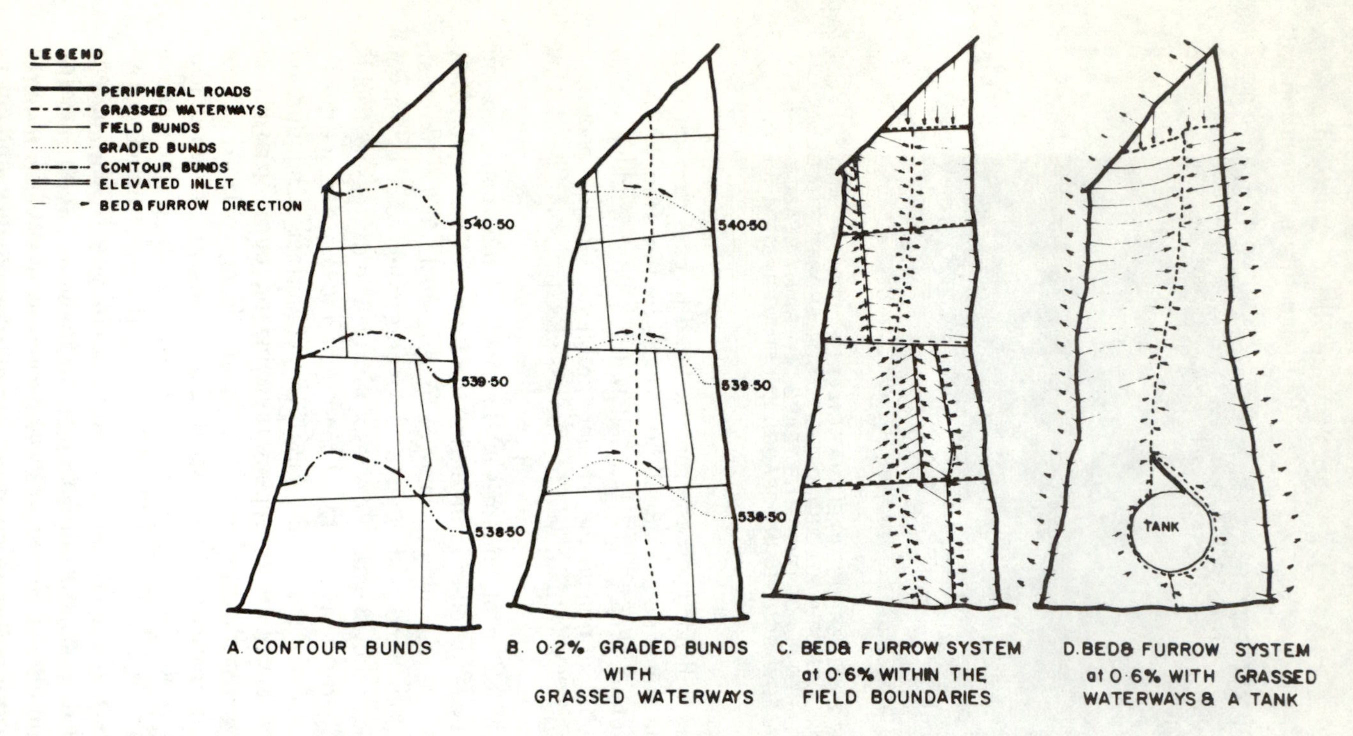

FIG. 17.3. VERTISOL WATERSHED (BW2) WITH FOUR POSSIBLE SOIL AND WATER CONSERVATION AND MANAGEMENT PRACTICES ILLUSTRATED IN THE SAME WATERSHED

(The system which was implemented 5 years ago and still exists is map "C" which consists of bed and furrow system at 0.6% slope within the field boundaries).

pea (ICRISAT Annual Reports 1973-1974 through 1976-1977). The authors have not been able to identify any results of controlled experiments which show crop yield increases due to moisture conservation by contour bunds in the SAT of India.

FIG. 17.4. STAGNANT WATER ABOVE A CONTOUR BUND IN LATE OCTOBER

Chickpeas planted in rainy season fallow vertisols in early October have been killed due to excess water. (Note sheet erosion across chickpea rows in the foreground).

It is apparent that past approaches to soil and water conservation to increase agricultural production in the SAT have not provided the break-through in soil and water conservation and utilization which is imperative for rapid development of agriculture in the SAT.

APPROACHES AND POTENTIALS FOR IMPROVED CROP PRODUCTION SYSTEMS

Strategy for Research and Development

The basic strategy for farming systems research and development at ICRISAT and in some national programs has been to investigate the natural, human, animal, and capital resources and to develop resource management, conservation technology, and crop production methods to optimize the utilization of these resources on a sustained basis.

The most severe constraint to the development of agriculture in the SAT is the lack of appropriate technology to make the best use of the available natural resources—climate and soils. Traditional systems of farming use only a small portion of the rainfall in any given year for crop production. Low effective rainfall combined with low water use efficiency results in a very low "rainfall use efficiency" in most of the present farming systems.

The farmer's actions are guided by experience with an undependable environment characterized by erratic rainfall which frequently occurs in high intensity and is interspersed with droughts. Therefore any new soil and water conservation technology must be associated with crop production systems which will provide immediate and clearly visible increases in yield levels and stability of agricultural production. Small gains or promises of future increases will not capture the imagination of the farmers and will not be rapidly implemented. Thus the two major strategies are:

(1) To provide technology for the improvement of the natural resource environment by better management so as to most effectively conserve and utilize the soil and water resource

(2) To provide technology for growing a continuum of productive and well-managed crops to facilitate optimum use of the improved environment from the onset of the rainy season for as long as soil moisture is available to plants.

Watershed-based System of Soil and Water Conservation

Since water is the first limiting natural factor in crop production in the SAT, improving the management of water and soil for increased crop production becomes the primary aim of farming systems research. In rainfed agriculture, the only water available is the rain that falls on a given area. Thus the watershed (catchment) is the natural focus for research on water management in relation to crop production systems and resource conservation and utilization.

Contour bunding (which has been used extensively in India) and watershed-based utilization systems employ distinctly different concepts of water conservation and management. In contour bunding, the excess water flows in concentrated fashion through minor depressions and collects at the bund and is forced to flow across the slope and out of the watershed where it is finally disposed of in roadside drains or gulleys (Fig. 17.3A). In cropped watersheds cultivated in graded beds and furrows, excess water is allowed to flow through furrows to the grassed drainage ways and is safely conducted to a tank and/or outlet. However, the velocity of flow of the water is controlled by the direction and slope of the bed and furrow system and no concentration in large overland flows is allowed. Since the 150 cm bed and furrow system can be a permanent feature, it can provide considerable protection against soil erosion on a year-round basis, even during the prolonged hot and dry non-crop season. The slope used should be one selected such that erosion during high intensity rain is minimized, infiltration is increased, adequate crop drainage is provided during prolonged rains, especially on deep Vertisols, and supplemental irrigation is facilitated as needed (Fig. 17.3 C & D).

ICRISAT's watersheds-based resource utilization research involves the monitoring of data on a wide range of activities which include inventories of natural resources, complete water balance studies (including measurement of precipitation, rainfall intensities, runoff, soil erosion, nutrient losses, consumptive use by plants, and deep percolation into shallow ground water), rainfall use efficiency, annual production, human labor utilization, bullock power use efficiency, the use of fertilizers, seeds, pesticides, and other material inputs, etc. Economic evaluations are made on all management systems throughout the year.

By investigating all facets of resource inventory, development, management, and utilization and all factors involved in crop production in a "system" approach, ICRISAT scientists expect to develop methodology which will attain alternative and economically viable systems of farming for the SAT. Although the research is focused on the small farmers of limited means, who comprise the majority of the population of the SAT, the large scale farmers who occupy most of the land and employ most of the landless labor are also expected to benefit.

PRELIMINARY RESULTS OF FARMING SYSTEMS INVESTIGATIONS

Studies on Bed and Furrow Systems

An effective soil and water management system should reduce runoff and soil erosion, increase infiltration of rainfall, provide surface drainage during the rainy season, and facilitate application of supplemental irrigation water. Systems involving graded beds (ridges) separated by furrows which drain into grassed waterways appear to possess most of the essentials for fulfilling these requirements. The surface drainage function of beds and furrows has been shown by Choudhury and Bhatia (1971) and Krantz and Kampen (1977). Preliminary research at ICRISAT indicates that the system can also be used, within limits, to manipulate the amount of runoff water and reduce erosion by altering the slopes of the beds and furrows. An example of the type of system envisaged is illustrated in a schematic drawing (Fig. 17.3D).

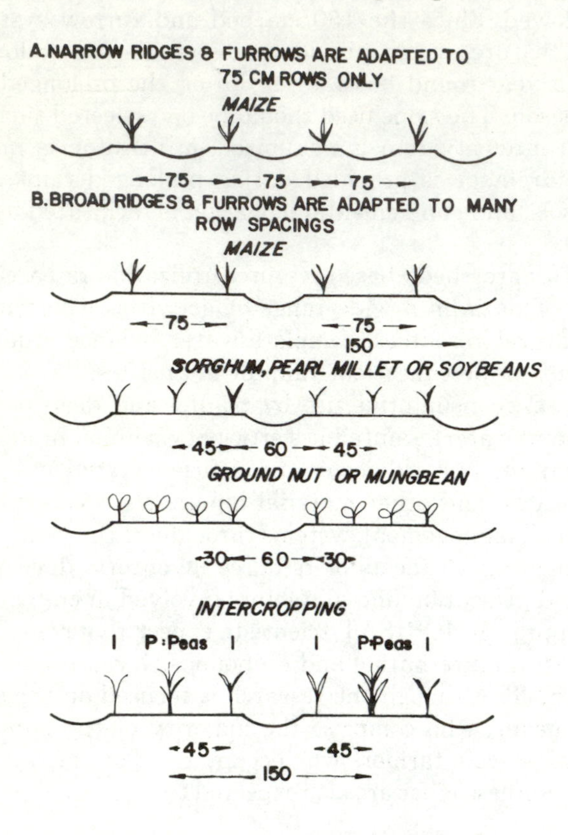

FIG. 17.5. SOME POSSIBLE CROPPING PATTERNS WITH 75 AND 150 CM BED AND FURROW SYSTEMS

During the 1975 season a broad (150 cm) bed and furrow system was compared to a 75 cm bed and furrow system on Alfisols. In earlier years, cross flow and erosion had been observed in the 75 cm bed and furrow system, especially in slight depressions due to instability of the 75 cm beds. This problem was overcome by the use of a 150 cm bed and furrow system. With the 150 cm beds it is possible to plant 2, 3, or 4 rows per bed at 75, 45, and 30 cm row spacing, respectively. The broad beds allow for planting pigeonpeas at a desired 150 cm spacing along with a wide range of plant geometries for the intercrop. An example of this is shown in Fig. 17.5. The 150 cm bed and furrow system can be easily prepared and maintained with bullock-drawn implements. The "bed-former" used has ridger units placed 150 cm apart with a levelling float in between the two ridgers (Fig. 17.6).

FIG. 17.6. RIDGER UNIT RE-FORMING BEDS FOR THE THIRD YEAR IN EXACTLY THE SAME LOCATION

The ridges or beds function as mini bunds, at a grade which is normally less than the maximum slope of the land. Thus, when runoff occurs, its velocity is reduced and infiltration opportunity time is increased. Instead of allowing runoff to concentrate in large streams, the excess water is carried off the land in a large number of very small flows. This tends to minimize soil erosion and provides more uniform infiltration over the whole watershed area. Some additional advantages of the bed and furrow system, observed in operational-scale research on watersheds, are:

(1) Only minor earth movement (smoothing) is required for implementation.

(2) Implementation can be executed by animal power or light tractors.

(3) The furrows provide a guide for draft animals to follow; this speeds up planting and facilitates uniform row spacing.

(4) While planting in between early monsoon showers, the top of the beds dries more quickly than flat cultivated areas, thus facilitating earlier planting on the beds.

(5) No land is taken out of production by the bed and furrow system (Fig. 17.3D).

(6) No further land development is necessary to facilitate application of supplemental water.

(7) The beds (mini bunds) can be permanent features which can be easily maintained with limited tillage; soil compaction in the bed (plant zone) is minimized.

(8) The 150 cm bed and furrow system can be implemented within farmers' field boundaries as shown in Fig. 17.3C. This system has been implemented and successfully operated for 5 years in a Vertisol watershed (BW2) (Fig. 17.3). However, more than 8% of the land area in this watershed was in field boundary bunds. By removing these bunds and aligning the farmers' fields with the permanent beds the land occupied by bunds could be added to the cultivated area of each farm (Fig. 17.3D). With the longer rows in this system, tillage and planting would be easier and faster and bund removal would also reduce weed and rat problems.

(9) Soil on the beds remains friable, facilitating initial primary tillage immediately after harvest of the last crop of the season. This plowing concentrates all organic residues and fertilizer in the plant zone and re-forms the bed and provides for erosion control throughout the year.

The Effect of Soil Management Upon Runoff and Soil Loss

Preliminary results show that runoff and soil loss can be greatly reduced by improved management in deep Vertisols (Table 17.1). In each of 3 storms during September 1974, the runoff from a rainy season fallowed watershed (BW4C) was about 5 times that of an adjacent cropped watershed cultivated to a bed and furrow system (BW3B). Soil loss from the former system was more than 12 times that of the latter. In addition to the soil loss observed at the outlet of the watershed, substantial interbund erosion was observed in the cultivated-fallow watersheds (an example of the long term effect of interbund erosion is shown in Fig. 17.2). Much more erosion was observed between field or contour bunds on flat planted watersheds compared to areas cultivated in a bed and furrow system. This is particularly evident if high intensity showers occur early in the growing season before crop canopy has developed. Cultivated fallow during the monsoon has shown no advantage in terms of moisture conservation or post-rainy season crop yields compared to the situation observed in watersheds cropped during the rainy season under the climatic conditions experienced at ICRISAT.

Water Balances and Effective Rainfall

Data collected in 1974 at ICRISAT indicate that improved systems of farming used a much higher proportion of the rainfall in crop production than the traditional practices. The percentage of monsoon rainfall (925 mm) used in crop production for watersheds with post-rainy season cropping only (BW6B), double-cropped on graded beds (BW1), and double-cropped on graded beds with recycled runoff water (BW5) was 30, 70, and 80%, respectively (ICRISAT 1974-1975 Annual Report).

TABLE 17.1. COMPARATIVE RUNOFF AND SOIL LOSS FROM A CROPPED WATERSHED WITH THE BED AND FURROW SYSTEM (BW3B) VERSUS A RAINY SEASON FALLOW AND POST-RAINY SEASON CROPPED WATERSHED (BW4C) ON DEEP VERTISOLS DURING FIVE STORMS IN 1974

Date	Rainfall	Runoff (mm/ha)		Soil loss (kg/ha)		Ratio of Soil Loss
		BW3B	BW4C	BW3B	BW4C	BW4C/BW3B
Aug. 9	51.9	1.74	3.00	182	455	2.5
Sept. 12	35.2	0.64	2.87	16	265	12.0
Sept. 24	29.2	0.68	3.81	37	445	12.0
Sept. 25	12.8	0.45	5.33	12	281	23.3
Oct. 23	56.3	4.57	10.00	420	941	2.2

In 1975-1976, on deep Vertisols, a double-cropped watershed with a 0.6% graded bed and furrow system had a total evapotranspiration of approximately 675 mm during the growing season with a seasonal rainfall of 1040 mm. Thus, the effective rainfall amounted to about 65% of the rainfall. Of the remaining portion of the total seasonal precipitation, approximately 15% was lost as runoff and about 20% percolated down to ground water. Rainy season fallow and post-rainy season cropping under traditional technology resulted in greatly increased runoff (25%), and a substantial portion (around 30%) of the rainfall was lost in terms of evaporation from soil and transpiration by weeds (ICRISAT 1975-1976 Annual Report). Compared to deep Vertisols, double-cropped watersheds on shallow to medium deep Vertisols were characterized by somewhat less total runoff and lower effective rainfall, but substantially greater deep percolation into ground water (30%).

On Alfisols in 1975-1976, a double-cropped watershed with graded beds and furrows had a total runoff of approximately 30%, an effective rainfall of 55%, and about 15% of the rainfall contributed to ground water.

Runoff Collection and the Use of Supplemental Water

During the 1974 season most of the runoff storage reservoirs were partially (50 to 70%) filled during the early part of the rainy season, thus providing water, if required, for "life-saving" irrigation during drought. Even the partially emptied tanks were completely refilled during the latter part of the monsoon and substantial volumes of water were lost across spillways. Evaporation and seepage loss measurements of water losses from various tanks indicated seepage losses ranged from 2.6 to 18 mm per day and evaporation losses varied from 3 to 5 mm per day. Since the water in these tanks is mostly used during the early part of the post-rainy season before temperatures start rising, the proportion of water lost by evaporation from deep tanks is relatively small.

The results of supplemental irrigation to crops on Alfisols during a 30-day drought in late August and early September of 1974 were quite spectacular. Yields of sorghum and maize were approximately doubled by the application of a 5 cm irrigation. At product prices prevailing at the time of harvest, gross rupee values of the average increase from a 5 cm supplemental irrigation at a critical growth stage in 2 watersheds were 3120 ($360), 2780 ($320), 1085 ($125), and 650 ($75) Rs per hectare for maize, sorghum, pearl millet, and sunflower, respectively. At these values the rupee value of a hectare-meter (8.1 acre-foot) of water (neglecting application losses) would be Rs.62,400 ($7171), 55,600 ($6390), 21,700 ($2490), and 13,000 ($1494) for the respective crops (Rs.9=$1).

During the 1975 rainy season, rainfall was uniformly distributed and no irrigation was required. In the post-rainy season, sorghum on deep Vertisols responded to supplemental irrigation at the grain filling stage. On Alfisols, tomatoes planted on beds in pearl millet stubble in 1975 responded to supplemental irrigation.

Alfisols which have a low water holding capacity may benefit from supplemental water during the rainy season every one to three years. However, even when supplemental irrigation is not needed, farmers with tanks can take advantage of yield increasing inputs at greatly reduced risks. During the post-rainy season supplemental irrigation may be beneficial on both Alfisols and Vertisols, especially for high value crops.

The Interaction of Various "Steps" in the Transfer of Improved Technology

To achieve the full benefits of watershed-based resource development, all facets of farming systems must be improved. In the development and implementation of improved technology there are many "steps" involved. The many facets were grouped into the following four phases—variety, fertilization, soil and crop management, and supplemental water—if needed.

In 1976, an experiment was conducted with sorghum, involving a comparison of "traditional" versus "improved" technology on an Alfisol. During the rainy season the rainfall distribution was adequate and fairly uniform and no supplemental water was required. Thus, only the first three phases were considered.

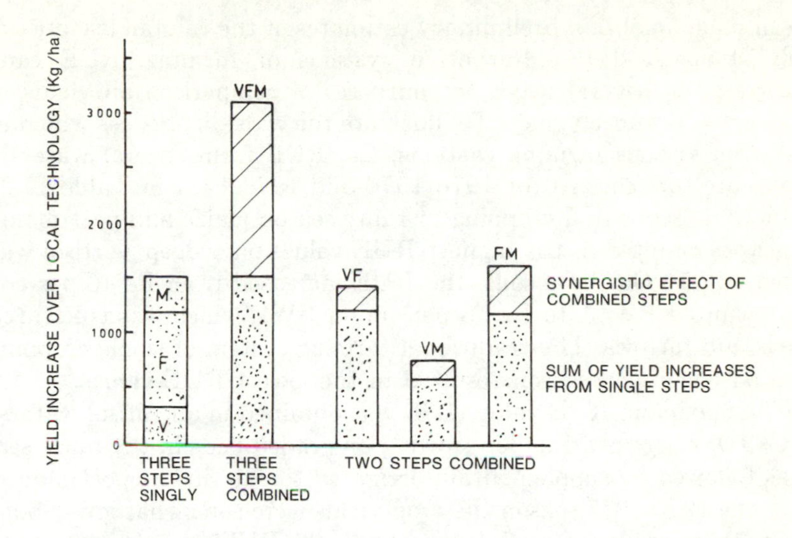

FIG. 17.7. SORGHUM GRAIN YIELD INCREASES FROM "IMPROVED" VARIETY (V), FERTILIZATION (F), AND SOIL AND CROP MANAGEMENT (M) SINGLY AND IN COMBINATION OVER "TRADITIONAL" TECHNOLOGY, 1976

(The lower portion of each bar represents the yield increase of improved over traditional for each factor applied singly and the upper portion the synergistic effect of the same factor applied in combination).

Figure 17.7 shows the yield increases of "improved" over "traditional" technology from the application of the three steps (Variety, Fertilization, and Soil and Crop Management) singly and combined. The yield increase from the 3 steps combined was double that of the sum of the increase due to the same 3 steps applied singly, thus illustrating the large synergistic effect when all 3 steps were applied together in a system. It was also noted that there was a slight synergistic effect when any two factors were combined; however, the magnitude of this effect was far less.

Similar results were obtained with sorghum on the same soil in 1975 and with maize on a Vertisol in 1976 and with wheat (Krantz *et al.* 1975).

Crop Production and Rainfall Use Efficiency

The basic objective of ICRISAT's farming systems research program is to attain the best possible crop production from the total seasonal rainfall by means appropriate for the farmer of the SAT. An attempt has been made to obtain preliminary estimates of the rainfall use efficiency (RUE) achieved by alternative systems of farming. RUE can be expressed in several ways; for purposes of comparison all yields were converted to rupee values. To illustrate the wide differences which exist between various farming systems, the RUE from several watersheds units are summarized for 1975-1976 and 1976-1977 in Table 17.2.

In 1975, sequential cropping of rainy season maize and post-monsoon chickpea resulted in the highest RUE values on a deep Vertisol watershed (BW1). On these soils, the RUE increased from Rs 46 per cm on flat planted BW4A to Rs 58 per cm on BW1, which was cultivated in beds and furrows. The traditional farming system of single cropping in the post-monsoon season resulted in the lowest RUE values.

The maximum RUE value (130) was obtained in an Alfisol watershed (RW1D) where pearl millet growing on graded beds in the rainy season was followed by supplementally irrigated tomato in the post-monsoon.

In the 1976-1977 season the crop yields were somewhat lower but the seasonal rainfall was much lower. Thus, the RUE's were higher on both soils in all cases where improved technology was used. Single-crop sorghum in the post-rainy season gave a RUE of only 9 Rs per cm of rainfall. The range of RUE values obtained on Vertisol watersheds (in 1975 from 12 to 58 Rs per cm and in 1976 from 9 to 69 Rs per cm) illustrates the production potential of this type of environment in the SAT, given the introduction of improved technology.

SOIL, WATER, AND CROP MANAGEMENT FOR VERTISOLS AND ALFISOLS IN THE SAT

After several years of operational research on a field-scale the following soil, water, and crop management systems have been evolved and found successful on the deep Vertisol watersheds.

(1) Establish a 150 cm bed and furrow system at about a 0.6% slope after minor smoothing to erase the microrelief. (These beds can be established on a permanent basis and maintained by limited tillage.)

(2) Till using left and right hand plow simultaneously on beds immediately after the harvest of the last crop of the growing season to kill all weeds and stubble and cover the cracks with a cloddy mulch, thus conserving residual moisture in the soil profile. By not allowing the soil profile to dry out to great depths, the soil profile is recharged more quickly during the early rainy season.

(3) After an early shower in the period February to June has moistened the clods and the soil has dried sufficiently to avoid compaction, use a ridger-cum-bed former to re-shape the beds. This completes the seed bed

TABLE 17.2. CROP YIELDS, GROSS MONETARY VALUES AND RAINFALL USE EFFICIENCIES[1] (RUE) OBTAINED IN SOME ALTERNATIVE FARMING SYSTEMS IN 1975–1976 AND 1976–1977

Watershed No.	Planting Method	Cropping System	Crop Yield Monsoon kg/ha	Crop Yield Post-monsoon kg/ha	Value Rs/ha	RUE Rs/ha-cm
1975-76						
BW1[2]	Beds	Maize, chickpea	4830	1790	6012	58
BW4A	Flat	Maize, chickpea	3680	1510	4756	46
BW4C	Flat	Fallow, sorghum	—	2190	1642	12
RW1D	Beds	Pearl millet, tomato + 5 cm irrig.	2750	2000	14362	130
1976-77						
BW1	Beds	Maize-pigeonpea intercrop	3360	720	4870	68
		Maize-chickpea	3310	600	3840	53
BW2	Beds	Maize-pigeonpea intercrop	2910	900	4960	69
		Maize-chickpea	3090	450	3410	47
BW4A	Flat	Sorghum-pigeonpea intercrop	2580	670	4130	57
BW4C	Flat	Fallow sorghum (trad. tech.)	—	600	660	9
RW1D	Beds	Groundnut-safflower	1480	180	4480	66
RW2B	Beds	Sorghum + ratoon	3080	550	3270	50
RW2B	Beds	Sorghum + ratoon + 10 cm irrig.	3080	1060	3730	57

Source: ICRISAT (1976).
[1]In 1975 the total seasonal rainfall in Vertisol watersheds was 1040 mm and in Alfisols 1100 mm. In 1976, the rainfall was 720, 680, and 650 mm for BW1-8, RW1, and RW2, respectively.
[2]BW refers to Vertisol watersheds; RW refers to Alfisol watersheds.

preparation during the dry season well ahead of planting time with minimal tillage and soil compaction. Each operation covers 150 cm at a time and the implement can be pulled by one pair of good bullocks. This also reduces peak power demands during the early monsoon.

(4) Apply a starter fertilizer of about 75 kg per hectare of 0-18-46 in. bands 8 to 10 cm below each plant row on the bed. This can be done a week or two before planting to further reduce peak power demands and speed up planting.

(5) About one week before the expected onset of the rainy season plant "dry" any crop which can be planted 5 to 7 cm deep. This includes crops such as sorghum, pigeonpea, maize, and cowpeas. In the past six years dry planting of such crops has been successful. This is much easier and less hazardous than attempting to plant after the rains start on these montmorillinitic clay Vertisols which become very sticky when wet. Because of the high water holding capacity of these soils and relatively deep planting, no germination will take place after small showers of 5 to 10 mm. When there is sufficient rainfall to germinate the seed at the 5 to 7 cm depth, there is sufficient soil moisture to keep the plant alive even without further rain for 10 to 14 days.

On Alfisols a different procedure has been followed:

(1) Establish the bed and furrow system on a permanent basis after land smoothing as in the Vertisols.

(2) In a single crop system, till the beds with left and right hand plow immediately after crop harvest to kill weeds and form a rough cloddy surface which is receptive to pre-monsoon showers and resists possible wind erosion.

(3) Where Alfisols are double-cropped and are very dry and hard after the second crop, the primary plowing may have to be delayed until after the first early rain which may occur at any time between February and June.

(4) Delay planting until after the rains have moistened the soil to a depth of 15 to 20 cm, because most Alfisols have a low water holding capacity. Since Alfisols dry out readily after a rain, there is less difficulty in planting after the monsoon has started than in the Vertisols.

Procedures which have been successful on both soils are:

(1) Cultivate and hand weed early, as soon after crop emergence as weather permits. In high rainfall areas where early weeding is difficult use a minimal application pre-emergence herbicide for early weed control in deep Vertisols.

(2) In intercropped systems involving a medium-duration cereal, re-

move the cereal as soon as possible after physiological maturity to remove light and moisture competition from the long-duration crop. In the case of an early maize intercrop on deep Vertisols, the tops can be broken over just above the ear to reduce shading of the long-duration intercrop and also allow the maize grain to dry in the field.

(3) Where ratoon cropping of sorghum is planned, harvest the heads and cut the stalks just above ground level at physiological maturity. A few days delay in harvest appears to reduce ratoon crop growth.

(4) Where double-cropping is practiced remove the monsoon crop as early as is feasible and cultivate and plant between the standing stubble. This technique has worked successfully and greatly reduces the energy and time required for land preparation. Saving energy is an important factor when all cultural operations are carried out by the use of bullock power; saving time increases the length of the growing season for the second crop and conserves soil moisture for germination.

REFERENCES

ANON. 1964. Study of soil conservation program for agricultural land. Program Evaluation Organization, Planning Commission, Government of India, New Delhi.

ANON. 1966. All India review of minor irrigation works based on statewide field studies. Committee on Plan Projects Irrigation Team, Government of India, New Delhi.

ANON. 1974. Draft fifth five-year plan, Vol. II. Planning Commission, Government of India, New Delhi.

ARNON, I. 1972. Crop Production in Dry Regions. Leonard Hill, London.

CHOUDHURY, S.L., and BHATIA, P.C. 1971. Ridge planted kharif pulses yield high despite waterlogging. Indian Farming *21*(3)8-9.

FAO. 1974. Improved productivity in low rainfall areas. Proceedings, FAO Committee on Agriculture, New Delhi, 1971. FAO, Rome.

ICAR. 1970. Soil and water conservation research 1956 to 1958. Indian Council of Agricultural Research, New Delhi.

ICRISAT. 1976. Annual Reports 1973-74 to 1976-77. Hyderabad, India.

JACKS, G.V., BRIND, W.D., and SMITH, P. 1955. Mulching Tech. Commun. *49.* Commonw. Bur. Soil Sci., Harpenden, England.

KAMPEN, J. 1974. Water conservation practices and their potential for stabilising crop production in the SAT. Proceedings, FAO Conference on Food Production in Rainfed Areas of Tropical Asia, ICRISAT, Hyderabad, India. FAO, Rome.

KAMPEN, J. *et al.* 1974. Soil and water conservation and management in farming systems research for the semi-arid tropics. *In* International Workshop on Farming Systems. ICRISAT, Hyderabad, India.

KRANTZ, B.A., and CHANDLER, W.V. 1954. Fertilize corn for higher yields. N.C. Agric. Exp. Stn. Bull. *366*.

KRANTZ, B.A., and KAMPEN, J. 1977. Water management for increased crop production in the semi-arid tropics. *In* Proceedings of National Symposium on Water Resources in India and Their Utilization in Agriculture. Water Technology Centre, IARI, New Delhi.

KRANTZ, B.A. *et al.* 1974. Cropping patterns for increasing and stabilising agricultural production in the semi-arid tropics. *In* International Workshop on Farming Systems. ICRISAT, Hyderabad, India.

KRANZ, B.A. *et al.* 1975. Agronomy of dwarf wheats. Indian Council of Agricultural Research, New Delhi.

PENGRA, R.F. 1952. Estimating crop yields at seeding time in the Great Plains. Agron. J. *44*, 271-274.

RYAN, J.G. 1976. Resource inventory and economic analysis in planning agricultural development in drought prone areas. Proceedings, Training Program for Agricultural Officers in DRAP Districts, Hyderabad, India, February. ICRISAT, Hyderabad.

TROLL, C., and PAFFEN, KH. 1965. Seasonal climates of the earth. *In* World Maps of Climatology, 2nd Edition. H.E. Landsberg *et al.* (Editors). Springer-Verlag, New York.

18

Crop Production Systems in Arid Tropical Zones

H. S. Mann and A. Krishnan

CHARACTERISTICS OF ARID TROPICAL ZONES

Location

The principal areas of the arid tropics are in North Africa and an area of about two million sq km in countries like Yemen, Saudi Arabia, Muscat, Mexico, India, and Australia (Fig. 18.1). Arid tropical areas in

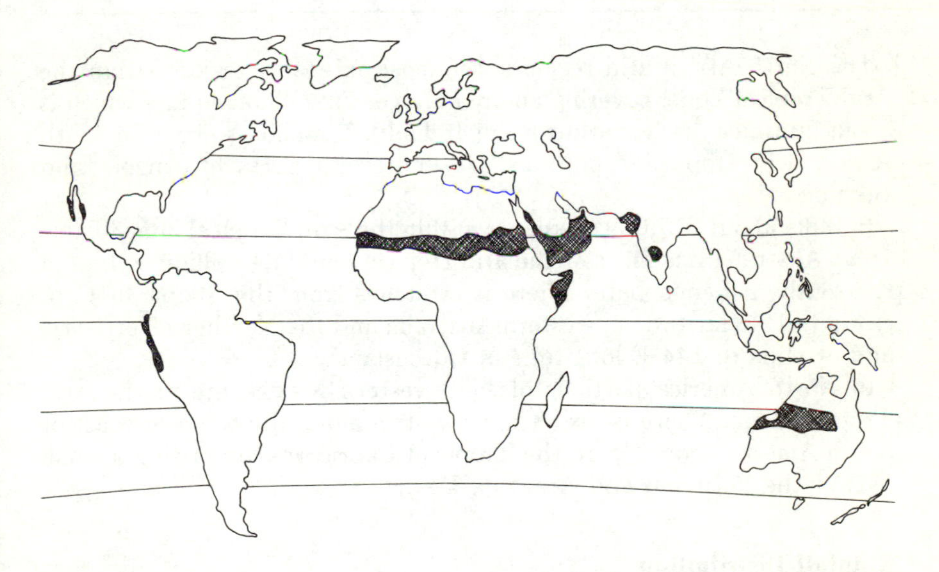

Adapted from Troll and Paffen

FIG. 18.1. MAP OF ARID TROPICAL ZONE

African countries as estimated by the method of Thornthwaite and Mather (1955) are given in Table 18.1. Assuming the Tropic of Cancer as the northern limit, only small southern portions of the deserts in Algeria, Libya, and United Arab Republic fall within the Arid Tropical Zone. The southern limit of the arid areas approximates with the 16°N latitude in Senegal, 13°N in Chad and Sudan, and almost the equator in Somalia.

TABLE 18.1. AREAS OF THE ARID TROPICAL ZONE IN NORTH AFRICA

Country	Total Area 1000 sq km	Area in Arid Tropical Zone 1000 sq km	%
Mauritania	1031	902	87
Spanish Sahara	266	79	30
Senegal	196	18	9
Mali	1240	808	65
Algeria	2382	355	15
Upper Volta	274	34	12
Niger	1267	1189	94
Nigeria	924	24	3
Libya	1760	228	13
Chad	1284	817	64
Sudan	2506	1412	56
United Arab Republic	1001	171	17
Ethiopia	1284	760	59
Somalia	638	438	69

The North Africa arid region is the most extensive region within the Arid Tropical Zone, covering an area of over 7.2 million sq km, which is about 20 times the hot arid zone of India (0.32 million sq km). In South Africa, arid tropical conditions are limited to parts of Angola and Bechunan.

In India about 317,000 sq km are within the Arid Tropical Zone (Table 18.2). Australia has most of the arid tropical land (1.4 million sq km) in the southern hemisphere. There it stretches from the Tropic of Capricorn to 19°S latitude in western Australia and the Northern Territory, and extends to 144°E longitude in Queensland.

In North America portions of southwestern Mexico are in the Arid Tropical Zone. There is also a narrow strip along the western coast of South America from 5°S to the Tropic of Capricorn, comprising a small part of the Peruvian and Atacama Deserts.

Rainfall Distribution

Seasonal distribution of rainfall for a few selected climatic stations in the Arid Tropics is shown in Table 18.3. In arid tropical Africa annual

TABLE 18.2. AREAS OF THE HOT ARID ZONE IN STATES OF INDIA

State	Area in Arid Tropical Zone 1000 sq km	Proportion of Total Arid Zone in India %
Rajasthan	196.2	62
Gujarat	62.2	20
Punjab	14.5	5
Haryana	12.8	4
Andhra Pradesh	21.6	7
Karnataka	8.6	3
Maharashta	1.3	0.4
Total	317.2	

rainfall decreases from south to north, ranging from more than 500 mm at 11°N latitude to about 200 mm at 15.5°N. Thereafter, real desert conditions occur with annual rainfall being less than 150 mm and generally decreasing to less than 50 mm beyond 18°N latitude. This is mainly due to the fact that the moist air does not penetrate beyond 18°N latitude. June to September rainfall is nearly 90% of the annual total in the 11° to 18°N latitude belt, and decreases rather sharply to less than

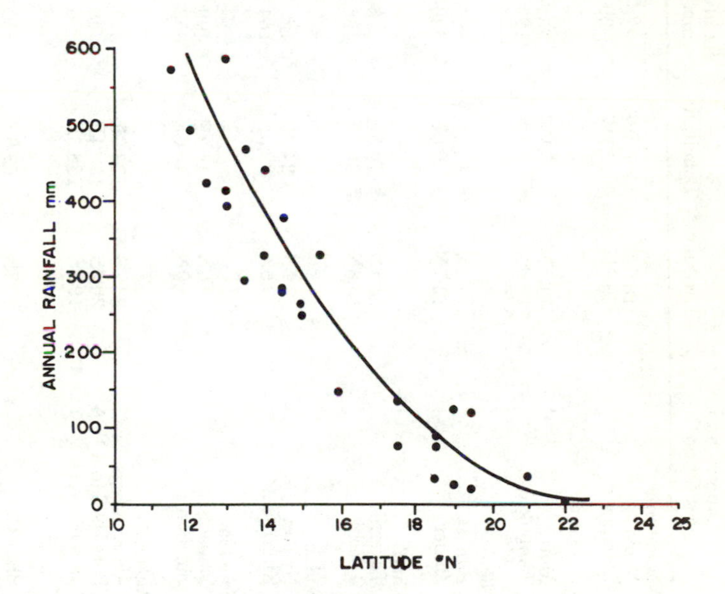

FIG. 18.2. LATITUDINAL VARIATION IN ANNUAL RAINFALL OF TROPICAL ARID ZONE IN NORTH AFRICA

20% beyond 18°N. The pronounced decrease in annual precipitation with increases in latitude in North Africa are shown in Fig. 18.2.

TABLE 18.3. SEASONAL DISTRIBUTION OF RAINFALL AT SELECTED STATIONS IN THE ARID TROPICS

Country and Station	°Latitude	°Longitude	Total Annual Rainfall mm	Seasonal Distribution of Rainfall % of Total			
				Dec.-Feb.	Mar.-May	June-Sept.	Oct.-Nov.
Mauretania							
Nema	16.6N	7.3W	288	1	5	89	5
Ettenna	20.9N	17.1W	29	17	10	39	34
Mali							
Mofti	14.5N	5.1W	551	0	8	90	2
Timbuktu	16.8N	2.9W	195	0	3	95	2
Niger							
Bilma	18.7N	13.4E	27	0	4	89	7
Zinder	13.7N	9.0E	520	0	3	94	3
Chad							
Tegerhi	24.4N	14.5E	5	40	40	20	0
Abechi	13.8N	20.8E	498	0	5	92	3
Sudan							
Nyala	12.1N	24.9E	497	0	8	89	3
Khartoum	15.6N	32.5E	164	0	4	94	2
Somalia							
Mogadiscio	2.0N	45.3E	498	3	27	52	18
Bulobunti	3.9N	45 6E	359	2	38	8	52
Australia							
Alice Springs, N.T.	23.6S	133.6E	245	56[1]	11[1]	11[1]	22[1]
Port Hedland, W.A.	20.3S	118.4E	288	62	20	17	1
Cammooweel, Q.	19.9S	138.3E	366	76	4	6	14
India							
Bhuj, G.	23.3N	69.8E	340	2	3	92	3
Pulivendia, A.P.	14.4N	78.2E	564	4	11	53	32
Bellary, K.	15.1N	76.9E	520	3	14	53	30

Source: Adapted from Thornthwaite (1962).
[1]In Australia the months included in the season columns from left to right are: June-Aug., Sept.-Nov., Dec.-Mar., and Apr.-May.

In Australia the rainfall decreases from nearly 500 mm in the North to about 250 mm at the Tropic of Capricorn. The principal rainy season is December to March, with its contribution varying from 50 to 80%. The decrease is not, however, as sharp as in North Africa.

The Indian arid tropics are characterized by considerably higher rainfall and most of this is in the monsoon months of June to September. For arid areas in peninsular India, the north-east monsoon (October to December) also provides some rainfall.

Recognizing the limitations of rainfall data alone for estimating crop production potentials, averages of Thornthwaite's (1948) aridity index were worked out for the three states of India that represent arid tropical areas (Table 18.4). Since an aridity index of 66.7 is considered the lower limit for the boundary between arid and semi-arid areas, it is seen that except for Kutch, the remaining areas are marginal for the arid zone.

TABLE 18.4. ARIDITY INDEX AND PERCENTAGE OF AREA UNDER TROPICAL ARID REGION FOR THE THREE STATES OF INDIA

State	District	Percentage of District in Arid Zone	Mean Aridity Index
Gujarat	Kutch	100	79
	Jamnagar	80	69
	Surendranagar	29	69
Andhra Pradesh	Anantapur	67	69
	Kurnool	31	67
Karnataka	Raichur	39	67
	Bellary	26	68

Land Use

Accurate and reliable statistics on land use, crop distribution, and yields of crops for different climatic zones within countries either do not exist or are not readily available. Consequently, the available statistics for whole countries having more than 30% of their area in the arid tropical zone are reported in Table 18.5, along with similar data for the principal arid-zone states of India. The proportion of land in crops is small and land use and crop yields vary from region to region and from year to year, based primarily on variations in rainfall and in lesser part on such related factors as low soil fertility and the incidence of diseases and pests.

The legal status of land varies from country to country and is often unclear in most of them. Individual ownership is the most common form of land tenure, but inheritance laws favor fragmentation of individual holdings and multiple ownership. Ownership of land is often concentrated in a few individuals and as a consequence a high proportion of farmers rent rather than own their farms. Rent is normally paid on a crop share basis.

TABLE 18.5. POPULATION DENSITY AND LAND USE IN COUNTRIES OR STATES WITH MORE THAN 30% OF AREA IN ARID TROPICAL ZONES

Country or State	Population per sq km	Percentage of Area		
		Crops	Pasture	Forest
North Africa				
Mauritania	1.1	0.3	38	15
Mali	3.8	1	30	4
Niger	2.8	9	2	12
Chad	2.7	5	35	13
Sudan	5.7	3	10	37
Ethiopia	18.2	10	56	7
Somalia	4.2			
India				
Gujarat	48.0	20	9	5
Andhra Pradesh	100.4	47	23	4
Karnataka	107.2	59	7	6

Sources: FAO (1968); Anon. (1976B).

Pastoralism is the age-old occupation and plays an important role because livestock (particularly sheep and goats) is capable of using almost all the land and is, therefore, an important economic activity. The percentage of area under pasture and forest varies from country to country and is normally much higher than the cropped areas. An exception to this generalization is in the arid tropical region of India where population density is high and the cultivated area is higher than for other countries (Table 18.5). Different forms of nomadic civilizations are supported in these countries.

Arid and semi-arid regions are estimated to contain more than half of the world cattle population, more than one-third of its sheep, and two-thirds of the goats. Animal husbandry, though a vital activity, is of low productivity—only 10 to 20% of that obtained in industrialized countries. Among the factors contributing to low animal productivity, besides low rainfall and its consequences, are "under equipment, under administration of the arid zones, the peripheral nature of their economy and. . .marginalization of their population" (Anon. 1977).

Because of the limited range of the available cultivated crops and their variable and low production, there appears to be a relationship between the percentage of area under cultivation and the density of population.

Both the population density and the percentage of the area under crops is less than ten in a large number of countries, indicating self-contained primitive economies. Australia is an exception, where development of the arid and semi-arid regions is impeded by labor shortages.

Cropping Pattern and Production

Despite water scarcity being the limiting factor, crop production is next in importance only to animal husbandry in the arid regions. Detailed statistics of the percentage area of the crop production and yields in different countries of the region are given in Tables 18.6 and 18.7. One characteristic of crop production in these regions is low productivity. Millet (*Pennisetum typhoides*), pulses, particularly mung (*Vigna radiata*), guar (*Cyamopasis tetragonodiba*), and moth (*P. aconitofolium*) are the main crops of the arid region of India. In Africa, maize, cotton, and groundnut are also extensively cultivated under dry conditions.

TABLE 18.6 PERCENTAGES OF AREA IN COUNTRIES OF ARID TROPICS IN PRINCIPAL CROPS

Crop		Mauretania	Mali	Niger	Chad	Ethiopia	Somalia
Total Area in Crops, 1000 ha		263	1,221	11,501	7,000	12,525	957
Sorghum and Millet	P[1]	30	82	21	17	39	37
Millet	Y[2]	8.1	9.2	5.4	7.1	6.8	2.6
Sorghum	Y	—	—	6.5	—	—	—
Pulses	P	8	1	6	2	4	—
	Y	2.5	5.7	1.6	5	5.8	—
Maize	P	7	6	0.04	0.4	6	8
	Y	6.2	10.7	5.8	10	9.6	5
Cotton	P	—	4	0.1	4	0.4	2
	Y	—	4.4	3.7	2.3	3.9	1.3
Groundnuts	P	—	13	3	2	0.3	0.3
	Y	—	10	8.4	6.8	5.8	10
Wheat	P	—	0.2	0.01	0.04	3	—
	Y	—	13.3	12.2	16	8.1	—

Source: FAO (1968).
[1]P is the percentage of the total area of each country in the crop indicated.
[2]Y is the average yield in q/ha.
Note: Crop areas and yield data are for each country as a whole.

TABLE 18.7. PERCENTAGE AREAS UNDER PRINCIPAL CROPS AND THEIR YIELDS IN ARID TROPICAL DISTRICTS OF INDIA

	Millet		Maize		Wheat		Groundnut		Sorghum		Cotton	
	P[1]	Y[2]	P	Y	P	Y	P	Y	P	Y	P	Y
Andhra Pradesh												
Cudappali	7.4	5.4	0.02	12.0	0.06	2.3	13.3	16.8	18.8	6.0	2.0	7.0
Anantpur	10.0	3.8	0.04	10.0	0.1	1.5	22.2	7.0	17.2	4.6	3.5	4.1
Kurnool	3.0	7.4	0.01	13.3	0.1	2.5	20.2	6.4	25.5	4.3	12.8	4.5
Gujarat												
Kutch	24.6	4.7	0.01	8.8	1.5	9.3	7.8	4.6	18.0	1.2	7.4	13.0
Jamnagar	14.6	5.5	0.06	10.0	4.7	11.7	53.3	5.1	13.5	1.1	3.2	15.8
Surendranagar	17.1	2.4	—	10.0	3.9	9.5	9.4	3.5	18.4	1.3	43.8	8.6
Karnataka												
Chitradurg	9.2	3.9	—	—	0.1	3.6	7.9	4.9	36.5	6.7	11.0	8.1
Bellary	4.4	5.7	—	5.0	0.08	3.0	10.0	4.7	27.7	5.3	16.9	8.1
Raichur	5.7	2.5	—	—	—	3.9	2.3	7.4	35.8	3.7	26.5	7.4

[1]Percent of total cropped area in specified crop.
[2]Yield of crop in q/ha.

Bullocks and camels are the main sources of power for field operations in India. One important factor in the production techniques used in the region is that crop production and animal husbandry are kept either totally separate or are inadequately associated. There is rarely an all pervading commitment to either agriculture or animal husbandry, and there are numerous examples of individuals and groups changing their methods of subsistence.

In Sahelian Africa, cattle are seldom used for field operations. Contracts are made with nomadic pastoralists for manure supplies. Oxen are used for ploughing in heavy soils; mules are used in mountain areas.

One crop a year or one in two years is the general pattern. The use of a rotation in which a crop and fallow alternate in a two years cycle is common in these countries. This practice is important for soil fertility restoration when there is no means of fertilizing the soil. Besides soil fertility recuperation, a further advantage of fallowing is that soils retain portions of the water from two rainy seasons for one crop. In saline soils, which are quite common in these regions, two years of fallow may result in leaching of excessive salts to lower horizons which otherwise would not permit good crops.

The analysis of the problems of crop production in arid tropical regions reveals that an actual ecological imbalance of the factors of production is responsible for low and unstable yields. Harsh and unfavorable climatic conditions, combined with wind blown soils, low in organic matter and poor in soil moisture retention are additional factors. Besides low, variable, and erratic rainfall, the solar incidence increases crop water requirements which limits cropping and reduces yields. In western Rajasthan, for example, solar incidence of 450 to 500 cal per cm^2 per hour results in a PET of 6 mm per day and a consequent high mean aridity index of 78%. Soils are mostly sandy and low in organic matter (0.1 to 0.45%), with poor moisture holding capacity (25 to 28%) and a high infiltration rate (9 cm per hour). Soil salinity and alkalinity extend to 45% of the area and surface crusts form after light showers. Both situations complicate crop production practices and reduce yields.

Variations in frequency and date of onset of limited rainfall are characteristic problems. In Rajasthan the early onset of the monsoon (occurring in one out of three seasons), and the late onset of adequate rains (probable in one out of four years) call for adequate and advance preparations. Also, drought combating strategies must be available for prolonged droughts (3 to 4 weeks) occurring at the seedling stage (during July) in one out of five years, and at flowering and grain-formation stages (August and the first fortnight of September), probable in two out of five years.

In above normal rainfall years, the moisture losses due to runoff and deep drainage are as high as 40% of the seasonal rainfall, each accounting for almost half of the total losses. It has been estimated that available moisture in the order of 284, 214, 173, and 124 mm could be obtained during cropping seasons experiencing good (>400 mm), normal (350-400 mm), below normal (250-350 mm), and poor (<250 mm) rainfall (Anon. 1976).

Existing cropping patterns on rainfed lands are traditional and subsistence-oriented. The following analysis of the cause-and-effect relationships will help with understanding the predominant maladies affecting crop production in these regions:

Causes	Effects
(a) Lack of moisture and restricted period of moisture availability	(1) Restricted choice of crops and varieties (2) Low cropping intensity (3) Inefficient utilization of inputs
(b) Lack of crops and varieties fitting into the existing rainfall pattern, appropriate soil and moisture conservation measures, fertilizer use, and plant protection	(1) Low and unstable crop yields
(c) The removal of natural vegetation as a result of uncontrolled grazing, the felling of trees and shrubs for fuel	(1) Inadequate vegetative cover, leading to a large scale soil and nutrient loss due to wind erosion
(d) Pressure of population	(1) The extension of cultivation to marginal and sub-marginal lands, leading to reduced productivity on the one hand and a greater risk of soil erosion on the other

THE APPROACH

The amelioration of arid regions falling in the rainfall range of 300 mm and below can be achieved best by a silva-pastoral system with suitable tree species. Crop production in marginal conditions in the region should be discouraged. Animal husbandry, range-management, and forestry should be the mainstays in these regions. India's National Commission on Agriculture has emphasized animal husbandry-oriented programs for improving the economy of such regions.

In areas receiving 300 to 400 mm of annual rainfall, improved cropping and animal husbandry technology for dryland crop production should be adopted. In order to make crop production a remunerative enterprise in these areas, the following measures should be adopted:

(1) Selection of crops and varieties having growth patterns matching the rainfall pattern and possessing high moisture-use efficiency
(2) Selection of suitable crops and varieties for conditions of delayed onset of the monsoon and other weather aberrations
(3) Adjustment of sowing time so as to ensure that the periods of flowering and grain-filling coincide with the periods of adequate moisture availability
(4) Tillage implements and practices conducive to good seed-bed preparation without much loss of soil, moisture, and nutrients; and techniques ensuring a proper placement of seed and fertilizers in the moist zone, leading to optimum crop stands
(5) Judicious fertilizer use commensurate with the status of soil nutrient needs of crops, plant population, and available soil moisture
(6) Adoption of appropriate water-harvesting techniques to ensure higher amounts of moisture for longer periods
(7) Moisture conservation through the use of partial or complete subsurface moisture barriers and organic mulches to cut down deep percolation losses and evaporation
(8) Weed control and adoption of effective plant protection measures
(9) Harvesting of crops, appropriate cropping systems, so as to impart stability in years of subnormal rainfall and to increase production in years of normal and above-normal rainfall.

Three basic approaches are suggested to cope with the problem of instability in crop production. These are: crop diversification, intercropping systems, and the recycling of runoff water or supplemental irrigation in the event of prolonged drought (Singh 1976). Castor and guar, among dryland crops, and *Cenchrus ciliaris* among perennial forage crops ensure stability of agricultural production. Nursery sowing of millets and its transplanting, although labor-intensive, is a valuable technique by which reasonably good yields can be obtained even in years of low or subnormal rainfall. In 1974 at Jodhpur (with 136 mm of rainfall) direct-seeded crops of millet failed, but transplanted 20 to 25 day-old seedlings gave 7 quintals of grain yield per hectare (Anon. 1974). In good rainfall years also, the transplanting of 24-day-old seedlings of millet gave more than 30% increase in grain yield over the direct-seeded crop (18.6 q per hectare) planted in the third week of July (Anon. 1976).

MANAGEMENT PRACTICES TO IMPROVE CROP PRODUCTION

Because of past inadequate efforts to dovetail plant-breeding programs with other improvements in dryland agriculture, individual practices developed by scientists for moisture and soil conservation have had marginal impact on productivity and have not found wide adoption. High yielding varieties of crops recently developed have given a silver lining to the otherwise dark picture of rainfed agriculture. However, despite their excellent yield potentials and favorable developmental and maturity patterns, these crop varieties need modification to conform to prevailing rainfall patterns. To obtain satisfactory production levels, crop varieties to be grown in such areas should possess tolerance to drought, should mature early, and should have high yield potentials. With these objectives in view, high-yielding varieties of cereals, pulses, and oilseeds have been developed in India and are listed in Central Arid Zone publications (Anon. 1976). Even with the availability of these varieties careful selection and management of crops are important. Thus, millet and sesamum failed one year owing to the late onset of the monsoon but crops such as mung, cowpeas, castor, and sunflower did give some production. Alternative crop planning must be adopted consistent with rainfall patterns.

Adjustment of Sowing Time

The sowing of crops should be so adjusted that flowering and fruiting coincide with periods of adequate soil moisture availability. In northwest India this period falls in the third and fourth weeks of August. As such, the appropriate sowing time is the last week of June to the first week of July, when the probability of receiving monsoon showers (30 to 40 mm) is usually high. Considering the staple dryland crops, viz, millets, sowing after the second or third week of July invariably results in either poor yield or total failure. Therefore, under the delayed onset of the monsoon, nursery-sowing and transplanting of millet are recommended. Situations as exist in other countries of the region must be studied and analyzed, and appropriate practices worked out.

Judicious Fertilizer Use

Fertilizer use in arid tropics is almost nonexistent, owing mainly to uncertain rainfall, the cultivation of non-responsive crop varieties, and inadequate scientific advice. As such, the earlier belief of farmers that fertilizer application was injurious to dryland crops was justified. With the availability of fertilizer-responsive varieties of dryland crops and the knowledge gained in some countries on the quantity, time, and method of nutrient application, a judicious fertilizer use offers possibilities for increasing dryland crop production under normal and above-normal rainfall situations and for stabilizing production in years of subnormal rainfall. Appropriate fertilizer-use technology assumes special importance in double-cropping systems feasible in years of high rainfall. These have occurred in 20 out of the past 75 years at Jodhpur. Research on fertilizer use on drylands has indicated beneficial effects from optimum nutrient applications with most crops. For instance, experiments conducted at the Central Arid Zone Research Institute (CAZRI), Jodhpur, have indicated in good rainfall years yield increases of the order of 35 to 150% with millet, 150 to 190% with sorghum, 22 to 100% with green gram, 11 to 77% with sunflower, and 180% with sesamum following application of 40 kg of nitrogen and 40 kg of P_2O_5, 60 kg of nitrogen and 40 kg of P_2O_5, 15 kg of nitrogen and 40 kg of P_2O_5, 40 kg of nitrogen and 40 kg of P_2O_5, and 40 kg of nitrogen and 40 kg of P_2O_5 per hectare, respectively. The response of millet and sunflower to nitrogen application was the highest, 10.2 and 7.0 kg per kg of nitrogen, at 40 and 30 kg nitrogen levels, respectively.

Water Harvesting Systems

Water scarcity problems can be partially overcome by adopting an appropriate water harvesting system which can provide additional quantities of moisture. Primitive societies in these regions have used these techniques as an art. For situations obtainable in the arid zone, two water harvesting systems, viz, inter-plot and inter-row systems, have been found to offer opportunities for *in situ* water harvesting. Studies at CAZRI, Jodhpur have shown that cropping only two-thirds of the field (leaving one-third for micro-catchments) with the use of runoff farming gives comparable production as obtained from conventional cropping on a flat surface. Runoff farming has been found to offer potentials for increasing and stabilizing yields, thereby lowering the risk of crop failure and saving inputs required for crop production (Anon. 1976). However, in case no land is to be sacrificed for water harvesting as catchment, the inter-row system of water harvesting is more practicable than the inter-plot water harvesting system. The modified inter-row water harvesting system developed at the CAZRI, Jodhpur induces an additional runoff in the trenches from the ridges and also from the micro-catchments provided in between the trenches. There was 13.4% more moisture in the soil profile under this system than under the flat system at the time of harvesting the millet crop—a distinct advantage conferred on the yield of succeeding crops. Recent studies have shown that there was a lower weed population in the modified inter-row water harvesting system than in the flat and inter-row water harvesting systems.

In the arid tropics it may not only be feasible but desirable to set apart a portion of the land for water harvesting and provide a tank to collect and store the runoff water. The stored water can be applied to the standing crops nearby as a crop lifesaving device during periods of prolonged drought either at the seedling stage or at the grain-filling stage of the crop. A good rainfall distribution during the cropping season, in some years, may not warrant the use of stored water for supplemental irrigation. In that event, the stored water can be used to provide a supplemental irrigation for long-duration crops, e.g., cluster bean and castor. The recycling procedure can also be used with advantage for giving one pre-sowing irrigation to supplement the moisture stored in the soil for raising two crops on the same field in the late-normal monsoon years.

Soil and Moisture Conservation Measures

Soil and moisture conservation programs are basic to any program of crop production in arid regions. More significant, however, is the wind action which may be particularly serious during summers. Stubble-mulching and wind strip-cropping are the basic remedies for the maladies caused by the action of wind. Misra (1964) reported that: (1) soil loss due to wind erosion increased with decreases in the size of the stubble of millet left at the time of harvesting. The minimum (761 kg per hectare) and the maximum (2088 kg per hectare) yields were recorded in the case of whole stubble and control plots, respectively. A stubble height of 15 cm was found to be optimum, both from the point of view of fodder yield and wind erosion control. (2) In studies on wind strip-cropping, average yields of 272 kg and 257 kg per hectare of *Phaseolus radiatus* (now renamed *Vigna radiata*) and *Phaseolus aconitifolius*, respectively, were obtained under unprotected plots. The yields increased to 484 kg and 372 kg per hectare, respectively, when the cropped strips were protected by perennial protective strips of *Lasiurus sindicus* and

Ricimus communis, established at right angles to the direction of the prevailing winds. Soil moisture in the protected cropped land was found to be higher (0.5 to 1.5%) than that in the unprotected land.

In arid regions, surface evaporation is as important as deep percolation, for once the surface soil (mostly sand to loamy sand) dries up, it acts as a natural mulch against the loss of moisture through evaporation. Nevertheless, the quick drying of the surface soil and excessive evaporation just after rainfall pose problems in seedling emergence and proper crop establishment if there occurs a spell of drought after sowing or during the establishment stage when the root zone is restricted to a soil depth of 15 to 20 cm. Under such conditions the use of surface mulches is advantageous in delaying the drying of the soil surface.

To cut down moisture losses from deep percolation in sandy soils, a technique of a partial moisture barrier (bentonite clay) incorporation has been developed at the CAZRI, Jodhpur. The placement of a bentonite barrier 75 cm deep in pits gave a yield of 160 q per hectare of *tinda* (*Citrullus vulgaris* var. fistulosus) against 100 q per hectare obtained from the control (no barrier) in 1973, a good rainfall year. Considering the performance of *tinda* with and without barrier over a period of 4 years, it was observed that the use of the barrier resulted in 35% average increase in production over that without the barrier.

This technique works very well when the barrier and runoff concentration systems are combined. In low rainfall years, the incorporation of the barrier alone may not help to increase storage of soil moisture. It is only in combination with the runoff system that storage of soil moisture is increased by 20 to 25% during rainy spells. In a low rainfall year (136 mm), the runoff concentration (with micro-catchment) in combination with a subsurface barrier of bentonite clay resulted in 43 q per hectare production against 19.7 and 13.4 q per hectare on a flat system with and without a barrier, respectively.

The use of organic mulches, such as millet husks, wheat straw, and grass mulch, helped to maintain a high moisture regime in the soil profile for more than 30 days after application and a uniform thermal regime conducive to plant growth in the soil-surface zone.

Timely Weed Control

Weeds compete with crops for light, moisture, and nutrients. Thus, timely control of weeds assumes great importance. Weed control is most important within the first 30 days in the case of millet and 20 days in the case of *mung*. Thereafter weed control is of little or no use, as the damage already done is irreversible. However, weeds should not be disturbed on lands lying fallow during winter until the first shower, which generally comes sometime in the second half of June. Here, a compromise has to be made between the conservation of soil fertility and soil protection. Obviously, soil conservation is the prime consideration, because fertility can be restored if there is soil, but soil once lost cannot be brought back. Moreover, if the weeds are incorporated into the soil with inversion ploughing, the nutrients they extract are brought back to the soil and may be available to the succeeding crop.

FARMING SYSTEMS

For ameliorating the existing conditions of low and unstable crop production in arid regions, an appropriate balance should be struck between crop production and animal husbandry. Perhaps silva-pastoral and agro-forestry developed on a scientific basis would provide an answer to the existing complex problems.

So far as crop production is concerned, any technology developed for these regions should be able to impart stability in years of subnormal rainfall and increase production in years of normal and above-normal rainfall. The role of crop diversification in stabilizing production has already been stressed. The intercropping of annual grain legumes between perennial grasses is yet another tool which can be capitalized on for stabilizing production.

Intercropping System for Drylands

Grasses are important in arid zones for soil improvement, fodder, and soil protection against wind erosion. Legumes can be grown with advantage as intercrops with grasses, the former benefiting from the improved physical condition of the soil induced by the grass sod, and the latter utilizing the improved fertility from the nitrogen fixation. Also, in the event of failure of rains, grasses may give a modest yield and avert a complete crop failure. For instance, at Jodhpur in 1974, with 136 mm of rainfall, though the legume crops failed, grass (*Cenchrus ciliaris*) as the base crop gave a yield of 49 q per hectare as green fodder (Anon. 1974). On the other hand in 1975, a year of very good rainfall (650 mm), the yield of guar (cluster bean) grain grown as an intercrop was 15 q per hectare, with a forage yield of 108 q per hectare from the base crop of *Cenchrus ciliaris*.

The Use of Non-monetary Inputs in Dryland Agriculture

The importance of management and other non-monetary inputs for a small farmer engaged in dryland agriculture needs no emphasis. For instance, simple practices such as soaking seeds in water and growing crops in paired-row systems have been found to result in good crop establishment and higher yields. The soaking of seeds of sunflower, safflower, and castor in water for 24 hours before sowing has been found to result in early seedling emergence and a good crop stand. In sandy soils of the region 5 cm as the sowing depth for sunflower and safflower and 15 cm for castor have been found to be optimum (Anon. 1976). Such depths may be important in areas where there is quick drying of the upper soil layer after sowing, and where the establishment of crops is a problem. Similarly, a paired-row system of planting has been found to result in higher yields and better moisture-use efficiency than a uniform system of planting in years of low and normal rainfall in the case of millet, grain legumes, and *Cenchrus ciliaris.* Recent experiments carried out at Jodhpur on systems of planting have shown that growing one row of millet in the paired and triple row systems of planting *mung* led to 2 to 2½ times more productivity than the planting of mung as a single crop. The yield of the principal crop (*mung*) is no doubt reduced (by as much as 50%) under the intercropping systems, but the decline in the yield is more than compensated by much higher total productivity, the major share being contributed by the companion crop (millet). For obtaining higher productivity per unit area per unit time, therefore, the planting of one row of millet in the inter-row spaces of *mung* may be rewarding. Recent studies have further revealed that the yield of millet is appreciably reduced (62%) if it is taken in the millet-millet rotation as compared with the *mung*-millet rotation.

GAPS IN KNOWLEDGE AND FUTURE LINES OF WORK

As the above sections indicate, research information on crop production in the Arid Tropics is available only for a few countries, and that for only the last few years. There is, therefore, a need for intensification of multidisciplinary research in these regions. Following are some suggestions:

(1) Make detailed analysis of the climatic data and related information for the entire arid tropics.
(2) Inventory natural resources (soil, land, vegetation) for scientific planning.
(3) Increase knowledge regarding plant types for intercropping and managing of intercropping systems in relation to nutritional parameters and radiant energy.
(4) Conduct tillage in relation to soil and nutrient conservation and crop establishment.

(5) Practice organic recycling.
(6) Collect and recycle runoff water.

Basic knowledge on the following points is still lacking:

(1) The crusting of soil and its control
(2) Soil-water-plant relationships, particularly relating to moisture stress
(3) The movement of moisture in arid zone soils
(4) Water balance, as influenced by different management conditions.

Although it is often said that the bulk of the arid lands is unfit for crop production, farmers will continue to grow crops on them during the foreseeable future. To put the economy of the arid areas on a sound footing, it is necessary to develop an appropriate crop-production technology dovetailed with variations in agro-meteorological conditions. Scientific crop production based on the principles of conservation and organic recycling, apart from enhancing the productivity of the arid zone ecosystem, can also in the long run appreciably check desertification. Research and development efforts should be directed toward achievement of this objective.

REFERENCES

ANON. 1974. Annual progress report. Dry Farming Research Main Centre, Central Arid Zone Research Institute, Jodhpur.

ANON. 1975. Annual progress report. Dry Farming Research Main Centre, Central Arid Zone Research Institute, Jodhpur.

ANON. 1976A. Improved dryland agriculture for western Rajasthan. Central Arid Zone Research Institute, Tech. Bull. *1*, Jodhpur.

ANON. 1976B. National Agricultural Commission Report. Government of India, New Delhi.

ANON. 1977. Development of arid and semi-arid lands: obstacles and prospects. MAB Tech. Notes *6*. UNESCO.

DALBY, D., and CHURCH, R.J.H. 1973. Drought in Africa. Report of the symposium, July 19-20, University of London. School of Oriental and Research Studies, School for African Studies, University of London.

FAO. 1968. Production Yearbook, Vol. 22. Food and Agriculture Organization. U.N., Rome.

KRISHNAN, A. 1968. Distribution of arid areas in India. Proceedings, XXI International Geographical Congress Symposium on Arid Zone, Central Arid Zone Research Institute, Jodhpur, Nov. 22-29. CAZRI, Jodhpur, 11-19.

MANN, H.S., and SINGH, R.P. 1977. Crop production in the Indian arid zone. Desertification and its control. Indian Council of Agricultural Research, New Delhi, 215-224.

MISRA, D.K. 1964. Agronomic investigation in arid zone. Proceedings, Symposium on Problems of Indian Arid Zone, Central Arid Zone Research Institute, Jodhpur, Nov. 24-Dec. 2. Ministry of Education, Government of India, New Delhi, 165-169.

SINGH, R.P. 1976. Towards stabilizing crop production on drylands of western Rajasthan. Farmer & Parliament II (7), 13-14.

THORNTHWAITE, C.W. 1948. An approach towards a rational classification of climate. Geog. Rev. *48*, 55-94.

THORNTHWAITE, C.W. 1962. Publications in Climatology. Laboratory of Climatology, Centerton, New Jersey.

THORNTHWAITE, C.W., and MATHER, J.R. 1955. The water balance. Publications in Climatology. Laboratory of Climatology, Centerton, New Jersey.

TROLL, C., and PAFFEN, KH. 1965. Seasonal climates of the earth. *In* World Maps of Climatology, 2nd Edition. H.E. Landsberg *et al.* (Editors). Springer-Verlag, New York.

19

Small Farm Cropping Systems in the Tropics

C. A. Francis

Throughout history, man's efforts to produce adequate food and fiber have been conditioned by his natural environment and the physical resources available for crop growth. More than 90% of all tropical farms have less than 5 ha. The national average is less than 3 ha per family in the Philippines, and between 1 and 2 ha per family in Bangladesh (Harwood and Borton 1978). It is further estimated that about 60% of mankind is involved in subsistence agriculture and related activities.

The proportion of crop production which comes from small farms is more difficult to estimate, due to the importance of produce for home consumption, barter, or trade. Estimates of 15 to 20% of total world food production have been advanced. For specific crops in the tropics this is much higher. Cowpea, the most important legume in parts of Africa, is grown 98% in association with other crops, presumably by small farmers. Ninety percent of the dry bean production in Colombia and 73% in Guatemala are in association with maize and other crops, while 80% of the crop in Brazil is grown in association. Through Latin America, an estimated 60% of the maize and 80% of the beans are grown in association (Francis *et al.* 1976). These multiple cropping systems are prevalent on small farms, due to the high labor requirement.

Through the experience of many generations of farmers, cropping systems have evolved which are specific to a local soil-climate complex. These systems often maximize production of a range of food crops for subsistence and for the market under the physical, economic, and social constraints which prevail in each situation. Limitations to production may include soil fertility, water availability, labor and power for farm operations, pests, low levels of education, distance to supplies and markets, lack of economic incentives due to government policies, and need to minimize risk. As diverse as are the plant species, soils, and climates of the tropics, the factors which limit production are many, complex, and interrelated.

A farming system as known by small farmers in tropical areas is a set of biological processes and farm management activities organized within the resources available to produce plant and animal products. Farm resources include light, land, and water (physical), plus cash, labor, power, and markets (economic and social) (Harwood and Borton 1978).

Other chapters in this book are focused on physical and biological factors which influence crop production. It is difficult to discuss increased production in the tropics, however, without some consideration of how the physical and biological potentials of the small farm interface with the economic, social, and political realities which influence the farmer's decision making process and his eventual success.

This chapter considers current and potential crop yields in the lowland tropics and current constraints to increased production. The contribution of research in cropping systems, agronomy, genetics, and plant protection is evaluated. New programs of the international centers and their contribution to development are discussed as an integral part of the present thrust to increase production in the tropics.

TABLE 19.1. CURRENT AREA AND PRODUCTIVITY OF MAJOR GROUPS OF CROPS (1974-1976), AND PERCENTAGE OF CHANGE SINCE 1961-1965

Crop	Region	Area (1,000 ha)	% Change	Yield (kg/ha)	% Change
Cereals (total)	World	743,112	9.8	1,873	28.3
	U.S.	71,443	15.6	3,309	20.9
	Developed	159,796	8.7	2,941	25.6
	Developing	303,892	14.2	1,346	23.8
Roots and Tubers (total)	World	51,723	4.1	10,763	9.1
	U.S.	596	−4.2	26,808	28.0
	Developed	3,589	−33.8	21,218	15.7
	Developing	20,074	16.0	8,507	14.8
Pulses (total)	World	69,823	3.1	684	9.3
	U.S	763	−1.3	1,382	−2.1
	Developed	3,488	−32.4	1,005	23.3
	Developing	47,654	12.1	505	−2.7

Source: FAO (1976).

CURRENT CROP PRODUCTION

Crop production in developing areas of the tropics has increased over the past two decades in approximate proportion to the increase in population. Table 19.1 illustrates the relative increases in area and productivity of major food crops which have accounted for these gains in total production over the past 12 years. Area under cultivation has been expanded in the developing world for cereals, roots and tubers, and pulses, while only cereals have increased in area in the developed countries. Further expansion of crop area in the tropics is possible, but becomes increasingly expensive as more marginal lands which require greater energy inputs for water, soil amendments, or land preparation are brought into cultivation. Major gains in the future will come from more intensive cultivation of existing lands. As shown in Table 19.1, increased yields have made a more important contribution to cereal production than increased area in developing countries, while the contribution to root and tuber production also has been significant. Only for the pulses, an important source of protein, have yields remained stagnant.

TABLE 19.2. CURRENT AREA AND PRODUCTIVITY OF THREE CEREALS AND ONE PULSE CROP (1974-1976), AND PERCENTAGE OF CHANGE SINCE 1961-1965

Crop	Country	Area (1,000 ha)	% Change	Yield (kg/ha)	% Change
Rice	World	140,044	12.8	2,413	18.3
	Thailand	7,972	14.8	1,822	12.3
	Colombia	364	24.2	4,313	119.5
	Philippines	3,560	13.1	1,715	36.4
	Brazil	5,344	40.3	1,479	−8.0
Maize	World	115,194	15.9	2,754	26.7
	Mexico	6,611	−5.0	1,270	19.9
	Kenya	1,250	20.2	1,163	9.0
	Guatemala	530	−20.3	1,285	44.9
	Thailand	1,170	177.3	2,356	21.9
Wheat	World	228,831	8.8	1,650	36.5
	Mexico	783	−2.4	3,802	82.4
	Morocco	1,840	16.6	1,015	19.8
	Tunisia	1,108	10.6	798	61.5
	India	18,902	41.0	1,306	56.4
Beans	World	24,029	12.1	519	12.6
	Brazil	4,140	41.0	517	−21.2
	Mexico	1,680	−8.1	616	48.1
	India	8,697	25.1	289	10.7
	Guatemala	103	35.5	715	10.2

Source: FAO (1976).

Data from individual countries (Table 19.2) show striking gains in a few unique cases, where a specific technology and favorable government policy have combined with local demand and export potential to expand the areas. Introduction of technology and local research have teamed up in the case of rice in Colombia and the Philippines to boost productivity. The greatly expanded area of rice production in Brazil has led to increased production, but the move to more marginal lands is indicated by a reduction in yields per hectare in the national average. Expanded area and increased production in Thailand of both rice and maize reflect the government's interest and stimulus in exporting these crops to Japan. The reduced area of maize production in Guatemala may be due in part to gains in productivity, and more likely is a response to higher prices for beans and other commodities which stimulated a shift to other crops. Finally, the striking increases in wheat production in several countries reflect the higher yields made possible by the research which was initiated in Mexico and carried to countries in the Middle East and Asia.

These examples of the green revolution are balanced by the limited increase of crucial subsistence crops basic to the diet in many countries: maize in Mexico and Kenya, rice in Brazil, and beans in Brazil and Guatemala. Production of these critical crops has not kept pace with population growth and demand. This is the challenge to increasing production of a range of crop species on small farms throughout the tropics.

AGRONOMY AND CROPPING SYSTEMS

Systems of cultivation are as diverse as the physical resources, economic systems, and cultures in which they have evolved. Classification and descriptions of the various cropping systems have been presented in previous chapters, and recently have been unified by Andrews and Kassam (1976). A number of references on tropical agriculture also have exhaustive detail on the systems which are used by shifting cultivators or small subsistence farmers (Duckham and Masefield 1970; Manshard 1974; Ruthenberg 1976). Among the criteria used to classify cropping systems are farm size, intensity of rotation and fallow periods, kind of cultivation practiced, water supply, tools and labor supply, cropping pattern and animal activities, and degree of commercialization. Recent symposia on tropical cropping systems (Papendick *et al.* 1976; ICRISAT 1974; IRRI 1976) have focused more directly on the constraints to production and how they can be solved. Current efforts in the international centers and the universities and ministries in the tropics are less descriptive and more action-oriented toward identifying and solving problems in an integrated manner.

Maintaining Soil Fertility

A principal problem in moving from shifting cultivation to a shorter or nonexistent fallow period, or to intensifying cultivation in other types of permanent agriculture, is maintenance of soil fertility. The stability of production which is crucial to the small farmer depends on replenishing nutrients removed from the soil by crops, maintaining desirable physical condition of the soil through addition of organic matter and proper tillage, preventing an increase in soil acidity and toxic elements, and minimizing or preventing erosion (Greenland 1975). A number of alternatives are possible for each of these problems, and some solutions require fewer or lower levels of purchased inputs than others.

Physical condition of the soil can be maintained and enhanced by the return of unharvested crop residues to the soil. Although the vegetative portion of some crops is used for livestock and small animal feed, the wasteful burning or other undesirable loss of organic matter should be minimized wherever possible. Crop rotations and choice of a particular sequence may enhance the buildup of organic matter and maintenance of soil tilth. Reduced tillage is another route to improved soil condition. Many of the fine textured soils of the tropics have a self-mulching capacity if left untilled. By employing more intensive cropping systems, use of small but judicial amounts of herbicides, and timing the sequence of operations in coordination with natural rainfall seasons, an economical cultural weed control approach can make the less intensive tillage systems desirable to improve soil physical conditions.

Use of cover crops as a source of nitrogen has not been popular in the tropics where food is scarce and available rainfall is used to grow food. However, the rotation of cereals with grain legumes is a sound practice for the restoration of some nitrogen in the soil, and the use of cereal-legume mixed cropping systems may be one of the small farmer's current practices which promotes low level but stable long-term maintenance of soil nitrogen.

Chemical fertilizers represent known technology of near-immediate application to solve many of the fertility problems of small farmers. The addition of artificial fertilizer is efficient, due to its nutrient concentration and relative ease of transport. Used in combination with other sources of nitrogen, the resupply of phosphorus and minor elements to many soils can provide a high economic return to investment in many situations. Complications to the use of fertilizers include cost, availability, proper recommendations for a crop which responds, and adequate financing to assure a timely application of the fertilizer. On low base soils which require the greatest concentration of added fertilizer, Sanchez and Buol (1975) and Spain *et al.* (1974) suggest the application of low levels of needed elements and selection of crop species and varieties which are more tolerant to nutrient deficiencies and toxicities. This leads to most efficient use of applied elements. FAO has conducted extensive fertilizer trials in the tropics, and has arrived at recommendations for many zones and soils. Examples of fertilizer response are many, and the new varieties of wheat and rice have shown that fertilizer used with varieties which respond can boost production in the tropics substantially.

Control of soil acidity through periodic applications of lime has been a common practice in the temperate zone. The relevance of this technology for the small farmer depends on cost and availability of lime, crop value to justify its use, and some source of finance. One traditional alternative to lime, the application of ashes from burned crop residues or cleared material from fallow, is not compatible with the return of this material to the soil nor with the elimination of the fallow period. Mulches of deep rooted species provide an additional alternative. The foliar or soil application of low levels of specific elements to overcome deficiencies in acid soils, rather than trying to correct the acidity, is another. Use of newly selected varieties, or species adapted to acid soils, is a route available to the small farmer at low cost.

TABLE 19.3. SOIL LOSS AND RAINFALL RUNOFF FROM FIELDS WITH THREE SLOPES IN IITA, NIGERIA

Slope (%)	Soil Loss (ton/ha) No Till	Soil Loss (ton/ha) Plowed	Runoff (mm) No Till	Runoff (mm) Plowed
1	0.03	1.2	11.4	55.0
10	0.08	4.4	20.3	52.4
15	0.14	23.6	21.0	89.9

Source: Greenland (1975).

Control of soil erosion is a final requirement of any sustainable cropping system. Reduced or zero tillage promotes better total cover during the year, especially with a mulch which can protect the soil surface against raindrop impact (Baeumer and Bakermans 1973). Data from Dr. R. Lal in IITA (Greenland 1975) indicate a dramatic reduction in soil loss and runoff on three different slopes (Table 19.3). Even with slopes up to 15%, there was minimal soil loss and less than one-fourth the runoff after rains in no-till systems compared to the plowed treatments. In conjunction with the use of more traditional tillage, erosion in sloping fields is minimized by cultivating across the slope, bunding, grass strips, or stopwash lines (Wrigley 1969). Although terraces are used in Asia and have been for centuries to irrigate rice and other crops, the construction of these devices today would be both difficult and expensive.

Water Capture, Conservation, and Use

Availability and control of soil moisture is intimately associated with crop yields. Irrigation is an ancient practice which has led to record yields with high inputs in temperate zones, and resulted in some of the most highly developed and densely populated regions in the tropics. Where labor and water are available, very efficient use of this resource can lead to increased yields on the small farm. Evaluation of the components of technology in bean production at CIAT showed irrigation to be more important than other inputs among those tested. In areas of low rainfall and intensive crop production in Mexico and Peru, modern irrigation technology is central to the system. Controlled irrigation is not available, however, to the majority of small farmers.

Recent research at ICRISAT in India has focused on water capture in the soil profile, trapping of additional runoff in surface tanks, and distribution of this moisture to the growing crop (ICRISAT 1975).

Mechanization and Land Preparation

Use of mechanization to increase production per man and ease the physical burden has been central to the modernization and development of temperate zone agriculture. This technology has been transferred directly to a limited number of regions in the tropics where the topography, crops, farm size, and capital were appropriate to available equipment and expertise. A smaller but still significant impact has been made in some areas with the introduction of small, hand-guided walking tractors and cultivators. Yet mechanization remains one of the most costly and least utilized elements of technology available to the small farmer.

Primary land preparation or tillage generally has been used to prepare a favorable seedbed which allows uniform mechanized planting and good early growth of the crop. Minimum or zero tillage as currently practiced by most small farmers presents many advantages, however. Unger and Stewart (1976) summarized the requirements for seed germination and early seedling growth, and compared conventional tillage systems with zero or minimum tillage. Adequate soil aeration to provide gaseous exchange in the seed and root zone probably presents little problem in either tillage system unless severe waterlogging or wet crusting occurs. Soil moisture generally is enhanced by reduced tillage systems, since surface residues absorb raindrop energy, reduce soil dispersion, reduce surface sealing and runoff, increase infiltration, and reduce evapotranspiration. Seed-soil contact after planting is a problem if conventional equipment is used, but the till planters which use straight or fluted coulters to cut through stubble and trash, open a slot, and provide positive coverage to the planter track are effective and available in some areas. This planting system enhances water content of the soil surface and promotes slower drying by evaporation, more rapid seedling growth, and the ability to withstand short drought periods. There is less soil crusting with crop residues on or near the surface.

Reduced tillage operations or restricting of traffic to the same narrow strips gives less soil compaction, better root elongation, and proliferation. The presence of residue on the surface also protects the young seedlings against wind and soil erosion. Early seedling growth requires a favorable light environment. Only in relay cropping with a tall and leafy associated crop already in the field would this be a problem. Residues on the surface generally reduce average soil temperature, depending on slope, residue levels, and soil water content (Unger and Stewart 1976). In Nigeria the temperature at 5 cm 2 weeks following planting sorghum was 10°C lower with 1 to 2 cm of crop residue compared to a clean tilled treatment (Rockwood and Lal 1974). Higher temperature in conventional tilled soil reduced germination and seedling vigor, and the no-tillage system produced 50% higher yield. In addition to saving time and energy, the no-tillage system can give more rapid emergence, early growth, maturity, and higher yields, under some cropping conditions. Under many conditions, yields are the same in no-till and conventional tillage treatments.

Additional evidence for the advantages of crop production under zero tillage was presented by Kannegieter (1967). In Ghana, maize yields following tropical kudzu were highest when the cover crop was killed with 2,4-D and maize planted directly into the residue. This was superior to clean culture, burning, or slashing and incorporating the residue. In IITA, Dr. R. Lal (Greenland 1975) compared monocrop cowpea to the cowpea/maize intercrop in four tillage alternatives (Table 19.4). Crop production was highest with zero tillage for both cropping patterns, compared to the plowed treatments. Weed, insect, and disease problems may be greater in the no-tillage system, and appropriate control must be exercised.

Mechanization for multiple cropping systems is complicated, especially in relay cropping when a second crop is planted into an existing one. Greatest efficiency in tillage is achieved when an entire area is prepared (Erbach and Lovely 1976). Little efficiency is lost if strip cropping is used, if the width of the strips corresponds to one or a multiple of the width of existing equipment. Alternatives exist to modify planting systems to fit present equipment, or to modify equipment to fit new and more efficient planting systems. There is no doubt that modifications of both will be necessary to arrive at compatible total systems, if some mechanization is to be incorporated in the development process.

TABLE 19.4. YIELDS OF MAIZE AND COWPEA UNDER FOUR TILLAGE SYSTEMS AND TWO CROPPING SYSTEMS

	Yield (kg/ha)			
		Intercropped		
Tillage	Monocrop Cowpea	Cowpea	Maize	Total
Plowed and Ridged	1185	665	1705	2370
Plowed and Flat Bed	1274	725	1675	2300
Strip Tillage	1538	1022	2337	3359
Zero Tillage	1649	941	2809	3750

Source: Greenland (1975).

A number of smaller items already have been widely adopted for tropical crops and systems, such as knapsack sprayers, rice threshers, maize mills, and cultivators (Duckham and Masefield 1970). The Intermediate Technology Development Group (9 King Street, London, WC2E 8HN) has a comprehensive program of design, modification, testing, and publications related to the use of simple and inexpensive machinery for all types of agricultural operations. Advancement of the slow but steady mechanization process reflects the growing awareness by farmers that these components of technology are important, but only a part of the process. The financing and infrastructure necessary to bring these implements to the farm also will promote the use of other innovations such as new seed, fertilizer, and pesticides.

Multiple Cropping Systems

Through the centuries of evolution of cultivated species, multiple cropping systems evolved at the hands of the small farmers. The complex mixtures of crops found on small farms in the tropics today are the modern legacy of this process. At first dismissed by modern and scientific agronomists as inefficient, low yielding, and without potential, these systems are currently receiving more serious attention from agriculturists in national programs and international centers seeking new routes to increased food production.

The pioneering work of Bradfield (1970) in the mid 1960s was instrumental in raising the consciousness of research leaders and policy makers in agriculture. He stressed that increases in area and yield per hectare were only 2 of the 3 routes available to increase production, and that increasing the number of crops per year from the same land area had perhaps more potential than either of the others. To exploit the time dimension, Bradfield argued convincingly to minimize the number of days in the year that land is idle. Several concepts he followed were:

(1) Plant on beds to facilitate irrigation and accelerate drying of the top layer where crops are planted and cultivated.
(2) Maintain volume of soil tilled and number of tillage operations at a minimum.
(3) Use early maturing varieties which produce high yields per day.
(4) Grow ratoon crops where possible to eliminate one or two planting operations.
(5) Transplant slow-growing vegetables to the field.
(6) Grow some crops to harvest green, such as maize and edible soya.
(7) Intercrop whenever possible, sacrificing yield of each component crop to give higher total production.

Researchers in Taiwan, India and tropical Africa also have followed some or all of these principles, to develop cropping systems which more productively exploit the resources of a specific region.

Many of the characteristics of multiple cropping systems as compared to monocrop agriculture have been described in the literature (Altieri *et al.* 1977; Ruthenberg 1976; Wrigley 1969). Multiple cropping is characterized by:

(1) Reduced susceptibility to diseases and pests through diversification; although weed control is more difficult, weeds are held in check by increased crop competition
(2) Adaptation of planting to changing soil conditions, with a better exploitation of the total soil nutrient pool and nutrient recycling
(3) Maximum utilization of light by a series of crops grown in proximity
(4) Varied food supply for family and a more balanced diet
(5) Soil covered throughout the year to prevent erosion
(6) Net production higher than in monoculture at the same levels of technology (LER >1)
(7) Economic stability higher due to fluctuations in crop yields and prices
(8) Social viability promoted, with long-term stability due to farmer

participation, and low ratio of land owning managers to agricultural laborers.

(9) Harvest spread over time in the year, a favorable situation to maximize use of family labor and provide food over the year where no storage facilities exist.

For a more in-depth treatment of multicrop-monocrop comparison, consult the above-mentioned references and others for specific examples.

Production potentials in multicrop systems, once thought to be low and with little possibility for improvement, are now known to be extraordinary (Figs. 19.1 and 19.2). Recent results from bean-maize intercropping in CIAT have shown consistent bean yields of 1500 to 2000 kg per hectare in association with a 5000 kg per hectare maize crop. Average yields of the bean and maize crops in tropical Latin America are about 600 and 1300 kg per hectare, respectively (Francis 1978). Intensive cultivation of monoculture climbing beans on small areas at CIAT produced more than 4000 kg per hectare in a period of 3½ months, opening the possibility for highly productive and diversified relay cropping or rotation schemes to greatly boost production of beans in the tropics. Crop yields, protein yields, land efficiency ratios and net returns from three cropping systems are shown in Table 19.5 (Francis 1978). The advantages of a bean-maize association in production of crop and protein are clear. The land efficiency ratio is variable, and ranges from a 21% to a 90% advantage for multiple cropping. Net income from the bean-maize intercrop consistently exceeded that from either monocrop, and represents little more investment than the monocrop maize. Further analyses of data from agronomic trials revealed this bean-maize system to be more stable in net return than with either monocrop.

In addition to the Bradfield (1970) reference, several recent symposia or workshops have dealt specifically with many aspects of multiple cropping and should be consulted (Papendick *et al.* 1976; IRRI 1976; ICRISAT 1974). The status of multiple cropping across the tropics and temperate zones to the year 1971 was summarized by Dalrymple (1971). Current annual reports of IRRI, ICRISAT, CIAT, IITA, AVRDC, and CATIE should be consulted by those interested in the latest results in this area.

Energy Component in Crop Production

Energy efficiency in agriculture is important in countries which currently import most or all of their petroleum. There is a significant drain on foreign exchange in countries where this is badly needed to

TABLE 19.5. EFFICIENCY MEASURES OF MAIZE AND BEAN YIELDS IN THREE SYSTEMS

	Monoculture Yield (kg/ha)		Associated Crop Yield (kg/ha)			Protein Yield[2] (kg/ha)			LER[3]	Gross Income[4] (US$/ha)			Net Income[5] (US$/ha)		
	Maize	Bean	Maize	Bean	Total[1]	Mono Maize	Mono Beans	Maize Beans	Maize + Beans	Mono Maize	Mono Beans	Maize + Beans	Mono Maize	Mono Beans	Maize + Beans
Bush Beans															
Trial 7501	6535	1738	7631	845	8476	654	382	949	1.65	784	834	1321	184	234	571
Trial 7501	8205	1738	8769	647	9416	820	382	1019	1.44	985	834	1363	385	234	613
Trial 7502	7221	2033	6926	1033	7959	722	447	920	1.47	867	976	1327	267	376	577
Trial 7511	5445	2165	6718	1443	8161	544	476	989	1.90	653	1039	1499	53	439	749
Trial 7516	3729	1531	3414	1083	4497	373	337	579	1.62	447	735	930	(153)	135	180
Average	6227	1841	6692	1010	7702	623	405	891	1.62	747	884	1288	147	284	538
Climbing Beans															
Trial 7509	5674	3635	6344	1480	7824	567	800	960	1.53	681	1745	1472	81	545	722
Trial 7511	5445	2688	4178	1275	8161	544	476	989	1.90	643	1039	1499	43	(161)	749
Trial 7515	5600	2688	4177	1275	5452	560	591	698	1.21	672	1290	1113	72	90	363
Trial 7517	4435	3696	4089	1732	5821	444	813	790	1.39	532	1774	1322	(68)	574	838
Trial 7518	4739	4307	4934	2075	7009	474	948	950	1.52	569	2069	1588	(31)	867	838
Average	5179	3298	5252	1601	5853	518	726	877	1.51	621	1583	1399	21	383	649

[1]Total yield (kg/ha) = sum of bean and maize yields in association.
[2]Protein yield based on maize with 10% protein + bean with 22% protein.
[3]LER = Land Equivalent Ratio = $\Sigma \frac{A_i}{M_i}$, where the yield of crop A in association is A_i, and the yield of the same crop in monoculture is M_i = LER of each monoculture is 1.00.
[4]Gross income based on maize price of US$120/t and bean price of US$480/t.
[5]Net income based on production costs: maize= US$600/ha, bush beans = US$600/ha, climbing beans =US$ 1,200/ha, maize/beans = US$750/ha.

import other inputs for development. Efficient use of energy is especially critical to the small farmer whose access to farm power is limited by distance and accessibility, or by lack of capital to acquire and operate machinery to "modernize" his system. Efficiency in the developed countries is of interest to cut costs of production on the farm, although the consumption of energy in the field production of crops is a very small amount compared to the total consumption in modern society (Hirst 1974).

From Francis (1978)

FIG. 19.1. ASSOCIATION OF CLIMBING BEAN VARIETY P589 (160,000 plants/ha) WITH HYBRID MAIZE H207 (40,000 plants/ha)

The efficiency of crop production in terms of fossil fuel energy is greatest in the primitive cultures where almost entirely human energy is used to collect and cultivate food in a relatively extensive fashion. Rappaport (1971) reports about a 20:1 efficiency ratio of energy output:energy input for the Tsembago culture of northern New Guinea. This extreme may be compared to reports of modern maize culture in the U.S. which has an output efficiency of about 3 (Pimentel *et al.* 1973). The increased efficiency of production per unit of labor has allowed production to increase substantially in developed countries and has permitted the exodus of people from the farm. However, there has been a substitution of energy-expensive inputs for the human input: larger machinery for land preparation and harvest, herbicides to replace

FIG. 19.2. ASSOCIATION OF BUSH BEANS AND MAIZE AT THE CIAT EXPERIMENT STATION

cultivation, etc. The relationship between traditional agricultural systems used to produce maize has been compared to human power systems, animal power systems, and highly mechanized systems (Heichel and Francis 1973; Heichel 1974). As production per hectare increased as a result of the greater fossil fuel energy input in mechanized systems, efficiency of output per unit input has decreased from about 16 to 20 to about 5. The increased efficiency of agriculture per unit of labor and per unit of land area has been an economically rational result of available technology and relatively inexpensive sources of fossil fuels in the developed countries.

Proposals for improving the energy efficiency of cropping systems (Pimentel *et al.* 1973; Steinhart and Steinhart 1974) in developed countries have a direct application to small farm systems in the tropics. Use of reduced tillage in primary land preparation already was mentioned in reference to the agronomic benefits which would accrue. This would reduce energy cost of production as well. The use of cover crops and animal manures is another route to improving soil fertility while reducing the cost of purchased fertilizer inputs. Control of weeds through crop rotations, cultural practices, an integrated program of cropping systems, low levels of herbicide, and mechanical or hand tillage can reduce costs in this crucial area. Hand application of herbicides, insecticides, and fungicides to small areas may reduce the total fossil fuel energy cost of these pest control measures. Irrigation costs can be reduced by selection of appropriate crops and choice of planting dates to coincide with availability of natural rainfall. Plant breeding approaches to insect and disease resistance, tolerance to drought, and early crop vigor to compete with weeds offer an exciting alternative to the small farmer who does not have the resources or experience to acquire and utilize effectively the other components of technology listed above.

The improvement of crop yields generally will require more inputs and greater investment of energy. The critical consideration is how to develop an appropriate technology and use that energy in the most efficient manner to increase production in a small farm context where both fossil fuel energy and other inputs to crop production are scarce and sometimes not available.

PLANT PROTECTION

The influence of pest management on crop production includes the effect of controlling insects, diseases, and weeds to increase production in existing systems, as well as the effect of changing agricultural systems on the occurrence and economic importance of these pests. Both aspects will be considered briefly.

Insects and Nematodes

Insect species have evolved in the tropics in conjunction with noncultivated and crop species, and man has controlled them in a number of ways to improve his edible crop yield. Crop rotations in the bush-fallow system and diversity in permanent cropping systems are but two of many approaches used by the small farmer to control the tremendous diversity of insects which attack his crops. Traditional varieties cultivated by the subsistence farmer also have been selected in each microclimate for a degree of tolerance or co-adaptation with prevalent insect species. One of the most common attempts to increase production by a direct transplant of technology from temperate zones to the tropics is the use of chemical insecticides, especially on commercial plantings of cotton, maize, beans, and vegetable crops. In many cases this has been successful in the short run, but non-selectivity of insecticides and lack of information and experience with conditions in the tropics have led to increased insect resistance, elimination of natural predators and parasites, and increasingly difficult control.

A recent and excellent review by Nickel (1973) outlines many of the problems inherent in controlling insects in changing agricultural systems in the tropics. He emphasizes that the diversity in traditional systems is closely related to stability, and that there is less probability of a massive outbreak of insect problems in shifting cultivation. This is also true of the multiple cropping systems still practiced by small farmers in many countries (Altieri *et al.* 1977).

Natural plant communities in the tropics have a wide range of ecological niches. When man introduces some stringent limiting factor into these communities, diversity of insect populations is reduced, although total numbers of insects may be unchanged or even increased. An example of this change is the introduction of monoculture systems to replace traditional multiple cropping systems. A further reduction in plant diversity occurs when clean cultivation and weed control eliminate these species from the fields. Unless research results prove that the farmer can deal effectively and economically with the predicted increase in insects which attack his crops, it may be advisable to seek more productive multiple cropping alternatives which maintain a degree of diversity throughout the cropping season (Nickel 1973).

Changes in plant parasitic nematodes also have been observed on newly cleared land by Caveness (Nickel 1973). Although population densities declined greatly after clearing, they increased rapidly during each successive cropping cycle. The species distribution changed drastically, from a small number of individuals of many species to large populations of two types which were specific to the crops being grown. Both these nematode species were present in very low numbers in the previous bush fallow.

The introduction of other components of technology such as irrigation, fertilization, and mechanization also changes the ecosystem in which a crop is planted, grown, and harvested by man in competition with insect pests. Irrigation can affect survival of insect species in a direct way by providing a more favorable habitat for their survival and reproduction (Rivnay 1964). Perhaps more important is the intensification of cropping sequence which results from available water and a potential for more than one crop per year. When the same crop is grown in sequence in the same year, there are usually increased insect problems during the second and succeeding crops. A relay crop system or crop rotation of carefully chosen species can minimize this problem for the small farmer.

Fertilization of tropical crops also can affect the severity of insect attack by changing the quality of the plant as a source of food, its attractiveness for oviposition, or even the susceptibility of the plant to specific insects (Nickel 1973). Although mechanization may increase insect attack due to the general shift to monocropping, there also may be better physical control of insects through deep plowing, shredding of residue, and ridging (Nickel 1973). Insect attacks also can be avoided by more timely planting and harvest, reducing pests both in the field and in storage.

The search for resistant crop varieties has been central to most plant breeding efforts (Painter 1958). The yield advantage of the "green revolution" varieties has been coupled with greater resistance to a range in insect pests. The reduction in crop diversity even within a species, however, could lead to severe insect problems if the breeding program cannot keep ahead of these pests. Smith reported that over 600 Bulu rice varieties in Indonesia recently were replaced by only a few varieties (Nickel 1973). This situation has been repeated in the tropics with wheat, maize, rice, sorghum, and other crops in many regions (Harlan 1975). Only continued efforts to recognize potentially serious insect pests and to incorporate new sources of resistance can provide varieties with higher yield potential which can be used on a broad scale in monoculture. The importance of preserving the genetic variability of original land races also has been stressed by Harlan (1975).

The consequences of indiscriminate use of pesticides present a unique problem to the serious researcher seeking long-term and sustained production of food crops. The effects of insecticides on predator populations are well documented (Newson 1967). The concept of pest management through integrated control is now central to most planning, and pesticides should be used as one component of these systems. Integrated control includes a combination of these elements: resistant or tolerant varieties, low rates and timely application of insecticides when needed, products specific for tropical conditions, high selectivity of materials, alternating methods of application and insecticides, crop rotations, and other methods including multiple cropping, trapping, and biological control.

Plant Diseases

Plant pathogens have evolved with host crops and other plant species. Similar to insect populations, pathogens are diverse in traditional cropping systems which involve a number of crop species in mixture, several varieties of each crop, small and scattered fields separated by some distance, and a mobile situation with shifting cultivators who do not replant in the same field until some years later. All these factors act as a defense against disease epidemics, and promote a stable endemic balance. Subsistence agricultural systems with a wide range of crop species and varieties, no fungicides, and low but relatively stable yields put little pressure on the organism to evolve new and more destructive forms. Varieties selected in these systems by farmers are well adapted to existing pathogens, and exhibit a wide spectrum of disease resistance in each area (Harlan 1976).

Changes in cropping systems lead to a greater emphasis on monoculture, larger fields with a greater degree of mechanization, more intensive cultivation of fewer species, less rotation, and more frequent repetition of the same crop (Harlan 1976). Success with new and high yielding varieties can lead to larger and contiguous areas planted to one or a small number of cultivars. Improved transportation, markets, and diffusion of seed lead to greater movement of crops over a larger area. The narrow genetic base of wide plantings of some crops raises fears of future susceptibility to some new or modified pathogen. The recent attack of southern corn leaf blight on maize with T-cytoplasm was a severe example of what a lack of crop genetic diversity can produce. Complicated by the rapid transport of seed and product, a potential now exists for rapid dispersion of a new and serious race of some specific pathogen.

An example of what changes in cropping system could do to the disease situation in one crop was summarized by Saari and Wilcoxson (1974). The new dwarf wheats in Asia and Africa were subject to an increased attack of rust in arid zones where irrigation was introduced. Additions of fertilizer to the system increased rust and other obligate parasites. Both problems can be alleviated by use of resistant varieties. Changes in tillage practices were found to have no effect on diseases. Crop rotation or multiple cropping practices reduced soil borne pathogens if the alternating crops were not alternate hosts for any specific organism. The researchers believe that the significance of some diseases has decreased, partly as a result of resistant varieties, and none has substantially increased as a result of the introduction of new varieties and practices.

Further evidence for the positive changes which accrue from changed cropping systems was summarized by Rotem and Palti (1969). Irrigation of crops allows greater flexibility in choice of planting date, and this opens the way to avoidance of some diseases through timely planting. Although the practice may provide a better environment for pathogens as well, plant species often are more vigorous and more resistant to pathogen attack under improved cultural and water conditions. The development of new varietal blends and multiple cropping systems has been suggested as an alternative which would reduce the probability of a severe epiphytotic. This route should be explored carefully, keeping in mind that the improved component crop varieties of any initial system have been developed for monocrop conditions, and may not give a fair test of the potential of the new system (Day 1973).

Weed Control

The diversity of weed species and year-round growing seasons in the tropics combine to magnify the problems of weed competition with crops. Although their effects on yields are usually less obvious than an epiphytotic disease or a devastating insect attack, weeds probably represent a more serious limitation to production in the tropics than any other single factor. Weeds also are hosts of insects and diseases, and can interfere with harvest.

Changes in cultural practices, cropping systems, or weed control measures have led to changes in weed populations and more severe problems in some cases. The move away from slash and burn agriculture eliminates the physical destruction and heat effects which accompanied the fire phase (Furtick 1969). A change from the traditional maize-groundnut scheme to cotton and tobacco in East Africa led to greater weed problems, and the need to develop, adapt, and test alternative methods of control (Sharman 1970). Erosion problems may be severe on some soils when clean cultivated and maintained without weeds. Improved technology is known and available, although its specific application in the tropics has been severely limited by a lack of guidelines and recommendations and the personnel to put them to use (Doll 1976; Furtick 1969). Adoption of part of an improved cropping system such as new varieties or fertilizer without consideration for weed control may lead to no net gain in production.

The technology needed for small farmer application of modern herbicides will require additional research to provide cheaper, simpler, and safer treatments (Parker 1976). This includes simple techniques for application, herbicides which are effective under relatively dry soil conditions, effective control of perennial weeds, crop varieties resistant to parasitic weeds, and weed control in systems with minimum cultivation. Although most weed control currently is done by hand, the increasing expense of labor, including the opportunity cost on the small farm for alternative activities, will make other weed control measures more feasible in the future. The possibilities of biological weed control also have been explored, and greater emphasis in research in this area may be justified especially for perennial crops (Goeden and Louda 1976; Wilson 1969; Alteiri and Doll 1976).

The fact that new technology has not been adopted by small farmers is due in part to a lack of appropriate technology, the means to buy and access to this technology, and low levels of education. It is necessary to teach farmers the sound agronomic practices used to increase crop competitiveness, the importance of preventive weed control, and the critical role that timely weeding plays in crop production (Doll 1976).

Use of chemical herbicides introduces another level of complexity which requires additional understanding and skills. Furtick (1969) points out that few trained people are available in the tropics to do this job. About 100 scientists are active in the entire tropics, while between 3000 and 5000 are involved with weed control research in temperate zones. Alternatives to the farmer's current hand weeding will be adopted only if they are appropriate agronomically, economically, and socially (Doll 1976).

VARIETAL IMPROVEMENT

Given the importance of small farm systems and multiple cropping in the tropics, little research has been directed at improving crops which would have specific adaptation to these systems. The only sustained effort to select crops for these complex systems has been the farmer's conscious attempt in each cycle to select and save seed for the next planting. The few efforts of agronomists in the tropics to introduce varieties into these traditional systems have focused on the improved cultivars which were developed originally for monoculture and relatively large use of other technological inputs. Even with these limitations, initial results have been favorable.

Varietal improvement already has been mentioned in regard to disease and insect resistance and greater competitive ability with weeds. Genetic tolerance to acid soils, ability to utilize lower levels of major and minor elements, and survival under stress conditions also were considered along with cultural solutions to these limiting problems. Whether the genetic route to solving each problem is the most appropriate depends on the cost and feasibility of modifying the environment to fit existing cultivars, the availability of genetic variability for the trait needed, and the probability that this quality can be included in a genetic package without sacrificing other important agronomic characteristics, especially yield and adaptation. A methodology for approaching this genetic improvement has been presented previously (Francis *et al.* 1976).

Central to the question of developing specific crop varieties for specific cropping systems is the magnitude of the genotype by system interaction in each crop and for the relevant range of cropping systems for that crop. If this interaction is small or nonexistent, then improvement and testing of germplasm in a breeding program may be conducted in any system which is convenient and economical. If the interaction is sufficiently large to indicate specific cultivar selection and adaptation to specific systems, then the breeding and testing scheme must be modified to most fully exploit the potential of each system. The decision to carry out two or more parallel breeding programs in one crop to select varieties specifically adapted to specific systems is an expensive one. A fixed amount of time and resources available to that crop would have to be divided, for example, between the efforts in monocropping and associated cropping, if these were the two predominant systems.

Several examples of this type of evaluation were summarized in a recent symposium article (Francis *et al.* 1976). Climbing bean varieties in Ecuador tested in association with 2 maize varieties as support showed a non-significant correlation for yield between the 2 systems, indicating a significant genotype by system interaction. Soybean yields in three systems reported from Tanzania likewise showed positive but non-significant correlations for yield between monocrop and associated cropping systems. Mung bean yields in monoculture and associated with maize in IRRI also showed non-significant correlations in two different seasons. Only one sorghum trial in Nigeria which compared 4 varieties under 2 contrasting systems had a highly significant correlation between yields in the 2 systems. These data suggest that the variety by system interaction may be of prime importance in a breeding program.

Recent data on bush beans (Francis *et al.* 1978B) and climbing beans (Francis *et al.* 1978A) at CIAT indicate that careful evaluation of the interaction in these crops may lead to different results from those reported previously. When 20 varieties of beans in each case were grown in monoculture and in association with maize over 3 consecutive seasons, positive and often significant correlations were obtained between yields in the 2 systems. An example for one season and test of climbing beans is shown in Fig. 19.3. Unpublished results on maize from three seasons in the CIAT program indicate similar correlations.

It was concluded from these trials that the risk of rejecting a genotype outstanding in association would be minimal if most early evaluations were done in monoculture, the most convenient system. The greater efficiency of testing in monoculture was shown also by greater expression of differences in yield potential, higher seed production in each cycle, lower CV's, and more precise observations of individual lines. Although the conclusion was to implement initial selection and screening

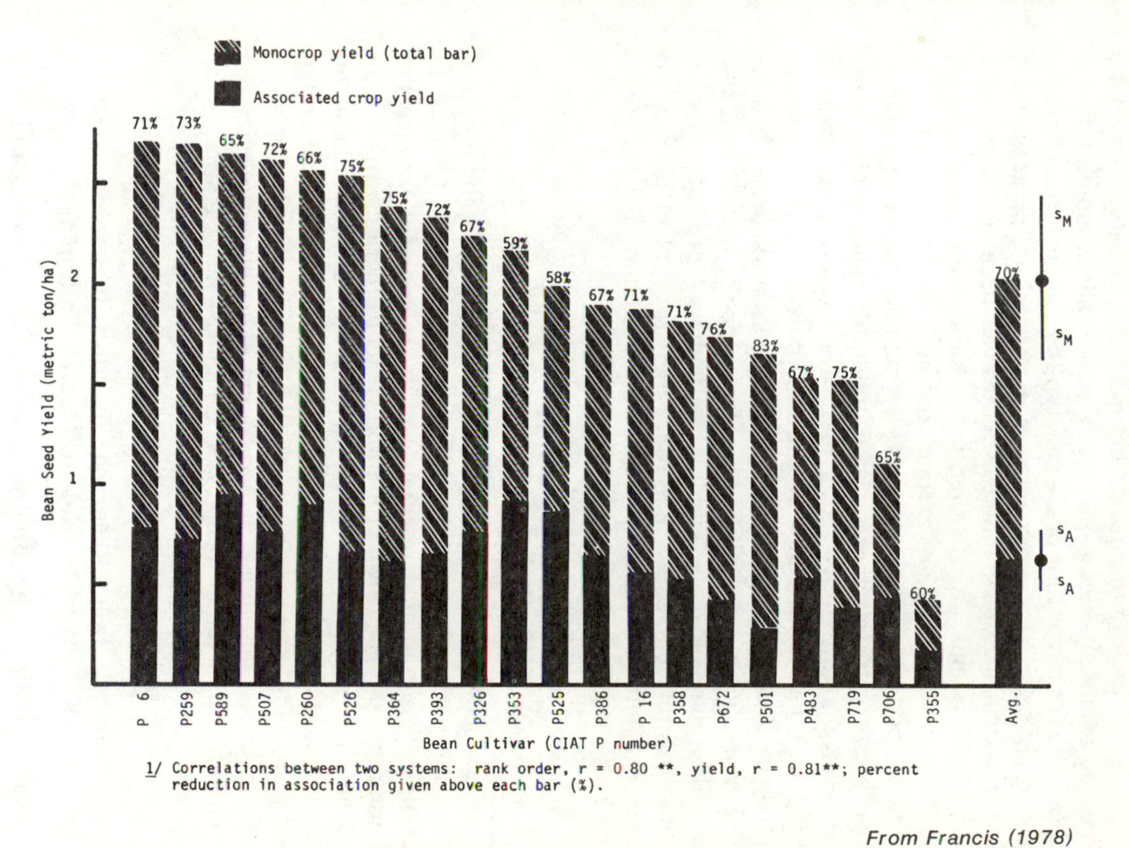

1/ Correlations between two systems: rank order, r = 0.80 **, yield, r = 0.81**; percent reduction in association given above each bar (%).

From Francis (1978)

FIG. 19.3. CLIMBING BEAN YIELDS IN TWO CROPPING SYSTEMS

in monoculture, advanced generation progeny must be evaluated in associated cropping systems. Concentration in the early generations on such qualitative traits as disease and insect resistance, seed color, and desired maturity can lead to rapid progress in these characters which are important to eventual success of a cultivar in either cropping system.

In addition to the magnitude of the genotype by system interaction, the decision to choose one or more systems in a breeding program must confront the circular problem inherent in the evaluation of germplasm in new cropping systems. When there are changes in fertility, plant density, land preparation, or cropping system, the selected materials with superior performance in previous systems may not be superior. It may be necessary to select germplasm under the new conditions. As systems evolve and improve, it is necessary to test each cycle of promising selections under current cultural practices. Conversely, optimum populations, planting dates, fertility levels, and species balance in mixed crops should be confirmed regularly using the latest available varieties. The decision on a breeding methodology depends on yield limitations in prevalent cropping systems, potentials of monocrop and intercropping, and the variety by system interactions for each crop (Francis *et al.* 1976).

PROGRAMS IN THE INTERNATIONAL CENTERS

Currently, much of the interdisciplinary work on developing new and improved alternative cropping systems for the tropics is organized in several of the international research centers. In conjunction with national programs and their own crop improvement programs, these integrative activities are directed at maximizing the use of available resources in systems which are similar to those currently used by small farmers to provide alternative systems which have greatly improved productivity. Since the approach is different in each center, they are listed separately with an address to which inquiries may be directed for more detailed information.

IRRI—International Rice Research Institute (Cropping Systems Program, IRRI, P.O. Box 933, Makati, Manila, Philippines)

The program initiated by Bradfield in Los Banos has expanded to include research activities in evaluation of cropping patterns, economics of cropping systems, weed science and plant protection, and varietal evaluation. The on-farm research approach leads to appropriate combinations of technological components for management of cropping systems under the constraints set by the resources available at each site.

Through work with individual farmers in a comprehensive outreach program in several countries, the IRRI team has developed a capacity to predict the adaptation of cropping patterns in situations which vary in soil, climate, landscape, and the availability of capital, labor, and power (IRRI 1976; Harwood and Price 1976; Harwood 1974; Zandstra 1978).

ICRISAT—International Crops Research Institute for the Semi-Arid Tropics (Farming Systems Program, ICRISTAT, 1-11-256, Begumpet, Hyderabad, 50016, A.P., India)

The primary constraint to crop production in the semi-arid tropics is scarcity of water and lack of dependability of rainfall from season to season. The program is resource centered and development oriented to achieve three goals: (1) to generate economically viable, labor intensive technology for improving and utilizing the productive potential of natural resources; (2) to develop technology for improving land and water management systems which can be implemented and maintained during the extended dry season to give additional employment and better utilization of animal power; and (3) to contribute to raising the economic status and quality of life for people in the semi-arid tropics by developing farming systems which increase and stabilize agricultural output (Krantz and Kampen 1976).

CIAT—Centro Internacional de Agricultura Tropical (Bean and Cassava Programs, CIAT, Apartado Aereo 6713, Cali, Colombia)

One of the objectives of the crop oriented improvement teams in CIAT is the integration of improved varieties into economically viable cropping systems for the tropics. For bean-maize and bean-cassava systems, in-depth research has explored the effects of manipulating agronomic variables on the productivity of the total system. The genotype by system interaction question has been explored in reference to bush beans, climbing beans, and maize. Current tests are being conducted in additional locations to confirm the initial results from research at CIAT. Basic work on the nature of competition between beans and maize in association is in progress (Anon. 1976).

CATIE—Centro Agronómico Tropical de Investigacion y Enseñanza (Department of Tropical Crops and Soils, CATIE, Turrialba, Costa Rica)

A comprehensive study of cropping systems for small farmers was initiated in 1973 by the Department of Crops and Soils. Over several years at Turrialba, crop patterns of beans, maize, rice, sweet potatoes, and cassava have been studied, including monocrop, crop rotation, and multiple cropping in the same season. Evaluations include agronomic effects of levels of technology and the economic feasibility of the best cropping patterns. An outreach program with personnel of CATIE working in cooperation with national research institutions of Costa Rica, Nicaragua, Honduras, El Salvador, and Guatemala is being implemented in several locations with small farmers. Using the farmer's present cropping system as the starting point, research is focused on improving food production through adapting and developing technologies appropriate to the socioeconomic and ecological conditions in each location (Pinchinat *et al.* 1976).

IITA—International Institute of Tropical Agriculture (Farming Systems Program, IITA, PMB 5320, Ibadan, Nigeria)

The objective of the Farming Systems Program is to seek ways to replace existing food crop production systems with more efficient ones that produce good yields of improved crop cultivars on a sustained basis. This is implemented by combining improved plant materials and aspects of their culture from the crop improvement programs with the environmental and management factors developed by specialists in the various aspects of cropping systems. Research is in progress to identify long-term sustained yield rotations, varieties adapted to these systems, and the soil management practices appropriate to the zone. Pest and disease limitations are being identified and solved in each crop improvement program, and mechanical techniques studied by the engineers. Economic studies of alternative cropping systems are comparing traditional agriculture with new systems based on technology (IITA 1975).

AVRDC—Asian Vegetable Research and Development Center (P.O. Box 42, Shanhua, Tainan, 741, Taiwan, Rep. China)

The broad objectives of the Asian Vegetable Center are to increase the yield potential of several important vegetable crops and develop cultural systems for the lowland humid tropics. Vegetables already are grown intensively in favored agricultural zones. AVRDC seeks to expand the adaptation of vegetables by improving the heat tolerance, disease resistance, and nutritional quality of these important sources of vitamins, minerals, and plant protein for the rice-based diet of Asians. Vegetables are a high-value component of cropping systems which small farmers now rotate with rice and intercrop with other cereals in some upland zones and near major population centers in Southeast Asia (Anon. 1976).

AGRICULTURAL DEVELOPMENT PROGRAMS

Although it is impossible to review current literature on development, a brief summary of how research in the tropics must interface with other factors which influence increased production is in order. Among other essentials for increased productivity, Mosher (1966) lists development of transportation and markets, access to supplies and equipment, adequate incentives to farmers, and production credit as critical to the process. The crux is to search out those factors which are causal and crucial to increased productivity, and not those factors which happen to be associated with development.

Agricultural research which has centered on the experiment station often has not been relevant to the surrounding country, and conclusions or recommendations based on biological and physical results have not stood the test of the economic realities of the small farm across the road (Mellor 1966; Hayami and Ruttan 1971). There need not be complex research or facilities to solve some of the most critical limiting problems in agriculture, and the early testing of possible alternatives on the farm is essential (Jennings 1974).

The IRRI Cropping Systems Program has moved the research-testing process further than those of other centers. A guide to this process has been published by Harwood and Borton (1978). An extension of the IRRI multiple cropping research pattern, this process involves both agricultural scientists and farmers in seeking to identify and solve the problems and limitations to production. A field testing program of possible alternative cropping patterns is designed based on agroclimatic data, the farmer's experiences, available technology, and evaluation of the farmer's resources and goals. Testing on these pilot or test farms can become the basis for an extension and education program.

The initial step in the development methodology is a classification of environmental factors and the location of target farms. The classification includes climatological and soils data, and proceeds with accumulation of data on crop and animal production under existing conditions and constraints. The farmer's goals and priorities are central to this preliminary analysis, and he participates directly with the scientist in the design of modifications in his system which are worth testing. Available technology known to the researcher is integrated with the experience and common sense of the farmer to plan the changes which both will implement together. Planning also includes the type of results to be measured and how goals will be evaluated at the end of the season. This "grass roots" approach to development provides a framework within which a series of disciplines from the research center can become involved in the analysis and problem solving on the farm.

The IRRI scheme has been developed over several years in Southeast Asia, and any evaluation of its effectiveness is too preliminary. It obviously is an intensive approach which will require a large number of technicians with scientific expertise and practical experience. It also is a unique and logical process which should provide the impact needed to move technology and stimulate change on the small farm.

Tropical cropping systems are complex, as indicated in the several sections of this review of the agronomic, genetic, and plant protection components of existing systems. Potentials for increased production exist through the application of available technology in some zones and systems. However, more emphasis needs to be placed on research which is designed to solve production limitations under the constraints most commonly encountered in each zone on the farm. The alternatives to present systems need to be designed and tested within the context of the small farmer's educational, technological, economic, and social situation. In addition to the technology needed to increase production, a wide range of constraints off the farm limit the availability of this technology and the incentives to adopt change in a cropping system. This is the challenge to the small farmer today, and to the researcher who is dedicated to increasing production in this sector of tropical agriculture.

REFERENCES

ALTIERI, M.A., and DOLL, J.D. 1976. Some limitations of weed biocontrol in tropical crop ecosystems in Colombia. Proceedings, IV International Symposium Biological Control of Weeds, Center for Environmental Programs, Institute for Food and Agricultural Sciences, University of Florida, Gainesville, 74-80.

ALTIERI, M.A., FRANCIS, C.A., DOLL, J.D., and VAN SCHOONHOVEN, A. 1978. A review of insect prevalence in maize (*Zea mays*) and bean (*Phaseolus vulgaris*) polycultural systems. Field Crops Res. *1*, 33-49.

ANDREWS, D.J., and KASSAM, A.H. 1976. The importance of multiple cropping in increasing world food supplies. *In* Multiple Cropping. R.I. Papendick, P.A. Sanchez and G.B. Triplett (Editors). Spec. Publ. *27*. American Society of Agronomy, Madison, Wisconsin.

ANON. 1976. Annual Report. Asian Vegetable Research and Development Center. Taiwan, Republic of China.

ANON. 1975, 1976. Annual Reports. Centro Internacional de Agriculture Tropical, Cali, Columbia.

BAEUMER, K., and BAKERMANS, W.A.P. 1973. Zero tillage. *In* Advances in Agronomy, Vol. 25. N.C. Brady (Editor). Academic Press, New York.

BRADFIELD, R. 1970. Increasing food production in the tropics by multiple cropping. *In* Research for the World Food Crisis. D.G. Aldrich, Jr. (Editor). AAAS, Washington, D.C.

DALRYMPLE, D.G. 1971. Survey of multiple cropping in less developed nations. For. Econ. Devel. Serv., USDA *FEDR-12*.

DAY, P.R. 1973. Genetic variability of crops. Ann. Rev. Phytopath. *11*, 293-312.

DOLL, J.D. 1976. Reaching traditional farmers with improved weed management in Latin America. Proceedings, 13th British Crop Protection Conference, Nov. 15-18, Brighton, England, 809-816. British Crop Protection Council.

DUCKHAM, A.N., and MASEFIELD, G.G. 1970. Farming Systems of the World. Praeger Publishers, New York.

ERBACH, D.C., and LOVELY, W.G. 1976. Machinery adaptations for multiple cropping. *In* Multiple Cropping. R.I. Papendick, P.A. Sanchez and G.B. Triplett (Editors). Spec. Publ. *27*. American Society of Agronomy, Madison, Wisconsin.

FAO. 1976. Production Yearbook, Vol. 30. Food and Agriculture Organization, U.N., Rome.

FRANCIS, C.A. 1978. Multiple cropping potentials of beans and maize. Hortscience *13*, 12-17.

FRANCIS, C.A., FLOR, C.A., and TEMPLE, S.R. 1976. Adapting varieties for intercropped systems in the tropics. *In* Multiple Cropping. R.I. Papendick, P.A. Sanchez and G.B. Triplett (Editors). Spec. Publ. *27*. American Society of Agronomy, Madison, Wisconsin.

FRANCIS, C.A., PRAGER, M., and LAING, D.R. 1978A. Genotype by environment interactions in climbing bean varieties in monoculture and associated with maize. Crop Sci. *18*, 242-246.

FRANCIS, C.A., PRAGER, M., LAING, D.R., and FLOR, C.A. 1978B. Genotype x environment interactions in bush bean cultivars in monoculture and associated with maize. Crop Sci. *18*, 237-242.

FURTICK, W.R. 1969. The importance of weed control in tropical and subtropical agriculture. Proceedings, 1st Asian-Pacific Weed Control Interchange, June 12-22, 1967. Univ. Hawaii, East-West Center, Honolulu.

GOEDEN, R.D., and LOUDA, S.M. 1976. Biotic interference with insects imported for weed control. Ann. Rev. Entomol. *21*, 325-342.

GREENLAND, D.J. 1975. Bringing the green revolution to the shifting cultivator. Science *190*, 841-844.

HARLAN, J.R. 1975. Crops and Man. American Society of Agronomy, Madison, Wisconsin.

HARLAN, J.R. 1976. Diseases as a factor in plant evolution. Ann. Rev. Phytopathol. *14*, 31-51.

HARWOOD, R.R. 1974. Resource utilization approach to cropping systems improvement. *In* International Workshop on Farming Systems. International Crops Research Institute for the Semi-Arid Tropics, Hyderabad, India.

HARWOOD, R.R., and BORTON, R.E. 1978. Small farm development: a research method for improving production with limited resources. International Agricultural Development Service, New York.

HARWOOD, R.R., and PRICE, E.C. 1976. Multiple cropping in tropical Asia. *In* Multiple Cropping. R.I. Papendick, P.A. Sanchez and G.B. Triplett (Editors). Spec. Publ. *27.* American Society of Agronomy, Madison, Wisconsin.

HAYAMI, Y., and RUTTAN, V.W. 1971. Agricultural Development: an International Perspective. Johns Hopkins Press, Baltimore.

HEICHEL, G.H. 1974. Comparative efficiency of energy use in crop production. Conn. Agric. Exp. Stn. Bull. *739.*

HEICHEL, G.H., and FRANCIS, C.A. 1973. Efficiency of energy use in producing maize with different technologies. Agron. Abstr. *43.* American Society of Agronomy, Madison, Wisconsin.

HIRST, E. 1974. Food-related energy requirements. Science *184*, 134-138.

ICRISAT. 1974. International Workshop on Farming Systems, Nov. 18-22. International Crops Research Institute for the Semi-Arid Tropics, Hyderabad, India.

ICRISAT. 1975. Annual Report. International Crops Research Institute for the Semi-Arid Tropics, Hyderabad, India.

IITA. 1975 Annual Report. International Institute of Tropical Agriculture, Ibadan, Nigeria.

IRRI. 1975. Annual Report. International Rice Research Institute, Los Banos, Philippines.

IRRI. 1976. Symposium on cropping systems research and development for the Asian rice farmer. International Rice Research Institute, Los Banos, Philippines, Sept. 21-24. IRRI.

JENNINGS, P.R. 1974. Rice breeding and world food production. Science *186*, 1085-1088.

KANNEGIETER, A. 1967. Zero cultivation and other methods of reclaiming *Pueraria* fallowed land for food crop cultivation in the forest zone of Ghana. Trop. Agric. *123*, 51-73.

KRANTZ, B.A., and KAMPEN, J. 1976. Farming systems research: 1976 program and summary of results. International Crops Research Institute for the Semi-Arid Tropics, Hyderabad, India.

MANSHARD, W. 1974. Tropical Agriculture: A Geographical Introduction and Appraisal. Longman, London.

MELLOR, J.W. 1966. The Economics of Agricultural Development. Cornell University Press, Ithaca, New York.

MOSHER, A.T. 1966. Getting Agriculture Moving. Praeger Publishers, New York.

NEWSON, L.D. 1967. Consequences of insecticide use on non-target organisms. Ann. Rev. Entomol. *12*, 257-286.

NICKEL, J.L. 1973. Pest situation in changing agricultural systems: a review. Entomol. Soc. Am. Bull. *19*, 136-142.

PAINTER, R.H. 1958. Resistance of plants to insects. Ann. Rev. Entomol. *3*, 267-290.

PAPENDICK, R.I., SANCHEZ, P.A., and TRIPLETT, G.B. 1976. Multiple Cropping. Spec. Publ. *27.* American Society of Agronomy, Madison, Wisconsin.

PARKER, C. 1976. Weed research requirements in developing countries. Proceedings, 13th British Crop Protection Conference, Nov. 15-18, Brighton, England, British Crop Protection Council. 801-807.

PIMENTEL, D. *et al.* 1973. Food production and the energy crisis. Science *182*, 443-449.

PINCHINAT, A.M., SORIA, J., and BAZAN, R. 1976. Multiple cropping in tropical America. *In* Multiple Cropping. R.I. Papendick, P.A. Sanchez and G.B. Triplett (Editors). Spec. Publ. *27.* American Society of Agronomy, Madison, Wisconsin.

RAPPAPORT, R.A. 1971. The flow of energy in an agricultural society. Sci. Am. *225*, 116-132.

RIVNAY, E. 1964. The influence of man on insect ecology in arid zones. Ann. Rev. Entomol. *9*, 41-62.

ROCKWOOD, W.G., and LAL, R. 1974. Mulch tillage: a technique for soil and water conservation in the tropics. Shell Public Health and Agricultural News *17*, 77-79, Derby, England.

ROTEM, J., and PALTI, J. 1969. Irrigation and plant diseases. Ann. Rev. Phytopathol. *7*, 267-288.

RUTHENBERG, H. 1976. Farming Systems in the Tropics, 2nd Edition. Clarendon Press, Oxford.

SAARI, E.E., and WILCOXSON, R.D. 1974. Plant disease situation of high-yielding dwarf wheats in Asia and Africa. Ann. Rev. Phytopathol. *12*, 49-68.

SANCHEZ, P.A., and BUOL, S. 1975. Soils of the tropics and the world food crisis. Science *188*, 598-603.

SHARMAN, C. 1970. Herbicide development in peasant farming areas. Proceedings, 10th British Weed Control Conference, Nov. 16-19, Brighton, England, British Crop Protection Council. 685-688.

SPAIN, J. M., FRANCIS, C.A., HOWELER, R.H., and CALVO, F. 1974. Differences between species and varieties in cultivated tropical pastures and soil acidity. *In* Soil Management in Tropical America. E. Bornemisza and A. Alvarado (Editors). North Carolina State Univ., Raleigh. (Spanish)

STEINHART, J.S., and STEINHART, C.E. 1974. Energy use in the U.S. food system. Science *184*, 307-316.

UNGER, P.W., and STEWART, B.A. 1976. Land preparation and seedling establishment practices in multiple cropping systems. *In* Multiple Cropping. R.I. Papendick, P.A. Sanchez and G.B. Triplett (Editors). Spec. Publ. *27.* American Society of Agronomy, Madison, Wisconsin.

WILSON, C.L. 1969. Use of plant pathogens in weed control. Ann. Rev. Phytopathol. *7*, 411-434.

WRIGLEY, G. 1969. Tropical Agriculture: The Development of Production. Praeger Publishers, New York.

ZANDSTRA, H.G. 1978. Cropping systems research for the Asian rice farmer. *In* Proceedings, Symposium Cropping Systems Research and Development for the Asian Rice Farmer. International Rice Research Institute, Los Banos, Philippines, Sept. 26-29, 1977. IRRI.

Index

Other AVI Books

AGRICULTURAL AND FOOD CHEMISTRY:
PAST, PRESENT, FUTURE *Teranishi*

AGRICULTURAL PROCESS ENGINEERING
3rd Edition *Henderson and Perry*

AN INTRODUCTION TO AGRICULTURAL ENGINEERING
Roth, Crow and Mahoney

BREEDING FIELD CROPS
Poehlman

CITRUS SCIENCE AND TECHNOLOGY
Vol. 1 and 2 *Nagy, Shaw and Veldhuis*

DRYING CEREAL GRAINS
Brooker, Bakker-Arkema and Hall

FUNDAMENTALS OF ENTOMOLOGY AND PLANT PATHOLOGY
Pyenson

HANDLING, TRANSPORTATION AND STORAGE
OF FRUITS AND VEGETABLES
Vol. 1 *Ryall and Lipton*
Vol. 2 *Ryall and Pentzer*

LABORATORY MANUAL FOR ENTOMOLOGY AND
PLANT PATHOLOGY *Pyenson and Barké*

PHYSIOENGINEERING PRINCIPLES
Merva

PLANT PHYSIOLOGY IN RELATION TO HORTICULTURE
American Edition *Bleasdale*

POSTHARVEST PHYSIOLOGY, HANDLING AND
UTILIZATION OF TROPICAL AND SUBTROPICAL
FRUITS AND VEGETABLES *Pantastico*

PRINCIPLES OF FARM MACHINERY
3rd Edition *Kepner, Bainer and Barger*

SEED PROTEINS
Inglett

SOILS AND OTHER GROWTH MEDIA
American Edition *Flegmann and George*

SOYBEANS: CHEMISTRY AND TECHNOLOGY
Revised 2nd Printing *Smith and Circle*

TREE FRUIT PRODUCTION
3rd Edition *Teskey and Shoemaker*